TEACHING AND LEARNING OF PHYSICS IN CULTURAL CONTEXTS

Proceedings of the International Conference on
Physics Education in Cultural Contexts

TEACHING AND LEARNING OF PHYSICS IN CULTURAL CONTEXTS

Cheongwon, South Korea 13 – 17 August 2001

edited by

Yunebae Park

Kyungpook National University, South Korea

World Scientific

NEW JERSEY • LONDON • SINGAPORE • SHANGHAI • HONG KONG • TAIPEI • BANGALORE

Published by

World Scientific Publishing Co. Pte. Ltd.
5 Toh Tuck Link, Singapore 596224
USA office: Suite 202, 1060 Main Street, River Edge, NJ 07661
UK office: 57 Shelton Street, Covent Garden, London WC2H 9HE

British Library Cataloguing-in-Publication Data
A catalogue record for this book is available from the British Library.

TEACHING AND LEARNING OF PHYSICS IN CULTURAL CONTEXTS
Proceedings of the International Conference on Physics Education in Cultural Contexts

ISBN 981-238-766-8

Printed in Singapore by World Scientific Printers (S) Pte Ltd

2001 International Conference on Physics Education in Cultural Context (ICPEC)

Organized by
Korean Physical Society

Sponsored by

International Union of Pure and Applied Physics(IUPAP)
International Commission on Physics Education(ICPE)
Korea Science and Engineering Foundation
Korea Research Foundation
PASCO EDUTEC Cheongju city
Chungcheongbuk-do Daekyo Co. Ltd
Kim Institute for the Korean Gifted
Hankook Chinaware Co. Ltd Physics Education Research Center
Dept of Physics Education, SNU Incheon Science Education Program
Center for Science Gifted Education, CNUE
Korea Institute of Curriculum and Evaluation
The Korean Federation of Science Education Societies
Institute of Science Education, Chungbuk Ntl University

Conference Advisory Committee

Chair & International Coordinator

Sung-Jae Pak

International Advisory Committee

Paul J. Black	R. Gunstone	E. W. Hamburger
T. Hyodo	Chang Sik Kim	Young Soo Kim
Keum-Hwi Lee	Xingkai Luo	E. L. McFarland
J. M. Ogborn	Edward F. Redish	T. Rossing
Tae Ryu	J. Sahm	Hee Sung Song
Vivian Talisayon	G. Tibell	Matilde Vicentini
Kaihua Zhao	Hwe Ik Zhang	

Local Advisory Committee

Hong Soo Choe	Don Hyung Choi	Byung Ghi Jang
Chan Soo Jee	Woo Soo Jeon	Jung Woo Kang
Beom Ki Kim	Hack Soo Kim	Sang Nyeon Kim
Sung Won Kim	Jae Kul Koh	Duk Gil Koo
Ki Jong Lee	Sung Muk Lee	In Ho Park
Hac Kyoo Park	Kwang Sub Soh	Sung Hyun Yoon
Keun Cheol Yuk		

Conference Organizing Committee

Director

Dai Shik Kim

General Planning

Ikgyun Kim
Byung Hoon Chung
Young Min Kim
Dong Suk Yang
Hey Gyoung Yoon

Domestic Affairs

Jaesool Kwon
Se Jung Jang
Jongwon Park
Jung Bok Kim
Jun Tae Kim

International Affairs

Jinwoong Song
Yunebae Park
Kyung Hee Choi
Sung Gi Kwon
Eunsook Kim

Pre/Post Sessions

June Hee Yoo
Myoung Hwan Kim
Jung-Ah Suh
Sang Woo Park
Sungmin Im

Paper Reviewers

(alphabetical order)

K. Akizo
M. O. Cambaliza
S. K. Cho
O. Choi
R. F. Gunstone
Sungwon Hwang
Se Jung Jang
Kwanghee Jo
Y. Jung
Hyoungseok Kim
J. Kwon
Keum H. Lee
A. P. Mazzolini
Sung-Jae Pak
Sang Tae Park
E. F. Redish
T. Rossing
J. Solomon
Jungah Suh
G. Tibell
L. W. Trowbridge
E. Bentillo
C. C. Chiaverina
K. Choi
H. Ferdinande
Ann Chin Hann
Sungmin Im
Kyung Ae Jang
L. Jossem
Eunsook Kim
Y. M. Kim
Jeongwon Lee
S. Markova
K. Nemoto
Churo Park
Yunebae Park
A. Rodrigues
Sangoh Seo
J. Song
V. Talisayon
I. Tokuya
M. Vincentini

Opening Address

On behalf of the Korean Physical Society (KPS), it is my pleasure to welcome you all to this International Conference on Physics Education in Cultural Context. The support from the IUPAP-ICPE and Korean Science and Engineering Foundation is gratefully acknowledged and let me commend the organizers and the invited participants making this event possible. Let us recognize Prof. Sung-Jae Pak in particular making this international event possible and the Organizing Chair Prof. Dai Shik Kim, who is also chairing Physics Education Division of KPS.

Education prepares the members of society for the future and hence should be carried out with feedback structure reflecting the societal change and need. Traditionally education was more of local nature. However, as transportation and communication becomes faster and cheaper particularly in the current information era, education has become more global and physics education is no exception. We are gathered here to share our ideas and experiences to improve physics education for the future with particular emphasis on the societal boundary condition called "culture."

We all know that active learning by motivated students is tremendously superior to passive education through the traditional top-down teaching. Fast bi-directional communication capability of current IT has opened a new horizon for improving education through possibility of multimedia education, resource sharing through internet, and prompt feedback of any change made.

Physics being foundation for modern science and technology, the Korean physics educators' contribution for national development was tremendous although it is hidden from public's view. You may wish to note that the Korean Physical Society has recently put physics education as its top priority and designated it as one of two main themes for its 50th anniversary in 2002. We strive for educational resource sharing at national as well as international level and any help you can give would be very much appreciated.

Please have a healthy and productive week. Thank you.

Hee Sung Song

President of Korean Physical Society

Preface

In August 2001, Korea National University of Education in Cheongwon, Korea hosted the 2001 International Conference on Physics Education in Cultural Contexts(ICPEC). The aims of the conference were to explore ways towards convergent and divergent physics learning beyond school boundaries, improve physics education through the use of traditional and modern cultural contexts, and exchange research and experience in physics education between different cultures.

In 5-day conference welcomed 211 participants from 14 countries, and the program included 8 keynote or invited lectures, 64 oral/poster papers, 4 symposiums, and 13 workshops/ demonstrations /exhibitions. In addition, various events and tours were also organized. The conference was certainly an enjoyable and stimulating experience for all participants.

The papers presented during the conference were collected and distributed to reviewers. Then, a total of 45 papers were finally selected for publication. Based on the subject matter of each paper, this book has been divided into three parts: Context and History, Conceptual Changes, and Media.

My sincere apologies for the delay in publishing this book, however, I hope everyone will enjoy reading this publication and that it will bring back many good memories.

Editor
Yunebae Park

Contents

Part II Conceptual Change

Part I

Context and History

PHYSICS EDUCATION IN CULTURAL CONTEXT:ISSUES, APPROACHES, AND PERSPECTIVES

SUNG-JAE PAK
Seoul National University
Seoul, Korea 151-742

Physics and physics education do not take place in a cultural vacuum, but rather within a geographical, historical, and social context. As "culture shock" is confronted when visiting other countries, students are forced to challenge a "cultural border-crossing" when they encounter the new culture of physicists while studying physics. Therefore, a more foundational and microscopic insight about culture is necessary when debating new issues, approaches, and visions related to the physics curriculum, instruction, and support system. The first cultural aspect is related to the renovation of the curriculum to include the 'physics culture' in balance and eventual establishment of a new culture. The second is related to learning and teaching in a cultural context, and the inclusion of "cultural validity" in assessment. The third is related to the issues, approaches, and future of a multi-cultural approach, as opposed to standardization, for the national curriculum and science education standards

1. Introduction

Physics research and education should both be regarded as cultural activities. As such, the linking of physics education with cultural activities could herald significant improvements in the curriculum, new ways of learning, teaching, and assessing, and new approaches to research, teacher education, and policies in physics education.

Accordingly, the current paper attempts to identify the origins of the reasons why students feel that physics is difficult and they do not like physics. Also the new frontier of physics culture education will be discussed, which may help overcome the limitation of Western culture-based science and form a new third culture.

The discussion will focus on the cultural aspects of learning and teaching in a cultural context and cultural validity in evaluation, along with the curriculum improvement. In addition, standardized and di-cultural approaches will be discussed in the context of previous research on national standards, international comparison studies, large-scale teacher education, and administrative support systems for physics education. Consequently, the resulting insights are based on a combination of established phenomena and the outcomes of recent research.

2. Cultural Aspects of Curriculum

Snow stated that the amazing development of Western culture-based science has widened the gap between the two academic cultures to the point where it is nearly impossible to bridge[1]. This gap may also be directly or indirectly related to the fundamental issues and problems facing today's our society.

2.1. *Physics Education for A New Culture*

Soh(2001) insisted that a new third culture, that can overwhelm the two existing cultures, could be formed through connecting activities between science and technology, science and philosophy, science and arts, etc. based on Connectionism, a traditional Eastern thought[2]. He also insisted that

> ' I connect, so I am', which is more than just imitating Descartes. No individual or unit can exist without connections. Life itself functions within a 'larger life', meaning that each individual is part of a larger network. Plus, Zhang's concept of a 'whole life', 'netting function of life in the world', or 'life space' can be interpreted as an ontological function.

Thus, Zhang insisted that the struggle for order or struggle for cultivation is essential to substitute the technology space in which the science and technology manifest with a culture space. Setting the science culture space in order means evolving into the order of new culture space with keeping and developing the inherited and respected values of the existing human culture.

The 'order of space' can distinguish and keep the admitted connection from the prohibited connection. Some serious problems could happen in the connection relationships. The first distinguished example of them could be derived from substituting the face to face connections with the face to computer connections. The second one is that no rooms for flexibility and creativity are allowed in the administration systems because the administration systems have become into automatic. The third one is prevailing reductionist prejudice in science. To form a new third culture, sloughing off the reductionist prejudice is important. Even though the fallacy of scientism, the view that science can substitute or reduce the liberal arts, should be thrown away, actually the science has just been recognized to pursue mainly the development of technology.

It seems that the characteristic activities of science, such as analytic, experimental and controlling of variables, have been totally isolated from the liberal arts, yet have strong power. Nevertheless, it does not appear that the atomism of science can understand, control, and solve all complicated problems. The new frontier of physics education with the aim of being nature-friendly, cooperating with individuals, harmonizing Western and Eastern cultures, and connecting the two cultures is necessary to overcome various issues, including extreme disasters.

The main premise of these arguments is that the ideology and practice of the education for a new third culture through combining two cultures would be the very important key.

2.2. *"Physics Culture" As A Content*

The discussions in constructing curriculum the cultural aspect can be started by a metaphor. Sometimes we say that "There is a tremendous number of cars, yet there is no car-culture." If there are no cars, we do not need to consider a car-culture. Although there are huge numbers of cars, we can say there is no car-culture when violating traffic signals, beeping the horn, not yielding and impolite behavior occur frequently.

The concept of physics and physics culture can be distinguished by this metaphor, though they are neither opposite nor isolated. Also the arguments for considering the culture into physics education is not the new rising one.

Generally, the considerations of culture have already been introduced into the physics education for quite some time. For example, one of the aims mentioned by the PSSC is to present physics as an intellectual and cultural pursuit which is the part of present-day human activity and achievement[3].

Also, Holton wrote "I agree strongly with I. I. Rabi, one of the members of Advisory Committee, who has said that physics is now at the core of the humanistic education of our time. This does not have to mean a 'soft' physics course or a course that involves poetry instead of physics. On the contrary, it can mean a physics course that accentuates just those elements which characterize the most thoughtful and fundamental achievements in physics itself from Newton to Bohn... introductory level should be physics taught from a humanistic point of view not only physicists' or physic teachers'. 79% thought that high school students stay away from physics because in their schools the course now given is too difficult to suit their abilities and desires, and yet 91% said also that a physics course with a cultural component is needed by nearly everyone."[4]

Recently, PTS/STS movements involved with issues related to energy and pollution have gained the spotlight in physics/science education.

Traditionally, physics concepts and conceptual schemes have been considered as the main contents of physics curriculum, but recently process skills have been included as the contents, in addition to the nature of science, the scientific history and the social construction of science. More recently, the various contexts of physical phenomena or learning activities have been rising to the surface. But all of them seem to be oriented to the "physics" education, with partial connections between the core physics and other areas on the premise that the physics, mathematics, technology and society can be distinctive and opposite each other.

When someone argues that the contents of physics and some aspects of culture are included in their curriculum, could it mean that physics were combined with liberal art rather than mixed? How can the biomechanics, electricity civilization, quantum epistemology, nuclear war, cosmology, etc. can be constructed and organized into the differentiated curriculum with meeting the students' level, interest and the demands of the flexibility and balance in the curriculum? Furthermore, how about including the physics cultures of the lives and wisdoms of the Ancients, in my and neighbors' contemporary lives, and in regional and global problems in the traditional physics curriculum? At this point, we need to investigate the theory and ways to include physics culture in the school curriculum.

3. Cultural Aspects of Instruction

It is common that physics education should mirror the physics itself when the aim is to bring up the future physicists. But when the aim is to bring up the future citizens, and the physics is taught for all as a subject or a part of science subject, some evidence which indicates that copying physics itself is not enough have been accumulated.

On the other hand, the educators of general theories have insisted that the psychology of students, instruction theories, and social studies are important. But how can they teach a particular subject such as physics with such general theories?

The opposition or conflict between the contents subject education and general theory education in the pre and in service teacher education could be interpreted as a part of the historical opposition and disconnection between two cultures. Also the origin of students' dislike to physics could be interpreted as being originated from the disconnection not only between two cultures, but also between every day life and the physicists' activities.

3.1. *"Cultural Border Crossing" As Learning Physics*

A student's mind is not empty to absorb the teacher's knowledge. For example, the students have experience with rainbows, laser pointers, and the thin layer coating the lenses of glasses before they learn about refraction and interference of light in physics class. Even though they are not familiar with the scientific terminology, they may have their own "light cultures". It might be the key point of physics teaching how to connect the students' light culture with the physicists' culture, which physics teachers intend to cultivate.

Cobern and Aikenhead (2001) addressed that earlier psychological perspectives on the individual learner, such as Piaget, Ausubel and personal constructivism, have expanded to encompass sociological perspectives that conceptualize learning in social settings, including social constructivism, science for specific social purposes, and situated cognition, then, the next stage in the evolution of their learning in a cultural milieu. The perspective from cultural anthropology can provide fresh insights into familiar problems associated with students learning science[5].

An anthropological viewpoint for science education was proposed by Maddock[6]. He wrote that 'science and science education are cultural enterprises which form a part of the wider cultural matrix of society and that educational considerations concerning science must be made in the light of this wider perspective'. In cultural anthropology, teaching science is viewed as culture transmission[7], and learning science as culture acquisition[8], where culture means 'an ordered system of meaning and symbols, in terms of which social interaction takes place'[9]. Aikenhead(1996) conceptualized culture according to the norms, values, beliefs, expectations, and conventional actions of a group[10]. Within every culture, there exist subgroups that are commonly identified by race, language, ethnicity, gender, social class, occupation, religion, etc. Cobern et al.(1995) continued to insist as follows[11].

"Transmitting a scientific subculture to students can either be supportive or disruptive. If the subculture of science generally harmonizes with a student's everyday culture, science instruction supports the student's view of the world, and the result is enculturation[12]. When enculturation occurs, scientific thinking enhances a person's everyday thinking[13]. If the subculture of science is generally at odds with a student's everyday world, as it is with many students[14][15], science instruction tends to disrupt the student's view of the world by forcing the student to abandon or marginalize his/her indigenous way of knowing and reconstruct in its place a new (scientific) way of knowing. The result is assimilation, which has highly negative connotations[15]. Students struggle to negotiate the cultural borders between their indigenous subcultures and the subcultures of science."[17]

The patterns of students' moving into the subculture of science: are (1) 'Potential Scientists' who cross borders into school science so smoothly and naturally that the borders appear invisible, (2) 'Other Smart Kids' who manage their border crossing so well that few express any sense of science being a foreign subculture. (3) 'I Don't Know Students' who confront hazardous border crossings but learn to cope and survive. (4) 'Outsiders' who tend to be alienated from school itself and so border crossing into school science is virtually impossible, and (5) 'Inside Outsiders' who find border crossing into the subculture of school to be almost impossible because of over discrimination at the school level even though the students possessed an intense curiosity about the natural world [18].

It would suggest that "students need a conceptualized approach to teaching science that draws upon the cultural worlds of students and makes sense in those worlds. We need to develop teaching methods that allow the incorporation of the content or aspects of another culture into a student's everyday culture (autonomous acculturation) and enables students to enjoy and construct meaning out of Western science without the need to assimilate science's cultural baggage (anthropological instruction)[19].

Therefore, one implication is that instructional methods and materials should: (1) make border crossings explicit for students, (2) facilitate these border crossings, (3) promote discourse so that students, not just the teacher, are talking science, (4) substantiate and build on the legitimacy of students' personally and culturally constructed ways of knowing, and (5) teach the knowledge, skills, and values of Western science in the context of its societal roles[20].

To deepen and widen the cultural aspects of physics learning would have enough meanings by itself, but further studies on the cultural aspects of the physics culture learning should be focused. For the physicists, the researches the physics culture might be more difficult than researches physics. But the reverse might be true for the students. The study in physics culture could be easier and interesting for the students.

3.2. *"Cultural Validity" in Assessing Achievement*

The teacher should diagnose carefully whether the students can solve the problems properly or not and feedback promptly. In some cases, the students distorted the intentions of examiners or solved a different problem or a different way from the traditional. Since there would be some discords of cultures between students and teacher, or between students. The validity can not be surely determined only from the viewpoints of professionals.

Solano-Flores et al. (2001) proposed the concept of "cultural validity" as a kind of test validity in science assessment on the premise that the culture and society shape an individual mind and thinking. It means that the assessment must consider how the socio-cultural context in which students live, which include the values, beliefs, experiences, communication patterns, learning and teaching styles, and epistemologies as well as socioeconomic conditions, influence the way in which students make sense of science items and solve them[21].

Greenfield(1997) administered conservation of quantity tests to unschooled children in Senegal as she studied the Piagetian stage of concrete operations. After the children transferred water from a short fat beaker to a long thin one, they were asked in their native language whether the quantity of water was the same, more, or less[22].

Questions such as, "Why do you think it is the same(or more, or less) amount of water?" or "Why do you say it is the same(or more, or less) amount of water?" yielded silence on the part of the children, until the question was changed to "Why is the water the same(or more, or less)?"

The researcher insisted that it is wrong to conclude that the students who had no answer to the first style of question were not in concrete operation. Different assessment process is needed for those who have different epistemologies. And the proper assessment process in considering students' epistemologies may prevent this kind of misleading.

The item asked the student to choose the picture that shows how the river and mountains look now as opposed to how they looked millions of years ago[23].

Figure 1. NAEP Erosion Item

A 4th grade girl chose the picture B, the wrong answer. During the interview, she did not remember learning mountains at school. Nor had she experienced climbing mountains, but she recalled one first-hand experience related to mountains: a relative said, "look" and she saw it and it was like that(B). So the researchers asked her again, and she replied that picture A was the past and picture B was today. They asked her explain why, and she said that this one looked familiar (B) and this one (A) she did not know.

Solano-Flores et al. identified five areas in which the reasoning derived from the notion of cultural validity can contribute to improving science assessment: (1) student epistemology, (b) student language, proficiency, (c) cultural world view, (d) cultural communication and socialization styles, and (e) student life context and values.

The notion of cultural validity implies the need for a paradigm shift in assessment. Considering the cultural minorities raise a serious problem especially about the notion of "standardization" because back translation no longer should be considered a guarantee for proper translation in the culture and the context in which students live is not properly considered. In the context of multicultural society and a global economy, item equivalence across culture over "standardization" imply the need for new developments in psychometric theory and radical changes in the ways in which assessments are developed and administered[21].

4. Multicultural Approaches for Innovation

The success of the education in a country largely depends on the systems and environments such as teacher education, researches, administration, finance, and inspections. As recently, the physics education has been regarded as a powerful source of national development, the above systems and environments have been focused on a large scale. On the other hand, the increasing role of international studies in such national plans and activities have involved in the increasing of the international affairs and interchanges.

In the relations to the above phenomena, there would be two issues in cultural perspectives: national standards and international frameworks. First of all, the standardized concepts of the “national curriculum" or "the national standards" could be hot issues in the united states because of minorities’ cultures even unitary state like Korea, which has stratified structure of sub cultures in general. Secondly, as a result of international standards comparison, the blind devotion of developing countries for the developed or the monopolizing by developed countries should be discussed. Nonetheless, the support system for the physics culture education will be emphasized in this paper.

4.1. *Standardization*

There is so called 'national curriculum' in Korea, Japan, Taiwan and so forth. In case of Korea, the first national curriculum was established in 1945, and up to now there have been seven times revisions. The national curriculum has clarified the aims of general education and established subjects and time allocations in addition to the subject statements about objectives, contents and instructions.

The restrictions have been relaxed in nowadays, but basically the contents of each units and grades are fixed without any flexibility. Thus the several textbooks which have been approved by the Ministry of Education can be identified their own characteristics in the concepts and the structure of concepts. Worse than is that most of teachers teach and evaluate their class along with a textbook without any consideration of their own students.

About forty students, one of them could be a ‘talented’ or a ‘retarded’ having various backgrounds, interests, and future careers, should learn the same physics in the standardized curriculum and classes. From the perspectives in the equity of opportunity and social justice, these over standardization would become issues. In principle, the multi-cultural approaches should be adopted.

The national standards in United States are in somewhat different context from the national curriculum in Korea. However the national assessments are carried out to get the ‘actual state of achievement’ and according to the results, the national policies of the teacher education, school support, and awards have been established and put into actions for the purposes of the world best achievement in science and mathematics. But the problems and limitation of standardized assessment should be discussed and reconsidered.

Solano et al.(2001) made a statement that it is difficult to consider the large scale evaluation in situation that sentences and patterns of present standardized concepts are different in each cultural society, with realizing of cultural validity. But in multi cultural society, the truth that equal assessment should be suggested to each cultural society exists prior to standardization in systematic economic situation[24].

Anyway many assessment are getting out of strict standards, for examples in the test computer applied there is assessments in which different problems are given to students according to their response of pretest. If it is, is it strange to consider cultural validity?[25]

Needless to say, sufficient conceptual and methodological researches are needed in order for cultural validity to be applied.

On the other hand, Sharon(2001) said that " In spite of our best efforts towards equality and reducing gap, innovation movement in science education has failed in responding to various students suitably. What is obvious one is that "science for all" does not mean "one size fits to all". More 40% of U.S. students have different a cultural origin, language and race. In addition to this, students of multi language and multi culture are taught in various countries but international comparison study like TIMSS already affects policy and action of mathematics and science education in any country of the world."[26]

Even if the cost is the crucial problems in establishment of educational polices, we should try to overcome the restrictions to force the standardization. The intentions of doing education properly in addition to the firm philosophy and the research results in education will throw the new vision.

4.2. *Deculturalization*

The teacher education and researches in physics education are the crucial factors of the success in embodying the philosophy of physics culture education. The current teacher education could not get out of the primitive state of adapting the cultural context to the physics education without the whole understanding the philosophy of physics culture education. The renovation of teacher education should be expected. Also the practical researches how to adapt the physics culture education in classroom involving the theoretical researches should be encouraged. The support system including the inspections, administrations and finance can activate the teacher education and researches in physics culture education.

One of the serious problems is the blind follow of the developing countries for the teacher education, researches and supporting system of the developed countries which have absolutely different cultural background. Also the monopolization and the expansions of the developed countries' system toward the developing countries should be avoided. Of course the communication and exchanges of information should be encouraged, and the pursuit of finding suitable system through the comparison studies with the cultural understanding are encouraged to be continued.

5. Concluding Remarks

A new reformed curriculum should aim to produce students that are physics culture literate and include a balanced content of physics culture to contribute to the formation of a new culture that is nature-friendly and based on connectionism related to Oriental culture. This could help overcome the limitations of modern physics based on the analytic and experimental atomism related to Western culture. In line with the curriculum, cultural approaches to teaching and evaluation that are aware of cultural border crossing need to be included in physics classes. Furthermore, supporting activities, such as physics education research, teacher education, policy making, finance and administration should also be managed based on a multicultural approach rather than a standardized situation. In this way, the vision of a new third culture can be realized beyond the horizon of "two cultures".

References

1. C. P. Snow, The Two Cultures and the Scientific Revolution, Cambridge Univ. Press, New York, 1959.
2. H. Soh, Science Culture and Oriental Connectionism, *Paper presented at the Science Culture Research Center,* 2001.(Korean)
3. Little, Friedman, Zacharias & Finaly, The Physical Science, *The Science Teacher*, vol. 24, no. 7, November, 1957.
4. Havard University, *Harvard Project Physics*, 1967.
5. W. W. Cobern and G. S. Aikenhead, Cultural Aspects Of Learning Science, International Handbook of Science Education, Part One, Kluwer Academic Pub: Dordrecht. pp. 39-52, 1998.
6. M. N. Maddock, Science Education: An Anthropological View Point, *Studies in Science Education*, vol. 8, pp. 1-26, 1981.
7. G. Spindler, *Education and Cultural Process: Anthropological Approaches* (2nd Ed.), Wavaland Press, Prospect Height, IL., 1987.
8. H. F. Wolcott, Prospect and the Acquisition of Culture, *Anthropology and Education Quarterly*, vol. 22, 251-273, 1991.
9. C. Geertz, *The Interpretation of Culture*, Basic Books, New York, 1973.
10. G. S. Aikenhead, Science Education: Border Crossing into the Subculture of Science, *Studies in Science Education*, vol. 27, pp. 1-52, 1996.
11. W. W. Cobern, A. T. Gibson, and S. A. Underwood, Everyday Thoughts about Nature: An Interpretive Study of 16 Ninth Graders Conceptualizations of Nature, *Paper presented at the annual meeting of the American Educational Research Association*, San Francisco, CA, 1995.
12. Hawkins, and R. D. Pea, Tools for Bridging the Cultures of Everyday and Scientific Thinking, *Journal of Research in Science Teching*, vol. 24, pp. 291-307, 1987.
13. W. W. Cobern, A. T. Gibson, and S. A. Underwood, Everyday Thoughts about Nature: An Interpretive Study of 16 Ninth Graders Conceptualizations of Nature, *Paper presented at the annual meeting of the American Educational Research Association*, San Francisco, CA, 1995.

14. V. B. Costa, When Science is "Another World": Relationships between Worlds of Family, Friends, School, and Science', *Science Education*, vol. 79, pp. 313-333, 1995
15. M. Ogawa, Science Education in a Multiscience Perspective, *Science Education*, vol. 79, pp. 583-503, 1995.
16. O. J. Jegede, Collateral Learning and the Eco-Cultural Paradigm in Science and Mathematics Education in Africa, *Studies in Science Education*, vol. 25, pp. 97-137, 1995.
17. W. W. Cobern, A. T. Gibson, and S. A. Underwood, Everyday Thoughts about Nature: An Interpretive Study of 16 Ninth Graders Conceptualizations of Nature, *Paper presented at the annual meeting of the American Educational Research Association*, San Francisco, CA, 1995.
18. V. B. Costa, When Science is "Another World": Relationships Between Worlds of Family, Friends, School, and Science', *Science Education*, vol. 79, pp. 313-333, 1995.
19. W. W. Cobern, A. T. Gibson, and S. A. Underwood, Everyday Thoughts About Nature: An Interpretive Study of 16 Ninth Graders Conceptualizations of Nature, *Paper presented at the annual meeting of the American Educational Research Association*, San Francisco, CA, 1995.
20. Ibid.
21. G. Solano-Flores, and S. Nelson-Barber, On The Cultural Validity of Science Assessments. *Journal of Research in Science Teaching,* vol. 38, no. 5, pp. 553-573, 2001.
22. P. M. Greenfield, Culture as Process: Empirical Methods for Cultural Psychology. In J. W. Berry, Y. H. Poortinga and J. Pandey (Eds.), *Handbook of Cross-Cultural Psychology(2nd Ed.), Vol. 1: Theory and Method*, pp. 301-346., Needham Heights, MA: Allyn & Bacon, 1997a
23. NAEP, The Nation's Report Card: Assessment Science-Public Release, www.ed.gov/NCES/NAEP, 1996
24. G. Solano-Flores, and S. Nelson-Barber, On The Cultural Validity of Science Assessments. *Journal of Research in Science Teaching*. vol. 38, no. 5, pp. 553-573, 2001.
25. R. Kopriva, A Conceptual Framework for the Valid and Comparable Measurement of All Students, Paper Presented at the American Education Research Association Annual Meeting, Montreal, Canada, 1999.
26. L. Sharon, Conclusion: "Science for All" Is not Equal to "One Size Fits All": Linguistic and Cultural Diversity and Science Education Reform, *Journal of Research in Science Teaching*, vol.38, no.5, 2001.

Appendix : Pilot exemplars of teaching physics culture.

1. Enquiry of toys (cf. ICPEC workshop on enquiry with toys)

2. Exploring science at historical sites (cf. ICPEC Symposium 1. Metal Printing Museum Field Trip, Workshop on Field Trip, 2001 NARST Presentation, http://peer.snu.ac.kr (Korean))

3. Physics Tournament of Group Enquiry
(cf. The 14th IYPT, SNU Report (Korean), http://physed.snu.ac.kr (Korean))

PHYSICS, TECHNOLOGY AND SOCIETY

JOAN SOLOMON
Open University, UK

Science Technology and Society education (STS) may seem to some people to have burst onto the stage in Europe during the 1970s and then to have died out, at least in Britain, in the late 1980s because of opposition from the government's new educational policy, and possibly also from transnational industry. In fact neither of these statements is completely true. The roots of STS can be traced back to two different movements in the 1930s and 1950s respectively, but the social group which backed it during the 1970s was, in many ways, unlike any which had gone before. This makes it a very interesting development. And it is also true that STS is now recovering its strength, although it has now taken on yet another new form to suit the present times. This paper will attempt to define the original sponsors of STS by their special purpose and identity, perceived adversaries, and goals. From this we shall try to see the subsequent fortunes of this social component of science education as a conflict between educational utopianism on the one hand, and a *realpolitik* which includes industry and economics, on the other.

1. More of The Same Science! Before The First Phase.

We could begin with the movement which emerged in Britain during the 1930s and is often known as 'The Scientific Humanists'. This early movement had aims for self-education in science and the development of a 'scientific education' for all the people. (Notice that the word used is not 'literacy' but simple 'education'). Despite the title of Hogben's book – Science for the Citizen - it was just a textbook in general science with numerical questions, answers in the back, and some added history of a paste-on kind.

In the same way Mathematics for the Million was a textbook in mathematics. Hogben thought nothing of contemporary writings on 'popular science'

> *The clue to the state of mind of which produces these week-kneed and clownish apologetics is contempt for the common man.*

So you can see that science was to be taught to *'the common man'* without any motive other than yet more authoritative education. Some of the other scientific humanists of the 1930s had a notion that citizens might want a different 'citizen's science' which addressed special questions which were socialist in origin and ignored others. But Hogben's ideas certainly did not die with him. In the 1950s Charles Snow propagated his notion of The Two Cutures in much the same spirit. To us nowadays the idea that understanding the Second Law of Thermodynamics, difficult as that may be, would somehow make a non-scientist as scientifically literate as a knowledge of the plays of Shakespeare would make a scientist humanistically literate, seems extraordinary. The whole question of why lay people who were not going to be scientists might want or need to learn science is totally ignored.

The supremely arrogant comments of Lancelot Hogben can even be compared to similar comments made by Lewis Wolpert (1993) today. Wolpert, and his American counterpart Morris Shamos tried to label scientific knowledge as 'unnatural' because it is, he argued, quite unlike common sense and can have nothing to do with ethics or any other interest of the non-scientists. So we may categorised them as

Type	Identity	Adversary	Goal
Scientific Humanists (1930s)	**Marxist educators**	**Social elitism in science**	**Relieving the oppressed (e.g Irish)**
Charles Snow	**Speaking for science as a culture**	**Humanism as the only culture**	**Recognition of two cultres**
Wolpert, Shamos (1990s)	**Scientific literacy is impossible.**	**Popularisers**	**Preserving elite science**

There are some very sharp and useful distinctions in the above table. What have been omitted are the common features of

a) ***huge pride in science as a way of thinking***, and a corresponding
b) ***contempt for most other kinds of knowledge*** and
c) no special interest for **the ordinary citizen**.

2. STS Arrives. The First Phase. 'Being A Responsible Scientist.'

The first wave of STS arrived early in the UK. It arose out of the anxiety of a whole generation who watched the dreaded mushroom cloud envelop Hiroshima at the end of the Second World War. In particular the scientists felt that they had a responsibility to explain the dangers of atomic bombs to lay people. A group of young university physicists in Britain, led by Dr Bill Williams of Leeds University, rented an empty railway-carriage which they furnished with educational materials, and travelled across the country teaching the public about radioactivity.

Like other contemporary movements such as The British Society for Social Responsibility in Science, Pugwash, and those who ran the Bulletin of the Atomic Scientists in the United States, the scientists who started the STS movement in the UK believed that their duty to be 'socially responsible', rather in the same way as doctors should treat their patients responsibly. Their aim was to educate lay people so that they could exercise their democratic decision-making rights more effectively. At that time the greatest emphasis in the UK was on nuclear power, the testing of nuclear bombs and new ways of generating energy using natural phenomena such as wind and wave-power. Many left-wing scientists, mostly physicists, belonged to SANA - Scientists Against Nuclear Armaments - and took part in the protest marches against the testing of nuclear bombs.

In the USA the focus of early STS was more on environmentalism than disarmament, partly because of the power of politically repressive McCarthyism, but it was of the same general kind. Rachel Carson, herself a scientist like the other originators of other first wave STS groups wrote eloquently about the damage that was being done to the populations of plants and animals, and of the inevitable subsequent effects on all forms of life. She wrote about the science of ecology which was a new scientific discipline at that time, originating partly in England and partly elsewhere. Her main point was that the science being applied by the agricultural industry was hopelessly out of date.

> *The 'control of nature' is a phrase conceived in arrogance, born of a Neanderthal age of biology and philosophy, when it was supposed that nature exists for the convenience of man. The concepts and practices of applied entomology for the most part date from that Stone Age of science. It is our alarming misfortune that so primitive a science has armed itself with the most modern ad terrible of weapons, and that in turning them against the insects it has also turned then against themselves.* (Carson, p261/262)

By 1970 the STS group in the UK had received enough funds from the Leverhume Foundation to start an organisation called Science-In-a-Social-CONtext, or SISCON for short. This encouraged the teaching of courses in STS in universities and polytechnics. Many of the SISCON teaching materials are still in existence, covering subjects such as Limits to Growth, The Atomic Bomb, Science, Technology and the Modern World, etc. It was a time for learning by those who were going to launch the STS movement in the universities and polytechnics of Britain.

3. The Second Phase of STS. The Teaching Begins.

So STS started in higher education, and only later spread to the schools. I was myself the originator of the first comprehensive school programme called SISCON-in-Schools. The first international conference was held the Netherlands, where STS in schools and universities had arisen directly out of a political need. The Dutch government had wanted to hold a national referendum on whether or not to invest in nuclear power. They decided on an eight-year public educational programme on the physics of the issue, and this was still underway when the conference took place in 1978. The logical but slightly naïve nature of this public education phase seems a remarkable and admirable example of 'long-termism' as we would call it today.

This second phase of STS, was tied to a strong popular belief in democracy. No longer was it the scientists who would guard the interests of the man-in-the-street. Now the aim was to help them to make up their own minds and to act on their own ideas. Of course science education would be required, but not of the type for training more scientists. Our young students would need to learn it in association with an understanding of how technological industry worked and what regulations would be required and also how democratic civic processes worked. There would be no room here for the elitist or patronising attitudes of Hogben and Wolpert.

As John Ziman, a physicist who was very supportive of STS, wrote in Teaching and Learning about Science and Society (1980) *'The STS movement belongs to our own times, and to our own form of civilisation'*. It was a reaction to what has been likened to a passing meteorite which changed the environment in science education in a way that even the third and very dangerous phase of educational change was unable to reverse completely.

To understand how the second phase of STS operated we could look at Donald Campbell' evolutionary model of learning which he called Blind Variation and Selective Retention (BVSR). It was based on Darwin's theory of Natural Selection. In biology one could speak of a kind of *'learning'* on the part of the organism which takes place by means of random variation through genetic mutation, followed by the brutal process of selective survival of some, or none, of the mutant individuals. The learning about, for example, what structure of limbs favours a quick turn of speed, or how to glide from one branch of a tree to another by means of a membrane of skin stretched from wrist to torso, is produced by the variations in the animals structure. Selection, which is like evaluation of learning, takes place in the physical environment which in this case was the classroom. This is similar to the success or failure of engineering artefacts, like the shape of aeroplanes' wings which are tried out in the environment of wind tunnels and rejected if they fail (Vincenti, 1990), or to what might be called *'the struggle of conjectures'* to solve problems within an open learning environment (Popper, 1972). Campbell wrote of BVSR that it was *"fundamental to... all genuine increases in knowledge, to all increases in fit of system to its working environment."* (1960, p. 380).

Donald Schon's famous notion of teachers operating *'reflection-in-action'* seems to be of this same sort of innovation. He worked out the theory of this continuing process in considerable detail, and just equating reflection with cognition, which is all there is time for here, certainly does not do justice to his argument. Teachers' action was creative and should be allowed freedom to move from one classroom to another. For fairly obvious reasons large-scale government-inspired educational reforms in Britain, or any other country, are not often guided by such a naive free struggle between a great number of individual teachers' different methods of teaching.

Nevertheless this is a very creative way of developing new curricula. Recently Handal and Lauvas (1987) have written eloquently about what they call teachers' '*practical theories*' which are dynamic, provide a basis for action, and yet are also subject to change as the teacher sees problems with what s/he has been attempted and tries another way of teaching, in a very similar way to Cambell's BVSR

> *'The practical theory (of teachers) refers to a person's private, integrated but ever-changing system of knowledge, experience and values which is relevant to teaching practice at any particular time.'* (page 79)

If we want forward looking physics teaching which changes when the times require it to do so, then this is a valuable way of moving forwards. Clearly each part is a small scale innovation with the students reactions providing direct feedback to the teacher, and s/he will then make a personal decision on whether to abandon the new idea, or to persevere with it and improve it in the same spirit. In spite of an almost complete lack of funds the 1980s were a very creative time in schools that, in spite lack of governmental support, the STS movement seemed to be flourishing. It was a time of energetic and inspiring small scale activity in STS education which depended on the existence of an energetic and autonomous force of science teachers.

There was just such a need for new ways of teaching if students were to see the social nature of science and technology. This was quite new for science. It would no longer be enough to teach about astronomy or radioactivity as 'wonderful' advances in scientific theory. Teachers had to show how the new ideas would affect people's thinking about the world and whether this would produce conflict with older and valued cultural ideas. They would also have to consider the risks of radioactivity and the rights of citizens to know how this might effect them, before it happened! Physics teachers needed, as we saw at the beginning of this lecture, to hear the voices and opinions of their students. How could this be done? There were several practical answers to this including Role-play (eg. The re-trial of Galileo.Solomon 1995) small group discussions, and commentaries on television programmes (the DISS project, Solomon 1992). All were very new and took time to learn both by the teachers and by the students. By 1983 the first public examinations in the subject were created and there were reports of students enjoying their lessons very much and valuing what they had learned.

In Britain, during the early 1980s, funding was raised from industry and from the Association for Science Education to back a new project, called the Secondary Science Curriculum Review, which was committed to teacher innovation and also, to some extent, to STS. That meant a network of local 'working groups' of teachers trying out new ideas of their own and of other teachers. So, in the mid-1980s, the teachers had seemed a powerful independent force for change. In retrospect it is easy to see that a really sound educational initiative which was nationwide and likely to last, would have required a whole team of different players from parents, through teachers, to politicians, which the SSCR did not have.

So this second very energetic phase of innovation in STS which tried to help our youngsters to learn how to act in a democracy (Carr and Hopkins, 1996) could not last. Two particularly heavy blows were struck at the early STS movement. The first was a hostile government report in 1978 on STS courses at tertiary level, to be followed at the end of a decade, by the arguments about raising educational standards in schools and going *'back to basics'* (which was reinforced by Prime Minister Margaret Thatcher's personal problem with understanding what the word 'society' might mean, once memorably announcing *'There is no society, only people!'*) The Thatcher government wanted to reform education and emphasise only the basic skills which they believed were being eroded, They began by breaking the power of the teachers' Trade Unions, as they had already done for the coal miners'. Then they announced that no social or economic issues were to be considered in the science classroom. The normally independent British science teachers could only wait for the next moves.

4. Third Phase: The Government Acts Enacts Social Darwinism.

The social Darwinism connected with the names of Herbert Spencer and Adolf Hitler, has been almost universally rejected as being intolerably elitist, brutal and racist. However that rejection was not because Darwinian theory should not be applied to human beings. Social Darwinism advocated direct actions that society should follow in order to reinforce the selection processes. If you find weaknesses amongst the poor or disabled or badly educated it was not enough to rely on evolutionary selection, - you had to help it to wipe out the disability. Here there was none of the respect for the creative potentiality of variation which is so essential in BVSR which, as we have seen, can get passed on through teaching to construct new ways of learning. The crude methods of social Darwinism used to reinforce selection are reminiscent of a more precarious human existence at earlier times when kindness to the weak was a very rare commodity. Now, in more civilised and affluent times it should have no place in either our general thinking or in our educational practice.

When the government imposed its own educational reforms the children themselves did not appear to be the targets of the new policy. Of course their knowledge, manners, dispositions towards learning, and mechanical ingenuity may all be outcomes of their education and would, effect what they retained of new instruction. But school children have no direct power in education. They have never yet closed down an educational system, even when it was both physically brutal to them and also élitist. College students did try to do this in continental Europe in 1968 but not very much was achieved. Indeed children still at school are almost the only group in society which usually has no voice at all in the movements for educational reform, but they do suffer the results of it, so we must always proceed with great care and caution.

Power for large-scale reform in education resides in a very range of different constituencies. This diversity makes it almost impossible to please everyone so it seemed to the reforming government that the only course was to compare one school with another, that is a selection by severe testing of outcomes, school by school. Then they could simply close down the schools which 'failed' and move children to a school which achieved higher examination results. At the same time the curriculum, and even the methods of teaching, came under strict control. There were to be 'no social or economic issues' discussed in the classroom. Although this mandate was not followed slavishly by science teachers, the Head teachers of the schools often felt that all available time should be used to teach the sort of science which would be tested. Head teachers have some remaining power to uphold the educational values which they most respect, in their schools (Douglas, 1987). They regularly speak about these as they try to define the school's social and moral purpose, implicitly advertising their schools to potential customers – by which they mean parents, not their children.

At first it was the Head teachers were most vociferous in their criticisms of the new 1989 Education Act which linked education to the operation of crude market forces. Schools were to be paid according to the number of students in their classrooms, and how well the students had performed in multiple choice examinations would be published for parents to read. They would select their child's school by consulting the list and the 'failing' schools would be closed down both by the inspectors and also by their lack of funds. There was to be no move to help those schools which were in the poorer regions of the country!

Many of us saw at once that this process would suppress any creative variation amongst schools and amongst their teachers. This meant that very little STS teaching, which was so new and innovative and whose aims were not to be found in the words of the new science curriculum, would survive. There would also be a blurring of the values on which the whole concept of STS was built. Any yet the ideal of an STS education which would support citizenship and democracy did not completely disappear.

During the 1970s and 80s Malcolm Skilbeck and some other general educational theorists had had been examining the aims of school curricula through a series of books and articles. He proposed that one of the main aims of all education was to provide the next generation with skills for critiquing their own culture.

> *A curriculum plan in a school can aim to foster critical reflective thinking; it can stimulate and provide opportunities for participation in practical projects by which the community betters itself:* ***it can encourage pupils to see themselves as the organisers of their own society.*** Skilbeck 1975 p34.

Skilbeck rarely mentioned science in his books, but he developed a concept of education as what he called *'a preparation for social reconstruction'* which clearly encompassed citizens' decisions about the new technologies just as STS did. We can see Skilbeck's discussion of education as a preparation for community action and citizenship, and so possibly for a new phase in STS education

5. Is There A Rebirth of STS Education?

Now, in our new Labour era with a frail but just surviving STS movement, the British National Curriculum in science for the year 2000 has given some cause for hope. The emphasis started moving towards considering contemporary cases of science-based 'controversies', 'evaluating evidence', and considering uncertain 'ethical issues'. The hard pressed teachers looked at this with some amazement and suspicion. Did the government really mean it?

Most recently some help was at hand and it came not from science educators but from a professor of political and social science. In November of 1997 the then Minister of Education set up a group to report on "Education for Citizenship and the Teaching of Democracy in Schools". Their published aim was as follows:

> *We aim no at no less than a challenge in the political culture of this country both nationally and local: for people to think of themselves as active citizens, willing, able and equipped to have an influence in public life and with the critical capacity to weigh evidence before speaking and acting.*

They left the details of the curriculum vague but were quite sure, as STS teachers had always been (Solomon, 1988), that

- the students should think about moral issues, and voice a personal opinion,
- that they should contribute to discussions and debates,
- they should use their powers of empathy to consider the problems of others,
- they should role-play the situations and opinions of others, and
- should take part in school and community activities, and reflect upon them.

Since the most salient and worrying situations of modern societies are precisely those of science and technology Professor Crick spoke to science teachers at their annual meeting in January and the results he got there were very encouraging. The science teachers knew that there was a lot to think about, but that if they did not teach about the controversial issues in science lessons, it would make for very dull teaching and low motivation for the students to learn. So what is wanted now is that STS teachers begin once again to show how role-play and group discussions about scientific problems can best be tackled and to develop new ways of dealing with new problems. As Professor Crick wrote, *"I am not advocating anything new, just a timely reinforcement.*" So it may that STS is back on track again.

6. Conclusion

At the beginning of this paper I promised to look at STS education as a conflict between educational utopianism on the one hand, and a *realpolitik* which includes industry and economics, on the other. I hope it is easy to see the high ideals for a new sort of utopianism permeated this movement. At first the scientists began, as it were, looking at their own responsibilities. The Second World War had been called the 'Physicists War' because it had, in the last stages, been won by the use of atomic power which few non-physicists understood.

This made scientists wonder whether their advanced knowledge gave them a special responsibility towards the rest of the population. It was as if they might have to be stewards of the world. But the population of lay people were not content with that and began to demonstrate for peace and against nuclear weapons even though they had little understanding of either.

Once the idea entered the schools it was no longer a question of physics students being stewards, the science teachers wanted to educate all the children. They saw at once that in a democracy the citizens should be involved in decision-making and that meant all of them should be taught the appropriate science. So now there was another set of high ideals which permeated STS courses.

However it takes a truly democratic government to allow their students to be given knowledge with which they might decide to oppose the government (not to rebel, but either to demonstrate peacefully or to vote against it). The Thatcher government was certainly not of that sort. At the same time industrial forces also began to worry about STS. Would it activate the students against their operations – against industrial pollution and against transnational companies which were upsetting the development of the poorer countries? The combination of Government and Industry was too hard for the frail educational forces behind STS and it very nearly died.

Now that STS is starting again it seems that a few lessons have been learnt, but we cannot be sure. However by putting Citizenship alongside Science in the new STS programmes, it may that a balance has been struck. We must work hard in our classrooms and hope for the best.

References

Carr, W. and Hopkins, A. (1996) Education and the Struggle for Democracy. Buckingham. Open University Press.

Campbell, D. (1960) Blind variation and selective retention in creative thought as in other knowledge processes. Psychological Review Vol 67(6) 380-400.

Crick, B (2001) Citizenship and Science, Science and Citizenship School Science Review. (In press)

Douglas, M. (1987) How Institutions Think. London. Routledge Kegan Paul.

Skilbeck, M. (1975) The School and Cultural Development. In (ed. M. Golby, J. Greenwald, and R. West.) Curriculum Design. Milton Keynes. Open University Press p7-19

Solomon, J. (188) Science Technology and Society Education. Tools for thinking about Social issues International journal of Science Education 10(4) 379-387

Solomon, J. (1992) The classroom discussion of science-based social issues presented on television: knowledge, attitudes and values. International Journal of Science Education. 14(4) 431-444.

Solomon, J. (1993) Teaching Science Technology and Society. Buckingham. Open University Press.

Ziman, J. (1980) Teaching and Learning about Science and Society. Cambridge. Cambridge University Press.

PHYSICS EDUCATION PAST, PRESENT AND FUTURE: AN INTERPRETATION THROUGH CULTURAL CONTEXTS

RICHARD GUNSTONE
Faculty of Education, Monash University

In considering "Physics Education into the 21st Century" I take the obvious strategy – I will begin by looking to the past to attempt to understand changes from this past to the present. Then I attempt to project to the future. In doing this I am limited by my cultural and language narrowness; I am only able to seriously consider developments in English language physics curriculum and teaching.
The first major and influential curriculum development in physics, PSSC in the late 1950s, was embedded in a cultural context (a very specific and limited cultural context). This narrow cultural context involved a very strong focus on the nature of physics, and derived from views in USA in the late 1950s about ways in which one could attract more high achieving students to the study of physics. Today, views of physics curriculum have not abandoned the central views underpinning PSSC. Rather, both the nature of physics and the ways in which one might attract students to the study of physics are now seen more broadly. The context of the classroom has become significant through greater understandings of the detail of student learning in physics and through concern with the "content context" in which physics is taught as a curriculum issue. The cultural context that is today considered has broadened, but is still narrow. In the future we need to expand much more our views of the extent to which cultural contexts should be considered in physics curriculum, and embrace a much wider range of socio cultural contexts in the curriculum. This will impact on both aspects of the content of physics programs and the ways in which this content is presented and assessed.

1. Introduction

1.1 *An Initial Comment on A Broader Issue of Cultural Context, And The Limitations*

I come from a substantially (but not exclusively) European cultural context. I engage with Asian cultures but cannot be part of Asian cultures. In this engagement, in common with many from European/western contexts, I believe I observe in Asia a broad cultural/moral issue that appears to pervade education in ways not often known in western cultures. There is in Asian cultures a greater valuing of education for its own sake, a stronger view that education is of worth because it is education. While this is sometimes implicit it appears to me to be very all-pervasive and influential. It is in stark contrast with at least some western contexts (including my own country) where much more instrumental views of education have become very common in the last few decades.

Of course this contrast in broad underlying positions about the value of education and how this should be considered is not in any way specific to physics. But its significance for a consideration of school physics education is that, by comparison with western cultural contexts, concern for justifying the learning of physics and making it more likely that students will chose to study physics has had less direct curriculum prominence in Asian contexts. As I attempt to show below, concern with recruiting students to the study of physics has been the main driver of trends in physics curriculum development in western contexts since the 1950s.

So, while my monolingualism results in my not being able to understand the breadth of cultural contexts and influences that I would like to in order to be comprehensive in my attempt to forecast "Physics Education into the 21st Century", it is not only my monolingualism that causes this. There is also the broad cultural/moral issue in Asian countries of the greater valuing of education for its own sake that I cannot understand well. Hence, there are two reasons why the following arguments are derived from western contexts only.

1.2 *The Structure of This Paper*

In order to consider "Physics Education into the 21st Century" I first consider physics education in the past and present. My 'past consideration' is of the program generated in USA in the mid to late 1950s by the Physical Science Study Committee (PSSC Physics) (PSSC, 1960; see also later additions), and my 'present consideration' focuses on the ways concerns with context and student learning have shaped school physics. In these considerations, and in my conjecture about the future, I will mostly be concerned with the intended curriculum (ie the curriculum documents, textbooks, etc on which teaching is based). Where appropriate I will also consider the implemented curriculum (ie the ways the curriculum documents etc are implemented in the classroom) and the attained curriculum (ie the consequences in terms of student learning of the curriculum). However my prime focus is on the nature of the intended curriculum – what forms and structures are in the intended physics curricula we teach, and how are these influenced by cultural context.

Before beginning these three considerations I briefly discuss how I am using the notion of "cultural contexts" in these arguments.

2. Cultural Contexts

One common interpretation of "cultural contexts in physics education" is the extent and ways in which cultural contexts and other cultural issues are included in the physics curriculum. If that was the interpretation I was using here, then there would be no point in mentioning PSSC. As Fensham (1992) has observed, "It has been said that the curriculum movement of the 1950s/1960s took place as if science and science classrooms were in a historical and social vacuum" (p 793). The inclusion of cultural, or other contexts, was not something that PSSC Physics attempted.

While the statement by Fensham just quoted is certainly a very reasonable one, and is supported by all the contemporary writing about PSSC at the time of its development and implementation (see examples in the next section), a different view of cultural contexts is central to understanding the nature of PSSC and other developments at that time.

My focus here will be on the broader cultural contexts that have played central roles in shaping the physics curriculum. That is, I will consider what aspects of the broader culture have been influential in, responsible for the general significant features of physics curricula in the 1950s/1960s and 1990s. In each case concerns with attracting students into high school physics are central features of this cultural context. These concerns are also, amongst other things, helpful for considering the question "Why does a society value physics being taught?"

3. High School Physics in The 1950s-1960s: PSSC

3.1 *Why Consider PSSC?*

The choice of PSSC, and the era in which it was developed, as my 'past consideration' of the ways cultural contexts have shaped the physics we teach is a very straight forward one. Of all the many substantial curriculum projects/reforms in physics, PSSC was at least one of the most influential. Given the times in which it emerged it seems a very reasonable assertion that it had more influence on physics teaching world wide than any other single program. The original PSSC program was translated into at least 15 languages other than English (Haber-Schaim, 1967), was adopted in at least 35 countries other than USA, had substantial influence on other programs developed around the same time (Schoene, 1965), and has left a lasting legacy. It is PSSC that was responsible for the development and/or popularising of ticker timers, trolleys, slinkies, stroboscopic photographs, etc. At more structural levels, PSSC transformed the ways in which laboratory work was conceived in high school physics, at least in western schools, by structuring that lab work so it was integrated with the remainder of the program, and then by developing equipment cheap enough to allow this to happen. PSSC, relative to the programs in existence at the time, was a remarkable shift from an encyclopaedic collection of essentially unrelated topics to a cohesive program in which all the content covered had links to the eventual central point of the program (wave-particle duality) (Finlay, 1966). PSSC had structure and cohesion that existing programs did not.

3.2 *The Origins of PSSC*

Contrary to popular belief, PSSC was not developed in response to the launch of Sputnik. PSSC had already begun when this launch captured the attention of the world. The consequence of the launch was to dramatically lift the profile and funding of PSSC (Gatewood, 1968; Lee Dow, 1971). The project was initiated by research physicists who were deeply concerned about the relatively small numbers of high achieving US high school students who were choosing to study physics at high school. These research scientists were intent on developing a program that would attract more of these high achievers to physics, at high school and then college/university (Zaacharias, 1960). Although the expansion of the project post-Sputnik led to changes in the project, research scientists remained by far the most influential group in the development of PSSC. (Note that at this time there were no academics and developers of the type we today call "physics educators" or "science educators", and the terms "physics educator" and "science educator" were unknown. There were no individuals with the research and development expertise and experience that today we assume in physics/science educators.)

3.3 *The Nature And Purposes of PSSC*

The nature and purposes of PSSC were focussed by a broad concern to 'induct into physics/science' those students who studied it. In this it was quite consistent with developments across many countries. The general view of the physics community at this time about the nature and purposes of teaching high school physics is well summarised by statements from an international conference on Physics in General Education held in Rio de Janeiro in 1963. The report of that conference (Brown, Clarke & Tiomno, 1964) includes in a "Summary of the Conference" the following about the inadequacies of physics programs of the time, other than PSSC. "They fail to present in any convincing way the magnitude and grandeur of the intellectual achievements represented by modern physical theory, they fail to convey understanding of the more important concepts of physics, and they fail to show to young people anything of the approach which the professional scientist necessarily brings to his (sic) work." (p 1) PSSC attempted to do just these three broad things: magnitude and grandeur of intellectual achievements, understanding concepts, the work of scientists. Physics was presented as a 'pure' enterprise, not embedded in a social or cultural context. (It is illuminating to look at all the references that appear in the index of PSSC to a topic such as electricity/electric current/etc – mentions of use of electricity are rare and tangential. This pattern was repeated in other contemporary senior high school science projects, eg the index of Chem Study contains about 15 references to carbon - not one of these indicates that carbon has any use!) And in doing this of course, PSSC, and other programs such as Chem Study, portrayed strong values about science and its place in the world by claiming to display no values.

Fensham (1992), in discussing PSSC and similar contemporary senior high school science curriculum projects, notes that "it is now clear that academic scientists exerted a crucial influence on the sort of science content that was included as worthy of learning. They were important legitimators of *what topics* and *what sort of account of them* were appropriate for school science. Their exercise of this role ....[was] more constraining in the physical sciences...." (p 791; emphasis added). Gatewood (1968) argues the consequences of this, particularly the impact of scientists on the forms of approach to topics, in strong terms. In considering the consequences of large sums of money invested in projects staffed largely by research scientists he concludes that this "has resulted in the regeneration of a modern up-to-date version of the elite science curriculum that existed during the latter part of the nineteenth century. Further, much of the excellent scientific material thus created is essentially irrelevant to the needs of a major portion of students..." (p 19).

3.4 *The Cultural Contexts Influencing PSSC*

There are a number of interrelated cultural contexts that impacted in fundamentally determining ways on PSSC.

• The views of a number of research scientists that the United States was in dire need of a larger number of high achieving high school students choosing to study physics was the catalyst for the establishment of the project. (At least some of the countries that later adopted or modified PSSC had similar concerns.)

• The subsequent launch of Sputnik caused this concern to be taken up by others with a focus on the economic and defence futures of USA, and for the amount of money available to the project to be dramatically increased. (At least some of the countries that later adopted or modified PSSC had similar concerns about futures.) Through this change in funding research scientists remained firmly in control of the ways PSSC developed.

• A crucial and shaping assumption by the scientists was that they 'knew best' for schools. It was for this reason that PSSC, and contemporary projects such as Chem Study, focussed so strongly on what was universally termed "the structure of science" (Roberts, 1982). In this PSSC represents a clear case of the important distinction between "logical meaning" and "psychological meaning" advanced by the unfortunately ignored educational psychologist David Ausubel (see, for example, Ausubel, 1968).

• The decisions about *what topics* and *what sort of account to give of them* were made, deliberately, in ways that led to a curriculum almost devoid of considerations of applications of the physics and other forms of use of the knowledge. This was, as already noted, "pure" physics focussed on induction students into physics. (This is the most obvious reason for many physics teachers, myself included, having found great professional enjoyment in teaching PSSC.)

3.5 *A Short Comment on European Developments in The 1960s*

It is sometimes asserted that PSSC had little impact in Europe, and that the Nuffield curriculum projects (including Physics and A level Physics) were of a very different form. These are not reasonable positions, at least in absolute terms. There was substantial influence of PSSC in some European countries (see for example Nordfors, 1965), and less influence in others. Nuffield did have clear differences from PSSC, but in many ways, including the fundamental nature of the curriculum, these were differences of degree rather than fundamental (Hobbs, 1974). For example, while Nuffield Physics was in the hands of expert school physics teachers rather than research scientists, this did not mean fundamental differences from the features described above for PSSC. Fensham (1982) has described these teachers as, in terms of research scientists "their faithful imitators, elite science teachers." (p 53)

4. High School Physics in The 1990s: Context-Based Approaches And Student Learning

My 'past consideration', the era of PSSC, was an era of commonness of purposes, as I have implied above. Any consideration of the present will not find anything near the same commonness of purposes or of curriculum thinking in high school physics. The broader cultural context is now very different that in the 1950s/1960s, and much more variable from one location to another. For example, physics departments in many western universities are under threat from declining student enrolments and declining research funds. Such an occurrence was totally inconceivable 40 years ago.

Hence my views of the central trends in physics curricula in the 1990s will be more open to question. This is necessarily a more idiosyncratic view, and I recognise that.

I see the central trends of the 1990s as involving acceptance of the significance of two forms of context: the context in which the physics content is embedded ("context-based approaches") as an issue for considering *what topics* to teach and *what sort of account to give of them*; the context of the classroom and the ways greater understanding of student learning of physics can shape *what sort of account to give of topics*.

4.1 *Why Consider Context-Based Approaches And Student Learning?*

While the influence of each of these two trends on physics curriculum is substantially less than that of PSSC in an earlier era, each trend has been building since the 1970s, and each has led to significant (if variable) impact in many countries. To some extent the two trends are independent of each other (although as I argue later there are important links), so I initially consider each separately.

4.2 *Context-Based Approaches - The Context in Which Physics Content Is Embedded*

The ways in which context has become a significant influence on at least some physics curricula is in some senses a complex matter. 'Context-based' curricula are profoundly different from teaching the applications of physics knowledge. I begin by indicating the meaning intended for 'context' in the term 'context-based physics curriculum', then elaborate the difference between context-based curricula and teaching applications of physics knowledge, and then consider two different examples of a context-based approach. After these elaborations I briefly indicate the nature and purposes of context-based approaches.

The meaning of 'context' in context-based approaches: Consider the two extracts below from two different high school physics curricula, curricula widely different in time of production (1960, 1989), location of production (USA, Victoria), and perception of the nature of the curriculum (the first is PSSC - physics to attract the highest achieving students, the second aimed to have physics seen as attractive by all students). Each is taken from the attempts in the curriculum documents to describe to students what the thing called "physics" is.

> Physics is the fundamental science of the natural world. It tells us what we know about the world, how men and women found out what we know, and how they are finding out more today.
>
> When a flash of lightening shatters the dark, the radio crackles and your eyes are dazzled for a few seconds. A moment later you hear the roll of thunder, and a loose windowpane rattles. Three states away in a storm-location center, a radiolocator can pinpoint the lightening strike; and downtown the weather forecaster, hearing a faraway rumble, nods as if he had been expecting a storm.
>
> Here is a chain of events, different events, taking place at different places and different times. They are all linked. How are they tied together, and just what is happening to eye, to ear, to the radio, and in the air itself? ....
>
> Physics enables us to answer such questions....(PSSC, 1960, p 1)
>
> It has always been part of the human condition to marvel at the world we live in - stars and rainbows, the apple that falls to the ground or the lodestone that always points north - and to ask why the world should be that way. In western culture, this way of speculating about the physical world became known as natural philosophy and later, as biology and chemistry took recognisably different paths, physics. At the same time as this separation into distinct sciences was occurring, physics developed its own particular methods and procedures, valuing precise measurement and

highly reproducible experiments, and developing a powerful and fruitful partnership with mathematics.

It is also part of the human condition to use knowledge to gain control. Knowledge of physics has led to developments in technology, some of which (for example, radio communication and electrical appliance) have had a profound impact on social structures. The social effects of such technologies may be positive or negative and, as has been the case in nuclear sciences, the use to which the knowledge is put may itself direct the course which physics takes....

(Victorian Board of Studies, 1994, p 7)

These two extracts seek to make the most obvious of points: physics involves the study of and search for explanations about natural phenomena. However there is a vast difference between the two in the extent to which this link between physics and natural phenomena is clear to students. In simple terms, the link has little real place in PSSC. While the physicists who developed the program and the physics teachers who taught it understood and applauded the meaning of the extract, students had no experiences in the program which reflected the meaning of the extract In the second example, a context-based approach, the link is the fundamental rationale for the curriculum.

The second paragraph in the second of the extracts argues that there is more to the context of physics than the natural world from which it has emerged. This is also shown by dictionary definitions of "Context", for example "**context** ... the parts of a discourse or treatise which precede and follow a special passage and may fix its true meaning; associated surroundings, setting." (Chambers Dictionary, 1977, p 280)

One of the most popular first year university physics texts of the last 15 years (Giancoli, 1980) makes these points clearly, again in a message intended for students using the book.

Unfortunately, many people think science is a mechanical process of collecting facts and deriving theories. This, however, is not the case. Science is a creative activity that ranks with the other creative achievements of the human mind.

....Even something as seemingly simple as "observing the pertinent facts" requires imagination, for scientists can never include everything in a description of what they observe....

The act of creation is most obvious in the putting forth of scientific hypotheses or theories. Theories are never derived from observations -

> they are *created* to explain observations. They are inspirations that come from the minds of human beings....
>
> Science is a truly creative human activity. In this book, we will explore the subject matter of physics. We will apply the principles of physics to other fields such as biology and architecture. And we shall spend a little time, at least, pointing out the humanity and creativity of science. (pp 1-3)

Here are other aspects of context that are part of physics - the nature of the epistemology that is fundamental to physics, the centrality of humans and creativity in the discipline, and so on. Context can be seen in very broad ways.

Contrasting context and applications: A very common feature of physics courses is the use of applications of concepts in the teaching of those concepts. At first glance this might seem to be an example of the inclusion of the context dimension of physics in the curriculum. However this is not the case. Seriously considering context in determining a physics curriculum is not the same as including applications of physics concepts in the curriculum.

The inclusion of the context dimension of physics in the curriculum involves seeing the context as central to the learning of relevant concepts. "Applications", on the other hand, is a term that describes the use of illustrations of concepts in the curriculum after the relevant concepts have been learned (or at least taught) - and the selection of applications for this purpose usually involves the use of those that are unproblematic illustrations (Hart, 1995, 1998). As an example of this contrast, consider the difference between teaching "Force and Motion" via a context such as "Everyday Motions" and using applications to illustrate this topic. To use the "Everyday Motions" context would imply beginning with a relevant example of reality and then using aspects of this reality to provide a launching pad for the teaching of the concepts of the curriculum. This clearly precludes the use of such non-realities as frictionless surfaces and point masses (although it may well be appropriate for a curriculum which embraces contexts to eventually introduce such things as part of fostering an understanding of the ways of knowing that characterise physics). To use applications however will mean that such non-realities are appropriate, even likely. The contrast is between having examples of the natural world that are central to the learning of the concepts of the curriculum and having examples to consolidate already learned material.

There are different extents to which a context can be emphasised. Possibilities here range from, on the one hand, having the understanding of physics concepts as the central purpose of the curriculum and seeking to use contexts as the vehicle for this learning to, on the other hand, having the context so central that the essential purpose of the curriculum is to learn about the context via relevant physics. These two possibilities represent very different approaches. The latter, a central focus on the context, even influences the physics content that would be selected. For example, in the case of "Everyday Motions", having the development of an understanding of this context as the central purpose means that the concept of friction would have a more prominent role than in any approach that uses contexts as a vehicle for the central purpose of developing an understanding of introductory Newtonian mechanics.

PLON - an example of the use of contexts in a physics curriculum: The Dutch project PLON (a Dutch acronym for Physics Curriculum Development Project) began in 1972 with concern for junior high school physics topics in science programs at this level. In its work over the last 25 years it has made extensive use of context in its production of curriculum and materials for a range of high school levels (Years 8 - 12) and purposes (including physics for senior students bound for science courses at university). It is chosen here to represent context-based physics curricula because it has been the leader in this focus.

The adoption of context-based approaches by PLON as a major feature of their work is also closely linked with other emphases. These include relating materials to the local environment of the students in the early PLON work, and, more recently, a strong Science (Physics)-Technology-Society focus. It is interesting in this discussion of context to note that the origins of PLON were with Dutch enthusiasm for Harvard Project Physics (eg Holton, 1967). Project Physics itself can be usefully seen in terms of context, particularly from the perspectives of student learning (see below). An introduction to the approaches used in PLON is given by Eijkelhof and Kortland (1988), who include Figure 1 (below) to show the general form of a PLON unit.

I illustrate this structure by outlining the unit for senior high school "Ionizing Radiation". (An English translation of this unit has been produced by the Education Faculty at Monash.) The beginning point of the unit is a *basic question*, of significance to the society in which the students live and centrally based in physics. For this unit the *basic question* is "How acceptable are applications of ionizing radiation to you?" Then the *orientation* phase introduces a number of everyday situations where the use of ionizing radiation may be an Figure 1. General form of a PLON unit (adapted from Eijkelhof & Kortland, 1988, p 288) issue, and considers the nature of "risk" in terms of a complex mix of probability and effects. *Basic skills and knowledge* comprises the nature, effects and sources of X-rays and radioactivity, and concepts such as half-life and activity. Then three *options* are available to students for further study of information and issues of risk and safety - nuclear energy, nuclear arms, uses of radiation for medical purposes. Students have the task of *reporting* to the rest of the class about the option they have undertaken. Throughout each of these phases the *basic question* acts as an organiser for the progress through the unit and provides a focus for the physics that is the fundamental purpose of the unit (including giving some basis for the selection of physics content for the unit). In the final phase of the unit, *broadening and deepening*, the *basic question* is a central feature again, with the physics concepts and skills previously introduced being broadened and deepened by their application to situations in which the basic question is prominent. In the case of this unit this phase presents procedures for analysing and evaluating risk in cases such as being prescribed a brain scan or the dumping of radioactive wastes in oceans

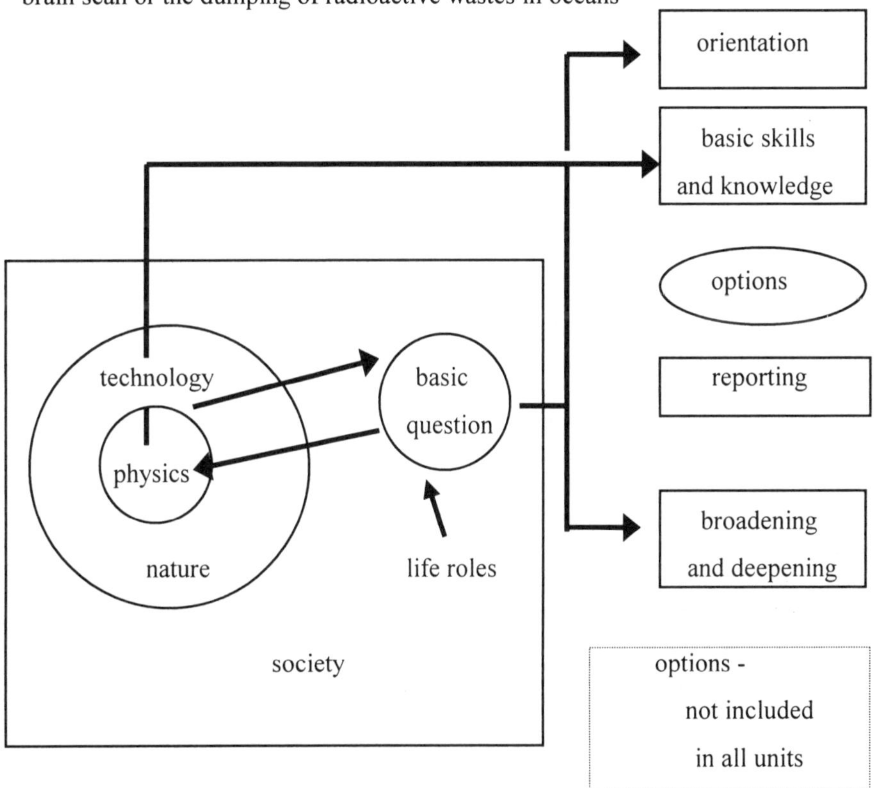

Researchers linked with PLON have investigated the nature of contexts that have greater appeal for high school students. Contexts related to the current interests of students were found to be of greater interest than contexts that were focussed on the nature of their intended future study (eg for students committed to the study of engineering at university, engineering types of contexts did *not* have greater appeal than contexts based on more immediate interests). In essence the students saw contexts that led into tertiary focuses as being able to be left until tertiary study began.

Extended projects as context-based approach: There are today a number of examples of senior school physics curricula that include an extended investigation task for students (eg A level in England, Year 12 in Victoria). This generally involves students in the planning, execution and formal reporting of an investigation of sufficient complexity as to require an extended period of time. Examples of investigations that have undertaken by students in this mode include Load and speed variations of a parachute, Compressibility of materials, Strength of human hair, Biophysics of long jumping, Performance of a pin hole camera, Natural frequencies of vibrating rods, The running down and recovery of a dry cell. The purposes of such extended experimental investigations are seen to involve learning about the nature and conduct of physics, and the development of experimental skills relevant to physics (as well as knowledge of the physics involved in the particular investigation, and affective issues).

Such investigations represent good examples of the use of context in the sense already described. However the purposes of learning about the nature and conduct of physics should also be seen in terms of context. In this sense, the context of "doing physics" is a central rationale for this approach to learning about physics. The context becomes a form of immersion in the processes of doing physics, with this immersion having the same potential advantages as immersion programs in the learning of second languages - the language to be learned (in this case the specialist language and methodologies of experimental physics) is learned through purposeful use and thus, it is argued, learned with greater speed and understanding. This is a much more valid representation of "doing physics" than is achieved by approaches such as PSSC where the same intention was approached though laboratory work whose nature and purpose was completely determined by the prescribed curriculum. In extended investigation tasks for senior high school physics students the nature and purpose of investigation is shaped by the phenomenon under investigation, and that is chosen by the student.

4.3 *Nature And Purposes of Context-Based Approaches – A Summary*

There are a number of positions that characterise context-based approaches. Aspects of the nature of these approaches has been elaborated above, and a summary is presented here. First I consider the purposes of these approaches.

• The fundamental purposes of context-based approaches have driven by concerns with declining numbers of physics students (particularly in some countries very small numbers of girls studying physics), and demonstrable lack of motivation for students to engage with physics. This has been seen in terms of a clear lack of relevance in existing curriculum.

• Hence context-based approaches have focussed on developing relationships between taught physics and students' everyday life experiences, on physics that relates to living in a late 20th century highly technological society, on physics-technology-society relationships as central to the study of physics rather than peripheral or absent. (A number of context-based approaches have in addition explicitly seen the relating of physics with living in a democratic society as a central shaping issue.) (See, for example, Eijkelhof & Kortland, 1982, 1988; Victorian Board of Studies, 1994; Wierstra, 1984).

I now summarise the nature of context-based approaches, as described by examples of these approaches.

• The physics that is addressed should be useful in everyday life in terms of social and personal decisions, and, at the same time, be that which is essential for those going on to the tertiary study of physics. In doing this an authentic view of physics should be presented. The term "authentic" needs careful consideration here. Consider again the extract from PSSC above that describes to students what physics is. As I have already argued, while the statement is laudable (particularly when seen in the broader context of the time at which it was written) there is a fundamental problem – the curriculum and materials that were produced for PSSC did nothing to represent the statement for students. An "authentic" view of physics, in the sense "authentic" is being used here, would have had PSSC actually focussing on the matters it raises in its introduction. The second extract above of a curriculum statement about what physics is (from Victoria) is from a curriculum that attempts to represent the statement in its materials and approaches. It is therefore, in the terms being used here, a much more authentic view of physics than PSSC.

• An authentic view of physics should also be presented by paying attention to the history, nature, and methods of physics. ("Methods" is a particularly important aspect here – see the discussion above of extended research projects for physics students.)

• Differences in student interests, abilities and plans should be recognised. (This is a necessary consequence of the basic concerns with developing curriculum of greater perceived student relevance, with a "physics for all" dimension.)

• A variety of activities that attract students should be involved. (Again a necessary consequence – see immediately preceding dot point.)

Both *what topics* to teach and *what sort of account to give of them* are, by comparison with previous approaches, substantially reshaped by a context-based approach. This is clearly illustrated by the ionising radiation example from PLON, given above. New topics (eg 'risk') are introduced to the physics curriculum, and traditional topics (eg the three forms of radiation) are approached in very different ways.

4.4 *Greater Understanding of Student Learning*

The most obvious feature of physics education research in the last 25 years is the enormous growth in studies of students' conceptions, in almost every content area of physics. There seems no need to describe this in detail here, so my comments are very brief. What is significant from this research for the present argument is the ways it has influenced physics curriculum, the ways it has led to much great understanding of the important distinction between "logical meaning" and "psychological meaning" (Ausubel, 1968).

Perhaps the most influential consequence of this area of research has been that it has made very clear that the content to be learned is a major variable in understanding that learning. This is what makes the research on student conceptions and its consequences for teaching physics totally different to and much more useful than the positions of "discovery learning" (Gunstone, 2000). This has led to data based new approaches to teaching and sequencing topics, and even to some reconceptualisation of content across a vast range of topics and levels of education (eg mechanics, electricity, optics, thermodynamics, quantum theory, etc). (For an example in introductory mechanics of such new approaches derived solely from this research see Gunstone & Mitchell, 1997.) This research and closely related development has also led to a range of new teaching and assessment approaches (eg concept maps, predict-observe-explain, relational (or Venn) diagrams, asking questions rather than answering questions, strategies for focussing discussions, and so on).

It is not possible to point to an example of a curriculum that has been totally influenced by this trend. Rather, there are many examples of curricula for which this trend has been one major influence. Many such examples are also curricula that have taken a contexts-based approach, for reasons briefly discussed below.

4.5 *Nature And Purposes of Approaches That Recognise What Is Known about Student Learning*

The nature and purposes of these approaches are completely consistent with the motivations of the researchers who have led the research. These researchers began and developed this research because they were quite dissatisfied with the quality of student learning of physics (in both their research studies and their previous experiences as physics teachers) (Gunstone, 2000). The central issue in these approaches is to have greater student *engagement with* physics. The central issue is then one of attracting students to physics, and having them learn with understanding.

4.6 *Important Links Between These Two Trends*

There are some significant links between the approaches of context-based and student learning/classroom context. It is clear that the nature of student physics learning is affected not only by the content to be learned, but also by the nature of the context in which the physics is presented. For example, in the extreme case of physics being presented as an essentially decontextualised entity, then students confuse the empirical aspects of physics knowledge (eg heavier objects fall faster) with the more powerful non-empirical generalisations of physics (for this example, all falling objects have the same acceleration in a uniform gravitational field).

More generally, the extensive research on student conceptual understanding has the consistent message that knowledge learned separate from any context is knowledge that students often cannot (or, sometimes, choose not to) apply to interpret contexts. It is also clear that being able to use knowledge to understand the context in which it has been learned does not mean that this knowledge can or will be used to interpret and understand a new related context. A major component of any physics curriculum that has concerns with student understanding of significant concepts must be consideration of a number of contexts in which these concepts are relevant. One issue central to this is also central to context-based approaches. It is that explaining the 'real world' is much more demanding of understanding than is explaining an artificial context (eg no friction) that exists for the purposes of learning physics concepts.

4.7 *A Brief Comment on The Influence of Research Physicists*

Relative to the development of PSSC, research physicists have had very little influence on the shaping of curriculum that has reflected context-based and student learning approaches. However they have had in some countries very powerful influence on the acceptability (or lack of this) of these approaches. For example, the context-based physics curriculum in Victoria referred to above has been considerably reshaped back towards much more conventional approaches over a number of years. (This includes major reductions in the form and importance of the extended research project.) It is solely research physicists in positions of power on committees responsible for decisions about the physics curriculum who have done this. Sadly, this conservative influence from physicists is not uncommon.

4.8 *The Cultural Contexts Influencing The Trends of Context-Based And Student Learning*

As for the earlier discussion of the 'past consideration' of PSSC, there are a number of interrelated cultural contexts operating in determining ways on these two trends.

• Broad concerns with numbers of students studying physics continues to be highly significant, but in much broader ways than was the case for PSSC. Today this is focussed on students in general (not just high achievers), with, in many countries, part of the concern being to increase the participation of girls in physics. Further, the broader western societies in which physics education is located today, by comparison with the era of PSSC, place much greater demands on education to engage students. Engagement of students with physics has been very important in both of these trends, and the need comes from broader shifts in the ways students and their parents see education.

• The central aspect of these approaches to attracting more students, greater engagement with physics, has been tackled through considerations of relevance. In a very limited way, relevance was also central to PSSC, in that PSSC produced a curriculum that was seen as relevant to very bright students who were planning to study physics at university. Relevance today is much broader – relating physics with the everyday lives of students and their futures as citizens in a society.

• One consequence of this focus on much broader notions of relevance has been, in these two trends, a much more explicit and realistic approach to developing an authentic view of physics among students.

• Greater understanding of learning has not been confined to physics (or science). Because this has had broader impacts, it has influenced the broader cultural context in which physics curriculum is formed. In essence, this broader shift has made it less difficult to have understanding of learning influence physics teaching.

• The influence of future study (which can validly be described as the influence of broader science culture) is also significant. In the era of PSSC, high school physics students who studied further were studying physics, chemistry or engineering. These fields had clear views about what was needed from high school by way of content preparation. Today this is much more complex, and is very different. While engineering and the physical sciences remain, biological areas in university education (particularly in the recent past biochemistry) have grown enormously in popularity and importance. Issues of pre-requisite knowledge are today much more complex, and much less certain. Physics is no longer the "automatic" path into tertiary science study that it once was.

• Decisions about *what topics* and *what sort of account to give of them* have been affected by most of the above, but particularly relevance and understanding of learning. The immediately preceding point (influence of future study) has had little or no impact on_*topics* or *sort of account*. This is most unfortunate I believe, as is argued in the final section of the paper.

5. High School Physics in The Future?

The beginning point of my crystal ball gazing is a simple summary of the things argued above about the forms of cultural context impacting on high school physics in the 1950s/1960s and today.

• At both times attraction of students to physics has been a very strong influence. However while this was seen in very narrow terms (a small subset of students only) in 1950s/1960s, it is today seen very broadly.

• While a form of relevance (relevance for studying more physics) drove PSSC, relevance today is seen very much more broadly and to be primarily based in the ways students see physics and their lives outside school.

• There is much stronger concern today with having students genuinely engage with physics.

• There is much greater concern today with explicit experiences for students that represent authentic physics (physics as it is practised and valued).

• Research physicists have much less influence on the creation of curriculum today (although in some places they have strong influence of a preventative form).

• The influence of further science study on participation in high school physics is much less today, because school physics is no longer an automatically accepted path into future science study.

In considering these points as a basis for suggesting the future, there is one obvious point to make first. That is that, at least in western countries, there remains great concern with the low numbers of students studying physics. The trends described above have *not* resulted in increased engagement with school physics. It is also clear that returning to traditional curriculum and structures (an approach actually argued by a small number of academic physicists) will not result in greater student engagement. My interpretation of the future is that concern with increased student numbers will remain a central concern.

I see two broad forms of future broad cultural context influencing high school physics, and shaping the nature of the curriculum. I conclude with a brief consideration of each.

5.1 *Influence 1 – Reconsidering And Expanding The View of Relevance*

One of the important characteristics of our approaches to creating relevance in the physics curriculum to this time is that the starting point has not yet been the world of the student (current world or future world of the student.) This far, we have sought relevance by looking for contexts that will both enable us to embed physics that we see as appropriate and draw links with the everyday experiences of students. This approach to relevance has been a relative failure in that we have not attracted more students to the study of physics in ways we hoped for, nor have we engaged those who do study physics in the ways hoped for. I see one important reason for this to be that it is us (physicists, physics educators, physics teachers) who have seen the relevance, not potential students. That is, I see that approaches to this point have not really adopted relevance. A rather different approach is to consider significant features of the current/future world of the student that have some physics base to them, and to use these to shape school physics. In such an approach there is a much greater chance that the student will see relevance, a obviously crucial requirement if relevance is to be used to attract students to the study of physics.

There are clear examples of possibilities at global, national, and local levels. At the global level the most obvious is the production, use, misuse and degradation of energy, and the closely related mater of global warming. It is clear that these are matters with a strong physics dimension involved in their understanding. (Put another way, how do you think the Kyoto protocol would have specifically been shaped and implemented if all politicians understood the underpinning physics?) Rather than seek contexts/relevance that enable us to teach, say, mechanical energy, I predict we will need to take issues such as production, use, misuse and degradation of energy, and the closely related mater of global warming and see these as the beginning point for high school physics. This would require a total reconceptualisation of both *what sort of account to give of* topics and *what topics* to include in the curriculum. For example:

(a) the teaching of mechanical energy would have to be abandoned (and, I assert, that would be an excellent thing even today; the strange assertion that energy is the ability to do work, with work being the product of force and distance, represents an extraordinary dislocation from the rest of science and from the world around the student);

(b) energy would become the central theme of the curriculum, with the corollary that the centrality (and even the place) of topics such as Newtonian mechanics would be questioned.

5.2 *Influence 2 – New Forms of Preparation for And Support of Tertiary Science Study*

Around the globe governments and other significant contributors to the funding of scientific research see the fields embraced by biotechnology as the appropriate focus for research efforts. These fields are developing rapidly, and are increasingly attractive to high school students moving to tertiary study. This represents a dramatic shift in the science-related cultural context in which high school physics is located. It will impact on the forms of physics that are seen to be appropriate as precursors to further study of science (and, to a lesser extent, on the forms of physics that are seen to be appropriate for the physics dimension of a general education).

This influence can already be seen in the teaching of undergraduate chemistry in Australia. A study by a research chemist at Latrobe University in Australia (Professor John Hill), and currently in progress, is exploring the ways in which undergraduate chemistry is taught across Australia. One of the unexpected findings is that there are a number of contexts where the requirements of the fields embraced by biotechnology have impacted on the nature of chemistry being taught in programs that are service subjects for biological majors. Sometimes this changed chemistry is being taught in the chemistry department, and sometimes it has been taken over by biological scientists who saw this as necessary in order to ensure that the forms of chemistry they wanted would be taught. The most consistent example of this changed chemistry is in introductory organic chemistry (Fensham, 2001). Functional groups are not a helpful structure for the teaching of organic chemistry if this content is to be useful to biologists – functional groups are not useful in understanding the ways in which these compounds interact with human organisms. Therefore organic chemistry is not taught from this beginning point where biologists now are controlling the content. This, in real terms, means different organic chemistry to that that has always been taught in the past.

It is a small step from this shift in undergraduate chemistry teaching to changes resulting from the same context in senior high school chemistry. And it is a small step to the same contextual pressures operating on physics, at both undergraduate and school levels.

References

Ausubel, D. P. (1968). Educational psychology: A cognitive view. New York: Holt, Rinehart & Winston.

Brown, S., Clarke, N. & Tiomno, J. (Eds.) (1964). Why teach physics? Cambridge, Mass: MIT Press.

Eijkelhof, H. & Kortland, K. (1982, July). The context of physics education. Paper presented at Second Symposium on World Trends in Science Education, Nottingham, UK.

Eijkelhof, H. & Kortland, K. (1988). Broadening the aims of physics education. In P. Fensham (Ed.) Development and dilemmas in science education. London: Falmer.

Fensham, P. (1982). Heads, hearts and hands – Future alternatives for science education. Australian Science Teachers Journal, 27(1), 53-60.

Fensham, P. (1992). Science and technology. In P. Jackson (Ed.), Handbook of research on curriculum. New York: Macmillan.

Fensham, P. (2001, May). Personal communication.

Finlay, G. (1966). The Physical Science Study Committee. In W. Martin & D. Pinck (Eds.) Curriculum improvement and innovation. Cambridge, Mass: Robert Bentley.

Gatewood, C. (1968). The science curriculum viewed nationally. The Science Teacher, 35(8), 18-21.

Giancoli, D. (1980). Physics principles with applications. Englewood Cliffs, NJ: Prentice-Hall.

Gunstone, R.F. (2000). Constructivism in the classroom. In D.C. Philips (Ed.) Constructivism in education: Opinions and second opinions on controversial issues (Ninety-ninth Yearbook of the National Society for the Study of Education Part 1). Chicago: University of Chicago Press, pp 254-280.

Gunstone, R.F. & Mitchell, I.J. (1997). Metacognition and conceptual change. In J.J. Mintzes, J.H. Wandersee & J.D. Novak (Eds.) Teaching science for understanding. San Diego: Academic Press, pp 133-163.

Haber-Schaim, U. (1965). The P.S.S.C. course. Physics Today, 20(3), 26.

Hart, C. (1995). The story of a curriculum document for school physics. Unpublished Ph D thesis, Faculty of Education, Monash University.

Hart, C. (1998). Addressing participation and the quality of learning through curriculum change: Some lessons from the experience of VCE physics. Australian Educational Researcher, 25(2), 19-37.

Hobbs, E. (1974). Physics courses X-rayed – A comparative analysis of high school physics courses in terms of basic assumptions. Science Education, 58, 153-159.

Holton, G. (1967). Harvard Project Physics. Physics Today, 20(3), 31-34.

PSSC (Physical Sciences Study Committee) (1960). Physics. Boston: Heath.

Lee Dow, K. (1971). Teaching science in Australian schools. Melbourne: Melbourne University Press.

Nordfors, B. (1965). The Swedish experience with a modified P.S.S.C. course. In E. Baurmann, W. Knecht & J. Lewis (Eds.) Teaching physics today. Paris: UNESCO Publications.

Roberts, D. (1982). Developing the concept of "curriculum emphases" in science education. Science Edcuaiton, 66, 243-260.

Schoene, H. (1965). Essential for a minimum course in physics. In E. Baurmann, W. Knecht & J. Lewis (Eds.) Teaching physics today. Paris: UNESCO Publications.

Victorian Board of Studies (1994). Physics - VCE study design. Melbourne: Board of Studies.

Wierstra, R. (1984). A study on classroom environment and on cognitive and affective outcomes of the PLON curriculum. Studies in Educational Evaluation, 10, 273-282.

Zacharias, J. (1960). The work of the American Physical Science Study Committee. In S. Brown & N. Clarke (Eds.) International education in physics: Proceedings of the international conference on physics education. New York: Wiley

YET ANOTHER PARADIGM SHIFT?: FROM MINDS-ON TO HEARTS-ON

JINWOONG SONG
Seoul National University, Seoul, Korea

Since science began to be first taught in schools maybe during the 18th century, school science education has experienced many substantial changes in its goals and nature over the period. The historical changes are usually referred by some key terms, like, mechanics' institutes, object lessons, heuristics, general science, inquiry, STS, misconceptions. To characterize these changes, science educators frequently use some slogan-like analogies referring parts of human body to indicate the movement of science education during a particular period of time: for example, 'Hands-On' for inquiry movement during 1960s-70s, 'Minds-On' for constructivist movement during 1980s-90s. In this paper, I briefly summarize the overall historical development of science education, then further expand the analogies to cover the overall process, that is, Ears-On → Eyes-On → Hands-On → Minds-On. In addition, to illustrate new directions of the 21st century, I propose a new analogy, 'Hearts-On' science education, and also discuss the meanings and implications of 'Hearts-On' analogy by illustrating how this analogy can be applied to reflect various current trends in science education.

1. Introduction

In 1963, following the first conference in Paris in 1960, a Conference on Physics in General Education was held to discuss the problems of the teaching of physics as part of the general education of all children in Rio de Janeiro under the auspices of the International Union of Pure and Applied Physics(IUPAP). In his address, Professor S. C. Brown of the MIT, then a president of ICPE (International Commission on Physics Education), pointed out the lack of interactions between scientists and the general public:

> This tremendous and rapid growth of science has resulted in a phenomenal involvement of our culture with science, but there has not been a corresponding acceptance of science as part of our culture. As physicists we are all aware of this. We are constantly confronted with evidence that the world at large looks on our field as one reversed for the narrow, mathematically endowed specialist, contributing only to our technological surroundings, talking a special language of his own, however, the public makes no effect to understanding. [4]

Brown criticised the traditional way of teaching physics in schools: "I believe that reason for the public's image of science in general and physics in particular derives from the attitude of our schools. ... We have been teaching physics in the framework of laws to be learned, of formulae to be brought out and applied to solve problems, and of routine laboratory exercises aimed at arriving at predetermined answers.[4]".

At the beginning of 1980s, Fensham[8] suggested a new sort of school science, what he called "hand science", emphasizing the mastery of practical skills based on scientific knowledge. He contrasted this approach, differently named "Tom Sawyer" approach, with the previous approaches, "head science" of the 1960s - represented by *PSSC* and *Chem Study* of the USA, *Nuffield* of the UK - and "heart science" of the 1970s - represented by *Nuffield Secondary Science* of the UK and *A.S.E.P.* of Australia. On contrast to his new "Tom Sawyer" approach, he called these approaches in turn "Wonderland" and "Disneyland" approaches.

> "Head science" with its emphasis on big ideas and the structure of knowledge; "heart science" with its more random, pleasure oriented approach without real rigour or structure have both failed in their turn to have significant impact on the majority of students. The intentions have been good and noble - so what can we do in the 1980's to achieve the scientific literacy we desire? [8]

In his analogy, Fensham categorized the post World War II time into three periods, i.e. 1960s, 1970s and then coming 1980s. He characterized the 1960s as a period of learning conceptual structures of science, requiring students' cognitive efforts. On the other hand, he saw the 1970s as a period of delivering students the pleasure of learning science, thus more emphasizing emotional aspects than cognitive ones. From the above analogy, it seems that he considered that learning science needs the both sides of learning science, cognitive as well as affective, and that the best way of to achieve this is through the mastery of practical skills based on scientific knowledge. For this reason, Fensham argued that the new kind for the coming 1980s need to be the learning of 'hand science'.

Apart from whether or not Fensham's characterization of the development of science education is the most appropriate to describe the historical process, it is always informative and useful to reflect the past and to try to draw a new picture for the future of science education with the historical reflection.

In this presentation, I will suggest a new framework with a different set of analogies to reflect the historical development of school science. The proposed framework consists of several analogies which in turn relate the parts of human body to the features of school science education of certain period of time. In addition, I will propose a new paradigm of science education, named here 'Hearts-On', for the coming 21st century and explain some key features of the new paradigm with examples.

2. A Brief History of School Science: From Ears-On to Minds-On

Here I will briefly summarize the historical changes of school science education during the last two hundred years, particularly in Britain which has shown most typical development processes during the period.

For centuries, education in England was the responsibility of the Church of England, and it was not until the end of the Victorian period that the clerical monopoly of schools and universities ended. It is for this reason that so much of English scientific activity took place within an essentially religious context. Scientific instruction consisted in the main of Aristotelianism imbedded in the curriculum of 'the seven liberal arts'.[3]

The increase of various technological activities at the beginning of the 19th century stimulated a demand for workers who were knowledgeable about how newly developed machinery worked and how to best operate the machinery based on relevant scientific principles. One of the responses to this need was the establishment of Mechanics Institutes for the working class in major cities and towns throughout the country.[5],[13] Lectures were given on the principles of areas of science such as Mechanics, Heat, Light, and Chemistry. Practical work in classrooms and lecture halls consisted of lecturers demonstrating the facts of science in front of the audience.[23] In addition to class teaching, lecture, and demonstration, the Mechanics Institutes also usually held exhibitions in which many educational items as well as new development from the industry were displayed. After the vast expansion between 1820s and 1840s, however, the movement of Mechanics' Institutes had been faded away since 1850s, and its original aim to teach scientific principles was replaced by a more broad and liberal ones to teach nearly all subjects, including social activities.[2],[30]

Science was a late starter among school subjects, as it was in the university. The teaching of science in elementary schools began to get under way in the 1840s, through the influence of remarkable men such as Richard Dawes, J. S. Henslow and Henry Mosleley.[17] However, the progress was quickly checked in 1862 through the introduction of a system of *Payment by Result.* Although this system was modified about ten years later to take some account of science, the situation of science teaching in training college and elementary schools were far from satisfactory.[14]

During the late 19th century, particularly compared with other major subjects, Elementary Science, together with Object Lessons, was the fastest expanding Class Subject, specially compared with other more traditional subjects, such as English, Geography, Needlework. During the second half, in general, there had been a steady growth of all major branches of science - i.e. Mechanics, Botany,

Chemistry and Magnetism and Electricity, while there had been a dramatic decrease of Latin subject.

Science teaching during the second half of the nineteenth century can be features particularly with the Object Lessons. The Object Lessons has its origin in the view of some educators, such as J. A. Comenius and Heinrich Pestalozzi, that children learn through their senses and should be led from the known to the unknown. The method was considered particularly suitable for teaching science, and consisted of bringing into the classroom either natural or man-made objects, each of which would form the basis of a lesson. This approach was introduced in Britain during the first half of the nineteenth century and further popularized during the second half of the century.[6],[18],[30]

In 1878 the London School Board issued a syllabus of Object Lessons for the younger pupils and of elementary science for the optional subjects and the more specialized subjects for the examinations of the DSA(Department of Science and Arts). In 1889, T. Twining, an enthusiastic teacher who played an important roles for vitalizing elementary science especially by introducing 'peripatetic science demonstrations' until the introduction of Armstrong's heuristic method, provided a *Suggestive Scheme* for science teaching in elementary schools. His scheme was basically consisted of two parts: the first part of simple Object Lessons covering Standards I to III, and the other part of the basic knowledge of natural sciences covering Standard IV and above - the Standard corresponds to today's grades. According to the Code of 1895, the Object Lessons became compulsory in Standards I to III of elementary schools. This led to a considerable increase in the number of children taking Elementary Science as a Class Subject in the upper standards .[30]

As the main value of Object Teaching, the following three principal uses were emphasized: (i) to teach the children to observe, compare, and contrast; (ii) to impart information and (iii) to reinforce the other two by making the results of them the basis for instruction in Language, Drawing, Number, Modelling, and other Handwork. Among the above uses, the first use was considered to be the most important. Some other educational advantages were also recognized, such as, to encouraging children to contact directly with nature, to make children by encouraging the exercise of brain, hand and eye, and to develop a love of nature and an interest in living things. While the educational principles of the Object Lessons were sound enough, in practice the Object Lessons frequently degenerated into boring rote learning. Together with the Object Lessons, the second half of the century was in large the period of the establishment of science as one of the core elements of school curriculum, by emphasizing the aspect of pure science as a means for mental training.

In the last decades of the 19th century there grew up a body of opinion on the value of science teaching and particularly of laboratory instruction in schools.

The main figures of advocating the importance of school practical work were the famous scientists, such as T. H. Huxley, C. Maxwell.[28] In Particular, Professor H. E. Armstrong who was a guiding spirit of the British Association Committee which placed a great emphasis on laboratory activities, advocated a system of teaching in which the pupil should discover things for himself. This was known as the heuristic method, and soon was considered to be the most desirable way to be pursued in school science.[30],[15]

According to H. E. Armstrong, however, the provision of sophisticated laboratories and their equipments alone could be the solution for practical work in school science. He emphasized the importance of using everyday things rather than ready-made apparatus and workshops rather than laboratories.

> Commence experimental studies at the very earliest possible moment, so that children from the outset may learn to acquire knowledge by their own efforts, so that observing and experimenting become habits.
>
> As to appliances, there is a very wrong idea that very special and expensive accommodation must be provided. This is not the case. There must, however, be space in which the work can be done; there must be a workshop, and much should be done with home-made apparatus. Use ordinary article, medicine bottles, jam pots, saucepans ...; invaluable opportunities are lost by providing everything ready made. (quoted in [22])

At the very beginning of the new century (i.e. January 1901), the Science Master's Association(SMA) - at first APSSM now ASE - came into being. The year 1902 is also considered to be the beginnings of important steps towards a unified system of education in England. The *Education Act of 1902* coordinated all forms of education under local education authorities and this swept away the old *ad hoc* system which had been inherited from the days of *laissez faire*.[1]

Despite the popularity of Armstrong's heuristic method, after the turn of the century, triggered by the experience of World War I, the growing appreciation of the impacts of science upon society and of the necessity of the teaching of science for wider audience gave a great impact towards two new main movements, i.e. *General Science*[7] and *Science and Citizenship*[15]. The later illustrates a typical example of the STS movement in school science during the first half of the 20th century, particularly driven by the socialistic ideas towards the relation between science and society.[27]

A growing realization that British education system needed to be changed in order to provide school science teaching to a wider audience with more emphasis on the relevance, industrial and humanistic aspects of science was echoed by a lecture series called 'Science and Citizenship' which lasted for almost twenty years from 1938 and which was reported in the *School Science*

Review, then the only nationwide professional journal for science educators and by a group of professional scientists who had socialist ideas toward society. Lancelot Hogben was one of the key member of the group [31] and delivered the second lecture of 'Science and Citizenship', titled "Biological Instruction and education for Citizenship".[11] Hogben's main idea, illustrated in this lecture as well as in his famous and million-seller science textbooks, *Science for the Citizen*[10], was that science education should be a way of teaching citizen for promoting democratic society and to achieve that science need to be taught in more integrated, utilitarian and humanistic manners, for example by showing the usefulness, relevance, historical and democratic aspects of science.[12],[27]

1960s and 1970s were the period of innovation of science education led by the USA and soon spread across the world. In Prof. Glenn T. Seaborg's, the director of CHEM Study, forward to *The CHEM Study Story* which contains a complete chronicle of the development of the project, the main purposes of the project were clearly outlined.

> The general objectives of the Study were to develop new teaching materials for the high school chemistry course, including a textbook, laboratory experiments, and films. The more specific objectives were to diminish the then current separation between scientists and teachers in the understanding of science, to stimulate and prepare those high school students whose purpose was to continue the study of chemistry in college as a profession, to encourage teachers to undertake further study of chemistry courses geared to keep pace with advancing scientific frontiers, and thereby improve their teaching methods, and to further even in those students who would not continue the study of chemistry after high school an understanding of the importance of science in current and future human affairs. It was decided from the first to have the course be strongly based on laboratory experiments and be applicable to all students who take high school chemistry.[20]

Like the USA, the UK witnessed the development of numerous academically-oriented programs, such as *Nuffield* projects, during the 1950-60s. However, during the 1970s, there had been growing criticism against the discipline-centered science education and some new noticeable approaches had been made to compensate the contemporary trend. For example, although its main focus was on the integrated approach in school science, the *SCISP* was quite successful to illustrate the importance of the relationship between science and society. Following this example, *Science in Society* and *SISCON-in-Schools* were more ambitious in developing genuine STS programs. These two projects were developed simultaneously and took the form of modules, rather than of textbooks. Nevertheless, *Science in Society* was more concerned with the applied and industrial aspects of science while *SISCON-in-Schools* was more inclined to the historical, philosophical and social aspects of science.[26]

During the 1980s, far more ambitious attempts had been made to develop full-scale STS programs, i.e. *Salters' Chemistry/Science* and *SATIS*. These two programs have been developed with the active corporation from the ASE and soon became the typical examples of the STS approach across the world. Besides the similarities between them, *Salters'* approach is more application-oriented, subject-oriented, and textbook-like while *SATIS* is more socially-oriented, issue-oriented and module-style.[26]

Since the 1980s, together with STS movement, there has been a vast expansion of researches on students' conceptions based on constructivism across the world. Britain was not the exception. The most important effort in the UK was the start of CLISP(Children's Learning in Science Project) led by Prof. R. Driver at Leeds University, and this project triggered a number of similar studies not only in Britain but also in many different countries. This constructivist approach soon became the main paradigm of science education, and became known as 'Minds-On' in contrast with 'Hands-On' of the 1960s and 1970s which particularly emphasized the processes of scientific inquiry activities. It is true that the idea of 'Minds-On' paradigm is still the main driving force of current research activities across the world.

The following shows a brief summary of British science education.

– 18c focused on reading the classics of natural philosophy from the past

1800 the 1st mechanics' institute, Anderson's Institute established at Glasgow

1840s pioneering science teaching by Richard Dawes & John S. Henslow

1851 opening of the Great Exhibition, London

1853 establishment of the Department of Science and Arts

1862 the Education Department's Revised Code

1871 introducing 'Payment by Results' (science as a specific subject)

1878 introducing 'Object Lessons' in elementary curriculum

1901 the 1st science teachers' association, APSSM, established

1903 H. E. Armstrong's *The Teaching of Scientific Method* published

1938 L. Hogben's *Science for the Citizen* published

1963 ASE(Asso. For Sci. Educ.) established

1960s *Nuffield physics / chemistry / biology* projects

1970s *Science in Society* & *SISCON-in-Schools* projects

1980s *SATIS* & *Salter's* Projects

1988 introducing Science in the National Curriculum

1998 *Science beyond 2000* published

So far I have briefly reviewed the historical development of science education particularly in Britain. The history of science education can be analyzed from many different viewpoints. Although it is very difficult to divide the past into a series of periods of distinctive features, it would be useful to do so because such time division could give us a grand picture through which we can review the history and forecast the future of science education. What I propose here is to categorize the overall period into five main periods: until the 18th century, from the beginning of the 19th century to the middle of the 20th century, 1960s and 1970s, and from 1980s until the end of the 20th century.

This categorization of the history of science education is basically based on some important historical changes or events, particularly in the UK's experience: for example, the start of mechanics' institute movement in 1801, the beginning of professional association for science education, APSSM(the Association of Public School Science Masters) in 1901, science curriculum innovation movements during 1960s such as *Nuffield*, *Nuffield Secondary Science*, the emerge of researches on students' misconception based on constructivism in 1980s.

In order to characterize and compare the periods of the historical development, I propose to consider six aspects to consider as shown in Table 1: period of science as..., background philosophy, essence of science learning, focus of science teaching, examples, and corresponding analogy.

It is true that the above characterization inevitably contains a considerable degree of over-simplification of the historical development. For example, the second period (19C-mid 20C), could be divided further into four sub-periods as follows: (a) the beginning of 19C - mid 19C : the period of mechanics' institute movement; (b) mid 19C - 1880s : the period of school subject and of object lesson; (c) 1890s - 1920s : the period of Armstrong's heuristic method; (d) 1930s-1950s : the period of *General Science* and *Science and Citizenship* movement. Nevertheless, I believe that the general characterization of the period as 'Eyes-On' as shown in Table 1 is somewhat valid and it is still meaningful, for science educators not for the historians of science, to characterize the historical periods according to some important events in the history.

3. A New Paradigm: 'Hearts-On' Science Education

So far I have reviewed with some degree of oversimplification the overall historical change of science education by employing slogan-like analogies referring the parts of human body. And the overall pattern of the historical development of the paradigms would be summarized as follows: Ears-On → Eyes-On → Hands-On → Minds-On.

Table 1. The Changes of Science Education Paradigm

Period / Aspect	until 18C	19C - mid 20C	1960s - 70s	1980s - 90s
Period of science as	natural philosophy	school subject(s)	discovery method	personal construction
Background philosophy	deductivism	empiricism	positivism	constructivism
Essence of science learning	logic & reasoning	knowledge & utility	concept structure & discovery process	prior experience & cognitive conflict
Focus of science teaching	philosophical arguement	demonstration of usefulness	scientific inquiry process	children's thinking process
Examples	Principia, Dialogue	mechanics' institutes, object lesson	Nuffield programs, Alphabet programes	CLISP etc.
Corresponding analogy	Ears-On	Eyes-On	Hands-On	Minds-On

Then what would be the next paradigm of science education for the beginning of the 21st century? The answer I propose here is 'Hearts-On' science education. The reasons why science education needs to be Hearts-On could be summarized as follows:

• Throughout the last two century, science education have secure its position as one of the major school subjects. Students, parents and the general public agree with the idea that science is essential for education not only for the personal development of individual students but also for the development of society and nation as a whole. However, since there is still a big gap between school science and public culture, science is usually practised only in the context of schooling and examination. In many societies, science is alienated by a majority of students and by the general public. People are only interested in the fruit of the development of science and technology, but do not want to participate in learning science. Unless science succeed to show itself as an essential part of general human culture, science would be isolated from the public and most students, and continues to be reserved only for the narrow specialists.

• In general we, science teachers, have emphasized only cognitive and psychomotor domains of learning science. Sometimes we have tried, with limited success, to motivate students to study science by showing how scientific activities can be joyful. Particularly during the last two decades researcher of science education have mainly focussed on securing the understanding of scientific concepts based on so-called constructivist philosophy. However, as we all aware well, science became too much important to all aspects of modern life. Society can not exist at all without science and requires a great deal of responsible management of science and technology. It is important to consider science always together with humanism, not focussing on the welfare of human being only but also emphasizing the harmonious interactions between the man and the nature. Science should serve for human development, but science without humanism could result in the catastrophe of the world. In this sense I believe that the general philosophical background of science education should be 'scientific humanism', although there should be further discussions on what this means exactly.

• In order to teach science as a culture and to teach it on the basis of scientific humanism, the essence of science learning need to be not only cognitive (i.e. understanding) but also emotional (i.e. appreciating) and behavioural (i.e. participating). To achieve that, we need to emphasize students' appropriate actions as well as scientific knowledge and responsible appreciation of the relationship between science and society.

• To achieve 'Hearts-On' science, it is necessary to pay more attention to the contexts of science and science learning. Until the end of 1980s when APU Science had been put into practice, the focus of science education usually was either on the content (e.g. conceptual structures, students' conceptions) or on the process (e.g. discovery, inquiry) dimensions or sometimes on the combination of the both dimensions. But a number of studies in the field of science education and of cognitive psychology have shown that the context dimension plays very important roles in various areas of science learning and cognition, such as, deductive reasoning, probabilistic judgement, controlling variables, inquiry skills, memory, consistency of misconceptions, selecting cognitive strategies, problem solving, interest, attitudes and so on.[25] It became now a general agreement that the context of science education is at least as important as the content or the process dimensions. Table 2 shows some features of 'Hearts-On' science education.

The main features of 'Hearts-On' science education can be best illustrated by the following goals which are hoped to be achieved by students :

to understand the core of science,

to enjoy the joy of science,

to feel the beauty of science,

to experience the usefulness of science,

to appreciate the responsibility of science,

to participate in the development of science.

Table 2. Features of the New Paradigm, Hearts-On Science Education

Aspect	feature
Period of science as	Culture
Background philosophy	scientific humanism
Essence of science learning	cognitive, emotional and behavioural
Focus of science teaching	the context of science and science learning
Examples	COPHY, scientific field trips
Corresponding analogy	Hearts-On

Table 3. Text vs. Context

if TEXT is ...	then CONTEXT is ...
- smell - trees - words that teachers say - contents in textbooks - Newton's laws	- aroma - forest - message that students hear - meanings and values of the contents - Newton's life & ideas and relations to Galileo's & Einstein's theories

The most important elements of Hearts-On science education would be how to consider and put into use the context dimension. Kwon[16] illustrated the extended meanings of the context dimension by comparing it with 'text'. (see Table 3) Although we try hard to teach students the text of science, what they learn is the context in which the text is embedded. If we are not successful in teaching physics, it would not be due to the text which has been little changed for the last 50 years, but rather due to the context coming with the text.

If so, how we deliver the right contexts while we teach the text of science? One of the answer to this question would be STS approach to science teaching. But what we have learnt from the experience of implementing STS approach for the last ten years or so in Korea is that the STS approach could not be directly applicable to ordinary classrooms due to many practical difficulties. For example, as long as we apply the STS approach it is always difficult to avoid the problem of loosing coherent conceptual structures of science, as we experienced with so-called life-centered approaches of the 1930s-50s. In addition, adopting the STS approach in science cause the issue of reshaping the whole school curriculum because the STS approach need a lot of humanistic and sociological consideration and this requires the reduction of the time allotment for science subjects, which has been already marginalized due to over-crowded school curriculum through a series of national curriculum reforms.

For these reasons, several physics educators in Korea are trying to develop a new context-rich physics curriculum, named COPHY(COntextual PHYsics), which adopts an extensive contextual approaches but does strongly maintain the conceptual structures of the discipline. Table 4 shows one of the framework of the classification of the context dimension for developing the COPHY.

Table 4. Classification of the context in COPHY(COntextual PHYsics)

1. Contexts of Physics
    - Theory & concepts (e.g. ideal conditions, thought exp.)
    - Practical works (e.g. inquiry, demonstrations)
    - Physics studies (e.g. history, philosophy)
2. Contexts of Natural Phenomena
    - Natural geography (e.g. geology, astronomy)
    - Natural disasters (e.g. flood, drought, earthquake)
    - Earth environment (e.g. climate, pollution)
3. Contexts of Engineering & Technology
    - Electric appliances (e.g. TV, microwave oven)
    - Hygiene & medicine (e.g. X-ray, ultra-sound)
    - Vehicles & traffic (e.g. ABS, tires, airplanes)
    - Architecture & safety (e.g. buildings, bridges)
4. Contexts of Society & Culture
    - Sports & hobbies (e.g. tennis, ski, fishing, photo)
    - Traditional cultures (e.g. tools, cultural assets, proverbs)
    - Media & cinemas (e.g. Sci-fi movies, DVD, IT)
    - Toys & theme parks (e.g. space tops, bungee jumping)
    - Arts & popular cultures (e.g. painting, sculpture, fairy tales)

Another movement currently undergoing in Korea, which could be regarded as an example of Hearts-On science education, is to teach physics as a culture. Recently there have been a number of attempts in Korea, specially led by Prof.

S. J. Pak at Seoul National University, to develop physics/science teaching materials which are to be used for scientific field trips to various traditional and modern cultural sites, including old fortress, temples, museums, relics, and newly built 2002 Korea-Japan World Cup Stadiums in Korea. Teaching physics as a culture also requires a wide and diverse consideration on why and how to teach physics. Table 5 illustrates from how many different viewpoints we can look at physics as culture.

Table 5. Issues of teaching 'Physics as Culture'"

1. As a Discipline
- What would be the characteristics of physics?
- Why do we teach physics, aims and values?
- How do we teach physics, a discipline of natural science?
2. As Educational Activities
- What would be intellectual features of students studied physics?
- Which aspects of citizenship do we wish to acquire by studying physics?
- How can physics learning contribute to individuals fulfillment?
- How can we cooperate with other subjects to achieve educational goals?
3. As Social Phenomena
- How has physics been developed historically?
- What are the interactions between physics and society?
- How can everyday phenomena be explained through physics?
- How would social phenomena be seen differently via physics?
4. As a means for Social Development
- How can we identify and foster the gifted with physics?
- How the principles of physics would be applied to technology & industry?
- What would be physics teaching suitable for future man power?
- How can we prepare for the meeting of physics and other area? Etc.

4. Conclusions

In this paper, to review the overall historical change of science education, I proposed a framework consisting of a series of analogies relating the parts of human body with the features of science education of certain periods of the time. The pattern can be summarized as follows: Ears-On → Eyes-On → Hands-On → Minds-On. In addition, as a new paradigm of science education,

Hearts-On, is suggested and some discussions and examples explaining the new analogy are given.

These days, the Korean Physical Society is facing a tough issue of loosing its traditional customers rapidly. At universities, introductory physics courses are largely ignored by a majority of students including those of engineering and medicine schools. Most of the departments of physics are being suffered from severe under-enrollment. At secondary schools, physics (and other science subjects) are classified as selective subjects with giving much more freedom of choosing subjects to individual schools and students, and very likely high school physics would be taken by only less than 10% of the population.

The situation is not limited to physics. Science in general is widely unwelcomed by students. Last year the number of student who applied for the science-stream in the national matriculation examination became less than a half of that for the humanity-stream. The ratio of the applicants for the science-stream changed from 42.6% in 1994 to 26.9% in 2001, while that for the humanity-stream did from 48.3% in 1994 to 56.4% in 2001 and that for the art/sport-stream from 9.1% in 1994 to 16.2% in 2001.

Although this rapid decrease of science-stream student population is particularly alarming in human power-driven economy systems like Korea, the situation seems rather universal across the world. In modern society, people are more and more inclined to cultural activities and they seem to be less and less attracted by science which used to have strange, untidy, isolated and dark images.[21],[29]

There is no one rigid way of doing science and science might be best developed with no external factors and constraints, as argued by Feyerabend(1975). Similarly there should be no the only way to help students to study science, and as many as possible ways encouraging them to study science need to be practised. My belief is that when students can enjoy the joy of science, feel the beauty of science, experience the usefulness of science, appreciate the responsibility of science, participate in the development of science, that is through Hearts-On science, they can best learn science. When this Hearts-On science is attempted, the gap between science culture and humanity culture, mentioned by C. P. Snow, can be practically reduced.

> Once or twice I have been provoked and have asked the company (highly educated) how many of them could describe the Second Law of Thermodynamics. The response was cold: it was also negative. Yet I was asking something which is about the scientific equivalent of: *Have you read a work of Shakespeare's?* (C. P. Snow, *The Two Cultures*, 1959, p.15)

References

1. M. Argles, South Kensington to Robbins: An Account of English Technical and Scientific Education since 1851. London: Longmans, London, 1964.
2. G. Bishop, Eight Hundred Years of Physics Teaching. Fisher Miller Pub., London, 1994.
3. W. H. Brock, Science education. In R. C. Olby, G. N. Cantor, J. R. R. Christie and M. J. S. Hodge (Eds.) Companion to the History of Modern Science (Vol.2). Routledge, London, 1990.
4. S. C. Brown, "Cultural values in science teaching". in S. C. Brown, N. Clarke, and J. Tiomno (eds.) Why Teach Physics? (based on discussions at the International Conference on Physics in General Education at Rio de Janeiro, Brazil, July 1-6, 1963). MIT Press: Cambridge, Massachusetts, pp.13-19, 1964.
5. D. S. Cardwell, The Organisation of Science in England and Wales. London: Heinemann, London, 1972.
6. B-H. Chung, Comenius' pansophism as a historical origin of science education. JKARSE (Journal of the Korean Association for Research in Science Education), vol.14, pp.379-392, 1994.
7. R. Fawns, "The democratic argument for science curriculum reform in Britain and Australia: 1935-1945". Research in Science Education, vol.28, pp.281-299, 1998.
8. P. J. Fensham, Heads, Hearts and Hands - Future Alternatives for Science Education, The Australian Science Teachers Journal, vol.27, pp.53-60, 1981.
9. P. K. Feyerabend, Against Method. NLB, London, 1975.
10. L. Hogben, Science for the Citizen. George Allen & Unwin Ltd., London, 1938.
11. L. Hogben, Biological instruction and training for citizenship. The School Science Review, No.23, pp.263-281, 1942.
12. A. Hogben, and A. Hogben, (Ed.) Lancelot Hogben Scientific Humanist: An unauthorised autobiography. Merin Press, Suffolk, UK, 1998.
13. J. W. Hudson, The History of Adult Education. Longman, Brown, Green & Longmans, Paternoster Row, London, 1851.
14. R. Ingle, and A. Jennings, Science in Schools: Which Way Now? Heinemann Educational Books, London, 1981.
15. E. W. Jenkins, From Armstrong to Nuffield: Studies in Twentieth Century Science Education. John Murray, London, 1979.
16. J. S. Kwon, Text vs. context. personal communication, 2000.
17. D. Layton, Science for the People: the origines of the school science curriculum in England. London: George Allen & Unwin Ltd., London, 1973

18. T. May, The Victorian Schoolroom.: Shire Publications Ltd., Buckinghamshire, 1998.
19. G. McCulloch, E. Jenkins, and D. Layton, Technological Revolution?: The politics of school science and technology in England and Wales since 1945. The Falmer Press, London, 1985.
20. R. J. Merrill, and D. W. Ridgway, The CHEM Study Story. W. H. Freeman and Company, San Francisco, 1969.
21. D. P. Newton, and L. D. Newton, Young children's perceptions of science and the scientist. International Journal of Science Education, vol.14, pp.331-343, 1992.
22. E. Nightingale, The teaching of science in Britain - a historical retrospect. School Science Review, No.150, pp.320-329, 1962.
23. M. Nott, "Keeping scientists in their place", School Science Review, No.285, pp.49-50, 1997.
24. C. P. Snow, The Two Cultures. (Canto Series, republished in 1998) Cambridge University Press, Cambridge, 1959.
25. J. Song, Review and analysis of the studies on contexts in science education. JKARSE (Journal of the Korean Association for Research in Science Education), vol.17, pp.273-288, 1997.
26. J. Song, The process of the quickening and development of science-technology-society education in the UK : during to 2nd half of the 20th century. JKARSE (Journal of the Korean Association for Research in Science Education), vol.20, pp.52-76, 2000.
27. J. Song, British movement of ;Science and Citizenship' during the 1930s-50s and L. Hogben's 「Science for the Citizen」. JKARSE (Journal of the Korean Association for Research in Science Education), vol.21, pp.385-399, 2001.
28. J. Song, and S. K. Cho, T. H. Huxley as a pioneer of British school science education. JKARSE (Journal of the Korean Association for Research in Science Education), vol.21, pp.38-58, 2001.
29. J. Song, and K-S. Kim, How Korean students see scientists: the images of the scientist. International Journal of Science Education, vol.21, pp.957-977, 1999.
30. D. W. Turner, History of Science Teaching in England. London: Chapman & Hall Ltd., London, 1927.
31. G. Werskey, The Visible College: The collective biography of British scientists and socialists of the 1930s. Holt, Rinehart and Winston, New York, 1978.

PHYSICS FOR THE LAY STUDENT

LESLIE W. TROWBRIDGE
Professor Emeritus of Science Education,
University of Northern Colorado, Greeley, Colorado

Public understanding of physics and physical principles in the United States is at a low point. The term "physics" appears to be frightening and intimidating to average individuals, perhaps because of its mathematical connotations or because of an aura of mystery about how the world works. Many individuals turn to occult, paranormal and metaphysical beliefs about phenomena encountered in their daily lives. The rational ideas of science appear to be less attractive to the minds of many than the superstitious or non-rational explanations promoted by many groups in the United States. Professional scientists, including physicists, have not succeeded in communicating effectively their task of explicating the workings of the universe to the average public. As a result, these scientists are frequently viewed as "gurus" or "priests" of the unknown, dealing in ideas far removed from the daily lives of average persons. Teachers of physical sciences in the elementary school in the U. S. are generally poorly trained in understanding science. Consequently, they avoid or poorly present information about science to their charges. Children thus develop a negative look about science that carried into adult life. Certain programs have been developed in the past decades to overcome the problems cited above. Usually, these are rather formal modes of presentation to children in the schools. In some cases, they are successful but to this point have not had widespread effect on solving the problem of poor understanding of physics by the public. Many problems still persist in developing scientific literacy among the general American public. With increasing awareness, the proliferation of technological devices in most American homes, and attention to early education in sciences, there is hope that these may become a medium for raising the literacy level in the sciences for average persons.

1. Introduction

We live in a physical world. We encounter physical relationships on a daily basis, While we may subdivide knowledge in the physical world into categories of chemistry, biology, earth sciences, space science and others, none of these categories can stand alone without the underpinnings of physics. In physics, we encounter the basic relationships between matter, energy, space, and time which are frequently expressed in mathematical terms.

In addition, our knowledge of physics is constantly changing as new relationships and explanations of the world come to light through research, development of theories, and observations. Underlying our understandings of the physical world is the belief that the universe is explainable, rational, and predictable. This does not mean it is easy, but that it is within the realm of realistic expectation that human intelligence can understand the relationships we encounter.

The purpose of this paper is to call attention to the general lack of understanding by lay students of the principles of physics and of the pervasive influence of these principles in one's daily life. It is important to realize that the physical phenomena observed are explainable. They are not mysterious nor occult. The science of physics is predictable and rational. To establish this point of view is a clear step toward understanding the world.

2. Discussion

What is the situation with respect to the lay students' understanding of physics? It is not good. We are beset with non-rational explanations for many perplexing observations and experiences. The occult, the mysterious, the metaphysical, the paranormal –all of these seem to have an unusual attraction to the general public.

What is the reason for this? It has always been true that the unknown gives rise to superstition. When events seem unexplainable, one's first reaction is to ascribe a non-rational explanation. This seems to take it out of the realm of responsibility for the individual to apply reason, critical thinking, or logic to the problem. This is understandable because to apply these processes requires energy and concentration, which is fatiguing.

Another reason for the reluctance to engage in rational thought when looking for explanations of physical phenomena is the inherent fear and antipathy toward mathematics present among many persons. Because the physicists' approach toward problem solution is often mathematical, his explanations quickly become stratospheric for the average person and result in a turn-off, not because the content is uninteresting, but because the symbolic and abstract explanations become unintelligible to the non-professional. This results in a permanent mental blockage against any kind of physical explanation. The very term "physics" becomes an anathema for the layman.

A third reason for the abysmal lack of understandings in physics by the average person is the pervasive influence of television, particularly in American homes. This influence is insidious because of three factors: it is entirely spectator oriented, requiring no mental agility on the part of the watcher to engage in critical thinking or problem solving, second, it is anti-intellectual and is aimed at about a 12-year old level of understanding, third, it is time-stealing and deprives individuals of the necessary concentration time required for serious consideration of real problems in the physical world. The message given is that knowledge is easy to obtain and requires little effort to acquire. It is simply a matter of absorbing information in a passive manner, with little attention to mental participation in learning.

There are a few exceptions to this condition, namely in certain special programs provided by public television and infrequent presentations on commercial television provided by National Geographic Specials.

A fourth reason one can cite is the absence of qualified teachers of physics for any level-elementary, middle school, and high school levels. The absence of clear understanding of physics among teachers at elementary and middle school levels is particularly devastation. Students' interests in physical sciences are not stimulated at an age when the young mind is most receptive to new learning. Thus, children pass through a crucial stage of development and obtains a negative view of physics which will persist throughout life. Other interests replace or preempt one's incipient interest in physical science understandings which is never fully nurtured.

A revealing body of research is reported by Driver et al.[1]. This book describes voluminous research in dozens of countries in Europe, Asia, and many other countries in which ideas of the natural world held by children at elementary and pre-school levels are reported. When reading this research, one is struck by the realization that young children without formal education in the sciences had already formulated ideas of the world at a very early age.

These ideas are natural, frequently non-rational, based upon personal observations and experiences, and invariably persistent throughout the life of the child. This reinforces the findings of researchers on constructivism, formation of misconceptions, and difficulties in bringing about conceptual changes that more accurately match the beliefs of scientists on how the world works. The condition exists in all cultures and reflects a phenomenon present among all children as they pass through early developmental stages.

Unless interventions are taken in formal education to correct the naïve beliefs held by children in their early years, it is inevitable that as adulthood is reached, the naïve ideas will persist. This may partially explain the phenomenon referred to earlier in which pseudo-science has gained much popularity throughout the world.

Let us consider how children approach their understanding of the world around them. All knowledge comes to us through our senses. All normal human beings have five sense organs that keep on bringing information that is somehow accepted and assimilated into our prior knowledge. This accumulation of knowledge forms the basis on which we respond to the world and to new events. Much of what we learn is based upon common sense. This forms an interpretation of events in which the information provided by our senses forms a strong and convincing argument for what we believe

Unfortunately, common sense is often uncommonly unreliable. Frequently information that relies on common sense may produce misinformation and misunderstandings. As long as these are temporary in the life of the growing child and are soon corrected by further experiences or a word from a parent, teacher, or friend, they may do no harm. It is when the misinformation persists into adulthood that serious problems may arise.

In schools, teachers are often perplexed and frustrated at the difficulty that they experience in introducing scientific explanations for common phenomena. The naïve beliefs held by children seem to form an effective buffer that prevents full acceptance of the new explanations.

I should like to suggest a plan for attacking this problem[2]. The procedure may be equally effective for older learners as well as for children in school. There are four steps in this procedure. The first could be described as "First the action, then the words". This is based upon the well-known belief that "learning by doing" is superior to listening to someone describe the phenomenon and hoping fully to understand the explanation given. Strictly verbal learning, which tends to foster a false sense of security about one's knowledge, might thereby be minimized. Students would be able to demonstrate the concepts under study, or state principles in their own words, or suggest ways to apply principles in other contexts. Frequently, students concentrate so intensely on memorizing a concept's exact words or wording that they fail to grasp the meaning. Unfortunately, the types of factual tests we frequently give them tend to reinforce this manner of study.

A second step would be "Talking through the new concept". On an individual basis, if possible, have students explain their interpretation of the phenomenon. Try to elicit reasons why students hold their particular point of view they do. Ask questions tactfully to identify the points where clarification is needed. Use the basic premise of science that every cause has and effect and every effect is produced by a cause. During this time use simple demonstrations where possible and examine each phase of the demonstration in detail. Present new information or new points of logical reasoning that are non-threatening and non-judgemental. Avoid casting aspersions on previous poor teaching or previous experiences that might elicit resistance from the students. Finalize the session by having students explain in their own words their revised understanding of the phenomenon.

A third step might be to "Teach about the concept to someone else". We have all experienced the truth of the maxim that "one really learns a subject when one teaches it." Use this truism to help reinforce the new understanding in the students' minds. On a one-to-one basis to start with, have students explain the phenomenon to each other. Have students ask questions of their pupils to see whether the explanation has been clear and logical. Later you might have students explain or demonstrate the phenomenon is a classroom environment.

Finally, "Don't let the concept die". Every teacher has probably experienced the situation of memory lapse when explaining a science phenomenon before a class. The more frequently you can refresh your memory through repetitions, the more likely you will remember the details. This is analogous to regular practice on a musical instrument. Without such practice, mistakes are certain to occur. This principle applies also in the case of restructuring science concepts. One successful performance does not guarantee success for all time. It has been found that under pressure, such as tests, students have a tendency to revert to former knowledge. Frequent repetitions of the newly learned information are necessary and use of the information in new contexts helps to reinforce it in the mind.

We can all attest to the fact that knowledge gained through personal experience seems to remain with us longer than second hand knowledge gained by only hearing, or reading. Personal involvement perhaps engages more of the senses and thus produces more synaptic connections in the brain. A recent model of teaching and learning developed by Rodger Bybee, is the Learning Cycle in the form of five steps, Engagement, Exploration, Explanation, Elaboration, and Evaluation[3]. These steps are based upon a constructivist approach to learning, in which the individual "constructs" his own view of the world that is tailored to his own experiences and perceptions. The Learning Cycle approach using the steps enumerated above provides frequent opportunities for the learner to become personally involved in recognizing problems, formulating hypotheses, constructing apparatus, gathering data, analyzing data, making inferences, drawing conclusions, and reporting results of problems under study.

Functions of the Engagement phase are to create interest, generate curiosity, raise question, identify previous knowledge, and seek out sources of information. In the Exploration phase, students are encouraged to work together, observe and listen to other students, ask probing questions such as "What cause this"?, "What is happening"?, etc. Sufficient time is provided to explore different pathways and make choices. Teachers serve as consultants rather than information sources in this phase.

During the Explanation phase, students are encouraged to explain concepts and definitions in their own words, give evidence for their beliefs, relate personal experiences that support their beliefs, and listen critically to teacher and student explanations. In this phase, also, stress is put on understanding of scientific principles, with an attempt to encourage students to express such principles in their own words rather than merely repeating standard statements they have heard or read in books.

The Elaboration phase encourages students to apply the concepts and skills in new situations, to identify alternative explanations, to make inferences and draw reasonable conclusions from evidence The Evaluation phase assesses students' knowledge and/or skills, searches for evidence that students have changed their thinking or behaviors, and provides opportunities for students to evaluate their own progress and knowledge.

During the decade of the '60's, several attempts were made to strengthen science understandings among teachers and students of elementary science. A project entitled Elementary Science Study(ESS) was developed at the University of California, Berkeley[4]. Included in this curriculum were several physical science topics such as Light, Sound, Floating and Sinking, Kitchen Physics, and others that promoted better understanding of physics in everyday life. This project was eminently successful for the grade levels for which it was intended. The teaching strategy was an investigative approach, using a learning cycle format. For a limited number of children in the '60's and '70's, physics became somewhat familiar and less intimidating. As these children grew to adulthood, presumably the experiences they had encountered with ESS gave them acquaintance with a physics that many other children did not receive. While similar efforts by the American Association for the Advancement of Science(AAAS) and other federally funded projects during this era were equally successful, the numbers of public school children so affected was small, and the lasting results uncertain.

One of the areas of physics that is most confusing to the average layperson is kinematics, which deals with motion of objects. The average person has difficulty differentiating between position, speed, velocity, and acceleration. In addition to confusion with respect to the terms above, the concepts of velocity and acceleration are frequently misunderstood. Even for adults who may have had high school physics, there is evidence of great perplexity about these two concepts. As an illustration of this confusion, suppose we imagine two automobiles traveling side by side on a straight road. One automobile begins from velocity zero, accelerates to a velocity of 30 miles per hour and then maintains a constant velocity along the road. A second automobile begins at a later time but accelerates at a higher rate and rapidly overtakes the first vehicle. If, at the instant of passing the first vehicle, a question is asked of and observer, "How do the accelerations of the two vehicles compare?", a common reply is "They are the same", indicating a lack of understanding of the concept of acceleration. A second point of confusion among many individuals is the units used to express velocity and acceleration. The unit for velocity, for example, feet per second, is usually understood as the rate of covering distance between two points. The unit for acceleration, for example, feet per second per second (or feet per second squared), frequently causes greater problems, not being understood as the rate at which velocity is changing.

Research conducted at the University of Washington, Seattle, WA by David E. Trowbridge, has elucidated this problem[5]. The research attempted to develop a new approach in curriculum development in which computer assisted instruction is used effectively to overcome common misconceptions about motion, and teaches elementary concepts of graphing position, velocity and acceleration versus time for one dimensional motion. The software developed is called "Graphs and Tracks" and is presently available from Physics Academic Software[6].

The Computer program described above was designed for use in schools and colleges. It provides a venue for manipulating graphics in ways that simulate motions of real objects in a laboratory. The program has proven to be very successful. The software program won and award for Best Physics and Best Integrated Software in the 1988 EDUCOM/NCRIPTAL Higher Education Software Awards program.

With the astounding proliferation of personal computers among the general populace, it appears the above program might be a highly desirable and effective tool for educating the general public about motion of objects. The software is very "learner friendly" and is amenable to applications of a competitive nature among youth and adults.

A project developed in 1985 entitled PRISMS(Physical Resources and Instructional Strategies for Motivation Students)[7] by Roy Unruh and Tim Cooney has many components that are suitable for learning physics by the general public. The thrust of the project is to use common, everyday materials and experiences to learn physics and develop understandings of physical principles. While the project is designed for classroom, the fact that materials are easily obtainable and may be manipulated in a non-classroom setting with suitable simple instructions makes it possible for lay persons to improve their knowledge of physics by systematically using these materials.

As examples of the kinds of activities designed for use in this project, one might cite an activity entitled "In Time an You Will see the Moving Plot". The purpose of this activity is to 1) introduce the skills of graphing and graphical analysis and 2) describe a common motion by plotting a distance vs. time graph. It requires a small windup car(or friction-powered car), a tape timer, meter stick, and graph paper. With simple instructions, participants will obtain data that can be plotted, graphed and analyzed to gain understanding of position, velocity, acceleration, deceleration, and other factors relating to moving objects.

Another activity entitled "Slippery as an Eel" is designed to investigate the coefficient of sliding friction. Simple materials such as a smooth hardwood block, a smooth board, several materials such as glass, steel, sandpaper, or rubber are studied using a small pulley and weights to provide force. Very little knowledge of mathematics is required(division of the force of friction by the weight of an object) to give a coefficient of friction.

The above examples and others used throughout the PRISMS project illustrate the factors of personal involvement, an attitude of curiosity, dependence on self-gathered data, care in making inferences, drawing conclusions, repeatability of experiments, and other attributes of scientific investigation that are frequently lacking in the understanding of science by the public. This project avoids many of the negative connotations sometime attributed to physics by lay persons by using simple materials, providing opportunities for person involvement in data gathering, and limiting the use of mathematical relationships to those within the capability of the participants. Many high school and college students who are rapidly approaching adulthood benefit from this approach to physics and will in all likelihood retain a certain confidence in their own abilities to understand physical relationships in the natural sciences throughout their lives.

Two recent programs designed for students or lay persons, which involve audio or video tapes, are available from The Teaching Company [8] in Springfield, Virginia.

The first of these is an audio tape series entitled an "Einstein's Relativity and the Quantum Revolution" consisting of lectures by Richard Wolfson of Middlebury College, Vermont. It is part of a series called "Modern Physics for Non-Scientists. This series is an excellent production by a superior teacher who makes abstract concepts very understandable.

The second is a video tape series entitled "Understanding the Universe" which is presented by Alex Filippenko of the University of California at Berkeley. This series, too, is a highly understandable presentation for non-science students or lay persons.

Both of the above series are interesting, non-intimidation, and non-mathematical explications of modern concepts in physics and astronomy that have potential for educating non-scientists about the world around them. I recommend them highly for consideration.

3. Conclusion

The world in which we live has revealed itself as a highly interesting place, the nature of which is slowly becoming understood. In such a world it behooves us to become knowledgeable of the forces, energy, and matter and relationships between them that tie all these entities together. Only then can we say that we have become somewhat scientifically literate and need no longer be subject to superstitions and mysterious happenings that we cannot explain. There is a certain degree of responsibility that must be assumed by all individuals. They are encouraged to learn all they can about the natural world so that they can have a degree of control over their lives. Perhaps the greatest hindrance to achieving this goal is the natural human reluctance to take steps into the unknown, and to apply the necessary time and energy to accomplish the task. The task is achievable and not beyond the comprehension of average individuals. It is not necessary to be a mental genius to gain an understanding of one's environment and to exert some control for a more satisfying existence.

Reference

1. Driver, Squires, Rushworth, and Wood-Robinson, Making sense of secondary science: Research into children's ideas, Routledge, London and New York, 2000.
2. Trowbridge, Bybee, and Powell, Teaching secondary school science: Strategies for Developing Scientific literacy, 7th Ed., Merrill-Prentice Hall, Columbus, Ohio, 2000, pp. 180-181.
3. Ibid, pp. 245-247.
4. Elementary Science Study(ESS), University of California, Berkeley, CA 1960.
5. David E. Trowbridge, "Applying Research Results to the Development of Computer Assisted Instruction" Proceedings of the Conference on Computers in Physics Instruction, North Carolina State University, Raleigh, N.C.(August 3, 1988).
6. Physics Academic Software, North Carolina State University, Raleigh, NC 27695-8202(Ph: 1-800-955-8275) (e-mail: pas@ncsu.edu).
7. Physics Resources and Instructional Strategies for Motivating Students (PRISMS), Iowa Physics Task Force, The Iowa Academy of Science, University of Northern Iowa, Cedar Falls, IA 50614, 1985.
8. Modern Physics for Non-Scientists, The Teaching Company, 7405 Alban Station Court, A-107, Springfield VA, 22150-2318 www.teachco.com.

PHYSICS IN ANCIENT CHINESE CHIME-BELLS

YANLAI YAN, LINGDA KONG AND HUAHAN LIANG
Department of Physics, Jiao Tong University, Shanghai, 200030, China
E-mail: ylyan@online.sh.cn

1. Introduction

The Set of Zeng Hou Yi Chime-Bells, which consists of 65 musical bells covering five and a half octaves in the age of Confucius, is one of the earliest and most extensive surviving groups of ancient musical instruments in the world. Behind such an ancient cultural miracle, physics is richly involved, and it has recently become one of the most interesting topics for multimedia lectures in physics education.

Music was one of the most important elements in ancient Chinese culture. Bells that were assembled into chimed sets as percussion instruments played the dominant role in the music of the Bronze Age China. Zeng Hou Yi chime-bells dating back to about 2500 years ago were excavated in 1978 from Marquis Yi's tomb in Hubei province, China.

Bells as musical instruments also developed in western countries during the seventeenth century but two millennia later. Although both Chinese and western musical bells were made of bronze, they were designed in different ways. Therefore their vibration mechanism was different from each other and resulted in different sound characteristics. In this paper we will discuss the physics behind the almond- shaped Chinese bell and compare its unique musical properties with the western carillon.

2. Acoustic Properties of Zeng Hou Yi Chime-bells

The set of Zeng Hou Yi Chime-bells were covered roughly five and half octaves, only a little less than the modern piano, and it produces entire 12 semitones, which were most like current C Major, on the other words, all the notes found on a piano keyboard.

As shown in Fig.1, the Chime-bells are almond-shaped and many nubs (called as Mei in Chinese pronunciation) distribute over the out surface of each bell. The lower fringes of the bell are not in a plane, but arched up a little between the side spine (called Xian in Chinese pronunciation). This special design results to its double tone property. When the bell is struck at the so called front drum point (or Gu), it sounds at one fundamental and a series of partials called as A-tone, two of those vibration modes with $m = 2, n = 0$ and

$m = 3, n = 1$ are shown in Fig.2(b) with maximum amplitude there around Gu, while when the bell is struck at the so-called side drum point (or Sui), it sounds at different fundamental and another series of partials called as B-tone as shown in Fig.2(c) with maximum amplitude there around Sui. That means the positions of the vibrating part of the bell depends on where it is struck. It should be stressed that the nodal lines go differently between A-tone and B-tone. The side drum point, i.e. Sui is just located on the nodal line of A-tone vibration modes, while front drum point, i.e. Gu is just situated on the nodal lines of B-tone vibration modes.

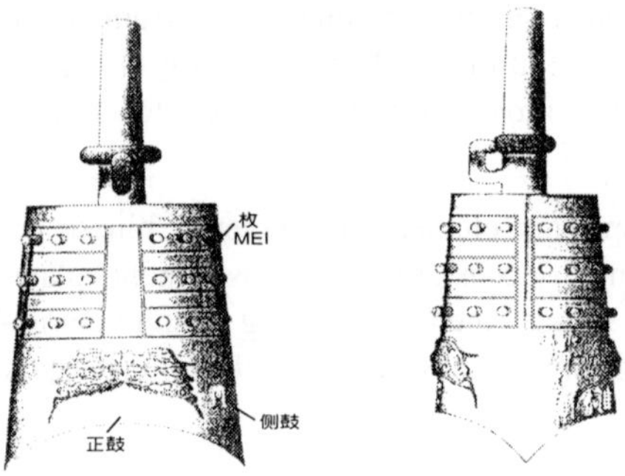

Figure 1. Zeng Hou Yi Chime-bells

For a round bell, no matter where it is struck, it gives one harmonic tone with many modes as shown in Fig. 2(a), because the same pattern will form symmetrically with respect to the striking point. The almond-shaped bell does not have much axial symmetry as the western carillons do, but have mirror symmetry. Such symmetry leads to two different sets of natural vibration modes, of which one is symmetric against the mirror plane while another one is anti-symmetric.

When the Gu is struck, all possible frequencies appear except those modes of which the nodal lines are along the Gu, on another words, all the B-tone modes keep silent. When the Sui is struck, all frequencies appear except those modes of which the nodal lines are along the Sui, on another words, all the A-tone modes keep silent. Therefore, almond-shaped bell produces two distinctly different tones. Holographic interferograms show the two-tone bell vibrates in fundamentals of A-tone and B-tone respectively as shown in Fig 3(a) and (b).

In addition to the intriguing double tone characteristics of the Zeng Hou Yi Chime-bells, one more property should be mentioned that the frequency difference between the two tones of the bell is always approximately either major third or minor third. In fact, the frequency of B-tone is always higher than A-tone by approximately either 400 cents or 300 cents. This fact demonstrates that in bronze age China not only advanced casting technique but also the tone adjust technique were already available.

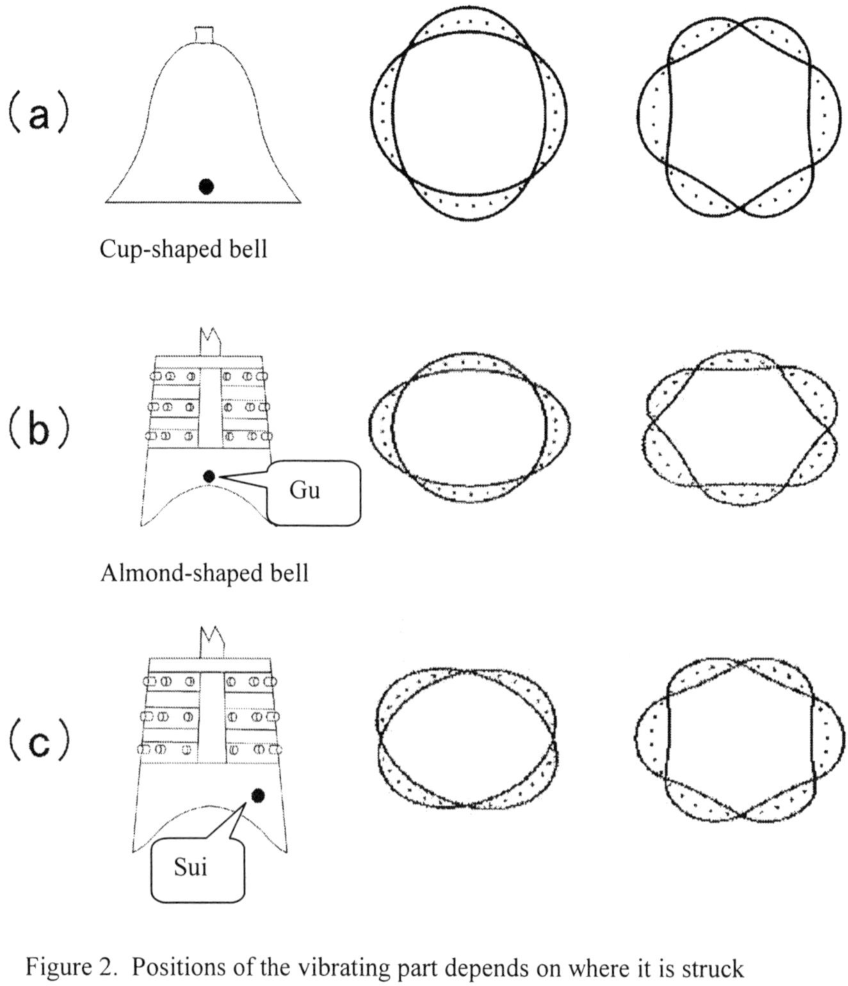

Figure 2. Positions of the vibrating part depends on where it is struck

Those nipple-like "Mei" (36 in total) around the bell brings more ancient atmosphere to the two-tone bells. However, what was the purpose to build them around the bell surface is still a puzzle. Our recent work shows that those Meis do have some effect to make acoustic differences on those ancient music bells besides of the design of decoration. We cast a special almond-shaped bell with longer Meis which is two times longer than usual for testing. We simply struck the testing bell and measure the frequency spectra of the bell and then we ground the meis to decrease their lengths step by step. After each grind, the results of the Fourier frequency analysis are shown in Fig 4.

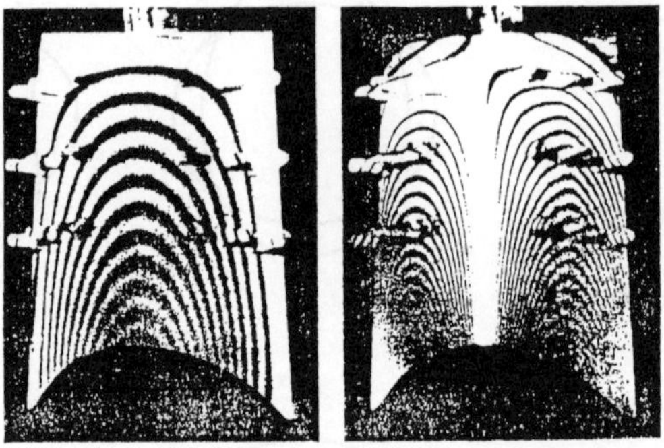

(a) (b)

Figure 3. Holographic interferograms of two-tone bell vibration

It is clear that the frequencies of higher than 4000Hz do not exist before the long meis were ground as shown in Fig 4(a). Along with those long Meis being ground to shorter and shorter, the partials higher than 4000Hz start to appear and the intensities of them vary with the change of lengths of Meis gradually. And the partials of high frequencies become stronger when Mei's were totally ground off. We believe that the Meis were designed not only for decoration of the bell surface, but also, possibly more essential, for filtering out those partials of higher frequencies and adjusting the audio tone quality as well. It has been noticed that those bells in the tallest row on the Frame of Zeng Hou Yi Chime-bells called Niu bell responsible for high tones do not have any Meis at all. While those bells on the middle raw of the frame called Yong Bell responsible for middle tones possess Meis with normal sizes. Whereas the bells on the bottom raw of the frame also called Yong Bell responsible for base tones all have the biggest Meis. It is so marvellous and obvious that we could not say that ancient Chinese music people who designed those Meis with different sizes around the bell out surface did not have any purpose for their acoustic results!

The third feature of the chime-bells is their short sound lasting property, i.e. the bell sound is damped relatively quick after the bell is struck, while the sound from the cup-shaped western carillon lasts longer. To give an explanation, we believe this acoustic property is originated from its unique almond-shape. Its two almond-shape curves meet at the bell's spine which is called Xian, and the top stem of the bell which is called Yong, are responsible for this short sound lasting property. These parts are the thick parts with large stiffness.

(a)

(b)

(c)

(d)

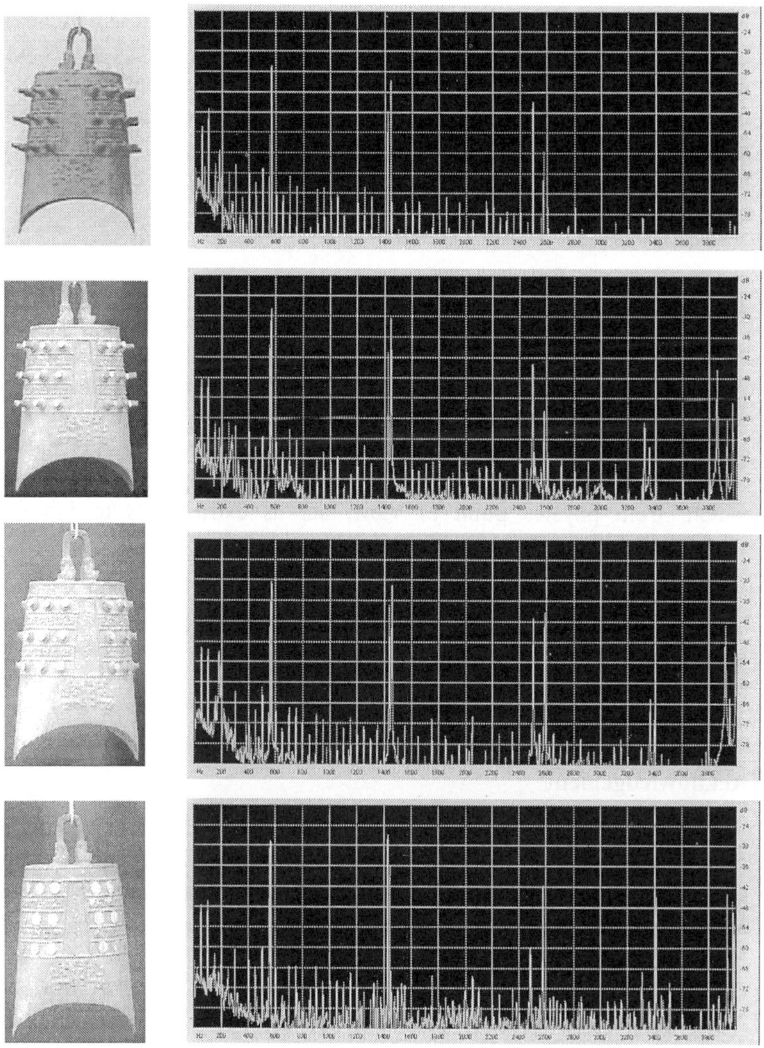

Figure 4. Fourier frequency analysis

The Xian damps the vibration modes of A-tone, because that the anti-nodal line vibrates along the Xian, while the Yong damps the vibration modes of B-tone, because that its heavy stem close to its anti-nodal line to obstacle the vibration at Sui. In short, the sound from a circular bell lasts longer than its counterpart from the flat one. Almost one millennium ago, a famous ancient Chinese scholar, Shen Guo (1031-1095) was the first one to reflect on the difference between bells with round as opposed to almond-shaped cross-section. He stated: "When a bell is round, its sound is long; when it is flattened, its tone is short. When the tone is short, it is abrupt; when the tone is long, it is undulating." For the set of chime bells, each individual bell should not vibrate for long, or it would interfere with the others. This acoustic property, therefore, makes it possible to play short melodic phrases by being struck with a mallet, however the western carillons being struck with clappers can only play slower music phrases in comparison

3. Conclusions

In conclusion, The Zeng Hou Yi Chime-bells brought us great inventions of two respects. One is its almond-shape, which generates not only two distinct tones but also short sound lasting property and at the mean time, makes the set of chime-bells covering broader octaves with less weight and volumes comparing with western carillons. The other is its nipple-like design of Meis, which, from physics point of view, act as extra loads to inhibit certain vibrations of higher frequencies of these chime-bells.

Inspiringly, in order to explore the acoustic effects resulted from those historic inventions in a cultural contexts referred to the ancient music instruments in bronze age China, physics is richly involved and some physics principals are well applied.

Acknowledgement

We would like to express our appreciation to Professor Sheng Zongyi*, who originally pointed out the basic effect of the nipple-like design Meis to us and also to the Chinese Bronze Company** for casting the chime-bell with longer Mei to make the above experiments possible.

* Sheng Zongyi, President of the Institute of Chinese Art, Jiao Tong University.

** Chinese Bronze Company (www.chinese-bronze.com), auxiliary to Jiao Tong University, Shanghai, 200030, China

PHYSICS IN KOREAN TRADITIONAL GAME, 'JEGI KICKING'

O. G. JO *& Y. M. KIM **

* *Mun Hyun Girls High School, Pusan, Korea*
** *Dept. of Physics Education, Pusan National University, Pusan, Korea*

Playing with a 'Jegi', which involves kicking a shuttlecock-like object with the inside of the foot, is an interesting traditional Korean game. The current study examined the motion of a Jegi using a video analysis method. Then, an equation for the motion of a Jegi was developed using a velocity-time graph based on data from the video analysis. The physics concepts that can be taught using the motion of a Jegi include the rising or falling time, distance, displacement change, speed, velocity, terminal velocity, acceleration, and net force, etc.

1. Introduction

Playing with a 'Jegi', which involves kicking a shuttlecock-like object with the inside of the foot, is an interesting traditional Korean game. Various physics concepts, such as gravity, free falling, air resistance, and so on, are involved in the motion of a Jegi.

When an object falls in air, the air resistance against the object's motion increases as the object's velocity increases, therefore, the object's acceleration will decrease until the air resistance equals the object's weight. The resultant force is then zero and the object falls with a constant velocity, called the terminal velocity, which depends on the size, shape, and weight of the object.

Conversely, when an object travels upwards in air, its speed decreases rapidly due to its weight and the downward air resistance. Yet, the air resistance decreases as the velocity of the object decreases, and at the top of the motion, the object weight becomes a downward force. Therefore, the decreasing resultant force also reduces the acceleration of the object.

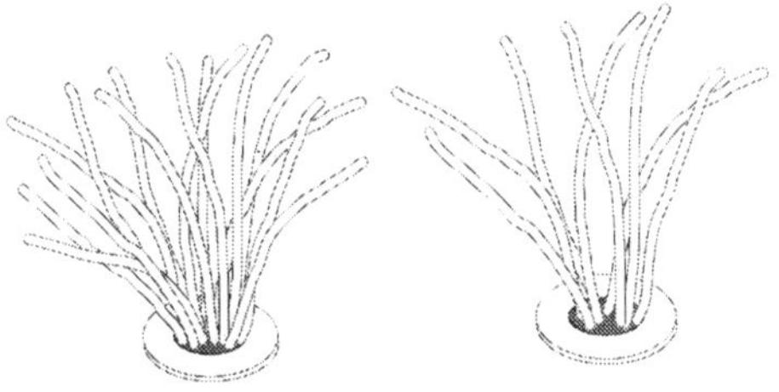

Figure 1. Figures of 'Jegi'

The current study observed the motion of a Jegi using a video analysis method, constructed a velocity-time graph for its motion, then created an equation for the motion of a Jegi.

Through playing with a Jegi, various physics concepts can be taught, including the rising (or falling) time, distance (or height), displacement change, speed, velocity, acceleration, force, impulsive force, and impulsive time, etc.

2. Equation for Motion of 'Jegi'

The direction of the air resistance acting on an object is opposite to the direction of the object's velocity. Plus, the magnitude of the air resistance varies according to the shape and velocity of an object.

Generally, for a solid object, the air resistance, R, is proportional to v^n, where v is the velocity of the object. There are various models related to the action of air resistance, R, on an object [2)] as follows:

a) constant (k)(will not reach a terminal velocity)

b) linear (k_1v)

c) quadratic (k_2v^2)

d) higher orders and their combinations,

where k is a proportionality constant that depends on the shape and size of the object and properties of the fluid.

When the velocity of the object is small, n equals 1. If the object's velocity in air is high, the resisting force is approximately proportional to v^2 (i.e. n=2). For a baseball, n is about 1.5~2.0.

In the case of the motion of a 'Jegi', let's assume that the resistance, R, is equal to kv (using the linear model), because the velocity of a Jegi is small. As such, the net force on a Jegi becomes mg+ kv when it is traveling upwards. Therefore, based on Newton's 2nd law, the equation of motion is as follows:

$$-(mg+kv) = ma$$

$$= m\frac{dv}{dt}$$

$$-(\frac{m}{k}g+v) = \frac{m}{k}\frac{dv}{dt},$$

$$\therefore \quad \frac{dv}{mg/k+v} = -\frac{k}{m}dt$$

After integrating each side of the equation, it becomes as follows:

$$\int_{v_0}^{v} \frac{dv}{mg/k+v} = -\int_0^t \frac{k}{m}dt,$$

$$\ln\frac{(mg/k+v)}{(mg/k+v_0)} = -\frac{kt}{m}$$

$$v+\frac{mg}{k} = (v_0+\frac{mg}{k})e^{-kt/m}$$

Therefore,

$$v = (v_0+\frac{mg}{k})e^{-kt/m} - \frac{mg}{k} \quad \text{..............................(1)}$$

The acceleration of the object is as follows:

$$a = \frac{dv}{dt}$$

$$= -\frac{k}{m}(v_0+\frac{m}{k}g)e^{-kt/m} \quad \text{...................................(2)}$$

Initial and terminal velocity

The initial velocity, v_0, can be acquired by inserting t = 0 into equation (1), while the terminal velocity, v = -(m/k)g, can be acquired by inserting t = ∞ into equation (1).

Magnitude of acceleration

From equation (2), the initial acceleration becomes as follows:

$$a_0 = -(kv_0/m + g)$$

In this case, the magnitude (an absolute value) is larger than the gravitational acceleration, g.

The acceleration of a Jegi at the highest point of motion can be acquired by inserting v=0 into equation (1).

$$e^{-kt/m} = \frac{mg/k}{(v_0 + \frac{mg}{k})} \quad \text{...(3)}$$

If equation (3) is then substituted in equation (2), the acceleration, a, is equal to -g, i.e., a = -g.

3. Experiment for Motion Analysis of 'Jegi'

If the mass of the Jegi is very large, i.e, a = -g, then the Jegi will basically free-fall. In contrast, if the mass of the Jegi is extremely small, then the exponential term on the right side of equation (2) becomes 0 and the acceleration a_0 approaches 0. In this case, the Jegi will fall with a constant velocity, called a terminal velocity.

Generally, a Jegi exhibits a motion of decreasing acceleration due to air resistance. This means its acceleration differs from every position.

The motion of a Jegi was examined using a video analysis method. A shooting machine was designed and built, as shown in Figure 2, and a scale was attached to the wall, as shown.

A Jegi was shot using the shooting machine and the time of motion measured using a video analysis method.

The motion was then analyzed by showing the video-tape in slow motion. Since the time for each frame was 1/30s, position data vs. time was obtained as shown in Figure 3 and velocity vs. time as shown in Figure 4. This data was then used to calculate the resistance constant, k.

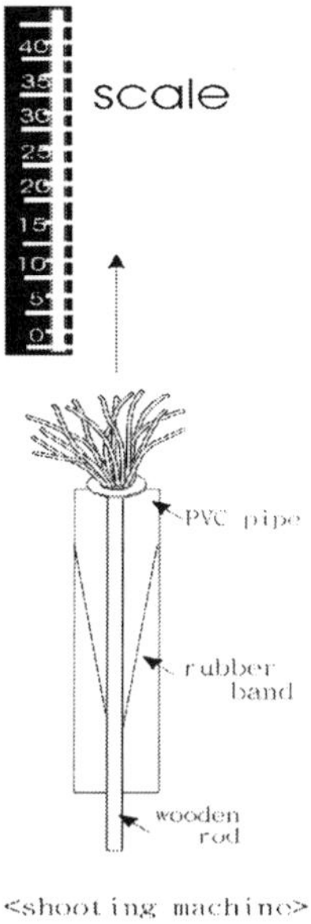

Figure 2. Shooting machine

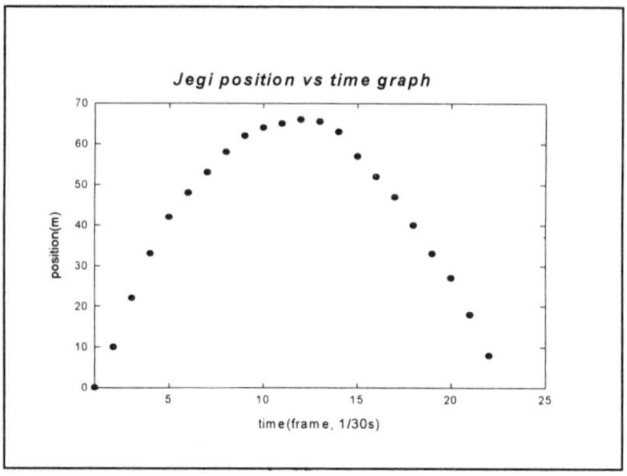

Figure 3. Position-time graph

formula ; v = A0 + A1 exp(-t/t1) ; A0=-3.54, A1=9.88,

In the current experiment, the air resistance constant, k, was about 1.26×10^{-2} kg/s.

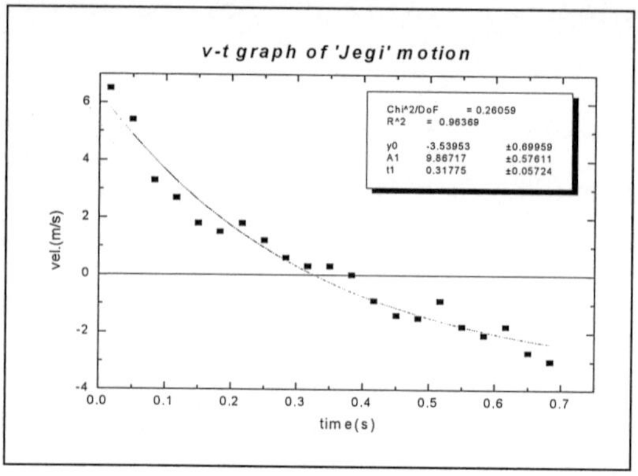

Figure 4. Velocity-time graph

Using the value of k, the time-dependent velocity of the Jegi was then calculated.

$$v = (v_0 + \frac{mg}{k})e^{-kt/m} - \frac{mg}{k}$$

$$= 9.11e^{-3.15t} - 3.11$$

References

1. Tom Duncan, *Advanced Physics,* John Murray (Publishers) Ltd., 1994.

2. S. Herbert and T. Toepker, Terminal Velocity, *Phys. Teach.* vol. 37, p96, 1999.

3. Hugh D. Young, *Physics*, Addison-Wesley Publishing Company, Inc., 1992.

CROSS BORDER QUALITY ASSESSMENT IN PHYSICS

GUNNAR TIBELL
Department of Radiation Sciences, Uppsala University,
Box 535, S-751 21 Uppsala, Sweden

1. Introduction

Quality assessment has become an important activity for the different organisations which take part in educational systems. In addition to more obvious reasons it is motivated by different arguments like a wish from the financial contributors, who could be the public, to ascertain that the resources are well spent. It could also be used as a means of securing the recruitment of students in higher education – many institutions are better supported if the student numbers are high. In the case of the physics discipline in universities there is certainly a need in many countries to promote the interest in studies. For instance, in some countries in Western Europe the number of physics students has fallen dramatically, and it is urgent to try and fill the places available, before the authorities decide to cut the budget. Naturally, it would be easier to require good support, if the institution is found to have a high quality in the physics programme offered.

In Europe there is the additional wish to facilitate the mobility of students between different countries. For this reason it is important to know if the study programmes are more or less equivalent, so that a period abroad does not automatically imply a loss of time for the students. There would be no strive for identity between the programmes – on the contrary the ambition could be to show local strengths in more specialised areas. On the other hand the basic programme should preferably be of the same high quality and not too different in scope.

2. Participators

In the pilot project in international evaluations to be described here the following institutions took part on a voluntary basis:

* Gerhard-Mercator Universität, Duisburg, Germany

* Universiteit Gent, Belgium, two faculties:

    Faculty of Science (Physics)

    Faculty of Applied Science (Civil Engineering in Physics)

* Universität Hannover, Germany

* Universiteit Twente, The Netherlands

The initiative was taken by the Ministries of Education in Belgium (Flanders), Germany (Lower Saxony and North Rhine Westphalia) and The Netherlands. The evaluation group consisted of six physicists from Austria, Belgium, Germany (2), The Netherlands and Sweden, namely

Professor Urbaan Titulaer, Austria (chairman)

Professor emeritus Jules Deutsch, Belgium

Professor Frans Habraken, The Netherlands

Professor Klaus Lüders, Germany

Professor emeritus Gunnar Tibell, Sweden, and

Dr Udo Weigelt, Germany.

The administrative responsibility was upheld by Zentrale Evaluations- und Akkreditierungsagentur, Hannover (ZEvA), chaired by Professor Hinrich Seidel. The present paper is a summary of the final report which was recently published by this agency.

3. First Document

The first document from the evaluators was produced after a preparatory meeting in Bommerholz, near Hannover in Germany, August 30-31, 2000. It was entitled “Minimum requirements for master’s and diploma study programmes in physics”, and covered Programme criteria, Implementation criteria, and Implementation of the minimum requirements.

The third section was split up into six parts:

* Formulating aims and objectives,

* Teaching and learning methods,

* Mathematics,

* Other sciences and non-technical fields,

* Recording teaching activities, and

* International contacts.

In the section "Programme criteria", it was stated by the evaluation group that a physics programme should enable graduates to:

Identify and solve problems in society that can be addressed with the methods developed in physics,

Appreciate the possibilities of research in physics and carry out independent research, e.g., in a subsequent doctorate study,

Understand the technical problems in society which confront the profession,

Comprehend the ethical characteristics of research and professional activity in physics and its responsibility to protect public health and the environment,

Maintain professional competence through life-long learning,

Work in an international context.

This first document was used like a set of rules, governing both the interviews made (according to check-lists) and, later, the analysis of the result of the hearings.

Even before this "Minimum requirements" document was produced the participating departments and/or faculties made self-evaluation reports which were published and sent to the evaluators as background material to be studied before the site visits took place.

4. Further Procedure

The evaluation procedure involved two-day visits to the sites, usually covering six hearings with different groups or individuals: teachers, students and administrative/technical personnel.

Some time after the site visits had been made, reports on each of the five programmes were written by the evaluators. These were then submitted to the faculties or departments which were invited to make comments, sometimes resulting in changes in the wording of the report. When the peer group considered the reports final, in February 2001, the responsible persons in the evaluated programmes had the possibility to add one page of comments which would then be included in the completed document. Only one university, that of Hanover, used this possibility to comment on the report. The final report which was mailed to the parties involved in July 2001, also contains a section on "General findings" as well as an "Executive Summary". As agreed at the outset of the evaluation no ranking between the universities was included.

5. General Findings

1) Organisational structure: The following citation from this section deals with the management structure in physics departments: "…we are thus seeing a shift in the way educational matters are handled, from committees towards an individual being ultimately responsible. In such a structure it may become clearer who can and must take initiatives." This person is sometimes called "Deputy dean" sometimes "Director of Education".

2) Aims and objectives: "Quality" is judged with respect to the aims and objectives set up. A general remark is concerned with the study time which, with one exception, namely Ghent, is "considerably longer than the nominal one". Another remark has to do with a lack of explicit preparation for professional life other than academic research. Again there was one exception: Twente.

3) Programme: the findings under this heading were subdivided into Structure, Contents and Methods, Theses, Examinations, Study Load, and Overall Assessment

With respect to **Structure**, there is a recommendation that the programme should preferably be five years; this is also the case everywhere, except in the Physics faculty at Ghent where the nominal time is four years.

Contents and Methods: *depth vs breadth* is discussed, in relation to the size and competence of the staff. The way *mathematics* is introduced varies quite a lot, from a common first year for physics and mathematics students, to specially adapted courses for those who want to go on with physics. Advantages and disadvantages can be noticed in both of these systems. *Non-technical courses* are discussed – they are more frequent in the more applied programmes!

Computer skills are taught in some of the programmes, in others not at all.

Theses: the number of hours programmed for the thesis varies widely: from 600 in Ghent to 2000 in Duisburg and Hanover. However the real time taken is probably more similar!

Examinations: again there is quite a large variation between the frequency and kind of tests given, oral or written.

Study Load: it is everywhere judged to be "very high", with the possible exception for Ghent.

Overall assessment: all programmes fulfill the "minimum requirements", but there are individual shortcomings, for instance, with respect to the possibilities for choosing courses outside the physics/technology sphere, for instance in social and economic sciences, the integration of mathematics in the curriculum and the preparation for professional life outside academic research.

4) Staff: there are comments on size, qualification, career possibilities, the lack of women among the staff, the lack of teaching records, and "in-breeding", implying a tradition to recruit staff to higher positions only from staff already present in the faculty.

5) Facilities and resources: they are quoted as being "adequate and reasonably up-to-date" everywhere. Details can be found in the site-reports.

6) Students: there are subsections on attainment level and selection, the evolution of numbers, counselling, duration of study and completion as well as drop-out rates, graduates and job prospects.

7) External relations: the comments are split up between "Industry and other sectors of the economy" and "International". Some criticism for some of the programmes concerning a lack of contacts in both of these is given, but also some praise. There are comments on the possibilities for incoming and outgoing exchange students, in one case pointing to unfortunate laws prescribing in which language university courses must be given, and in other cases on lacking efforts to help incoming students with accommodation. Both of these obstacles naturally tend to lower the number of incoming students.

8) Internal quality assessment and management: there are comments on the various ways these issues are handled. Sometimes, like in the German universities, there has not been any assessment before the one described here. In Ghent and Twente we saw student questionnaires and so called response groups which were used to give teachers an indication of the success of the interaction with the students. Recommendations were also made to try and find out from recent graduates and potential employers if they thought the studies had been adequate for the future career.

6. Executive Summary

The final report which comprises 160 pages, also gives an executive summary, stating, to begin with, that in general the minimum requirements from the first document of the evaluators, are fulfilled. Its contents can further be summarised in the following way:

A five-year study programme seems optimal, there is only one exception among the five institutions visited,

We found good balance between obligatory and elective courses,

There tends to be insufficient preparation for employment outside academic research,

Contacts with graduates should be improved which would imply learning from their experience,

Contacts with schools and the public should be improved and the rôle of physics in culture should be stressed,

Active participation of students should be encouraged as well as the participation of active researchers also in teaching basic courses,

The reasons for dropout should be investigated,

More training in information retrieval and in problem-solving skills, facilitating life-long learning, is needed,

Career possibilities of junior staff should be improved,

When hiring staff more attention should be paid to teaching, and it is recommended that "teaching records" be established,

Training programmes for academic teachers should be encouraged.

7. Conclusions

For the European scene evaluations like the one described here will probably be very important in the near future. The European Ministers of Higher Education have started to meet with regular intervals to try and make university education in the different countries structured in such a way that it becomes practical for young people to travel and have their academic degrees recognised across the borders. In Bologna, Italy, in 1999 and recently in Prague, Czech Republic, the ministers have made statements which give indications of a wish to make the academic careers more similar. In particular the so called Bologna declaration gave guidelines for the introduction of a two-cycle degree structure, Bachelor and Master/Doctor. Studies for the first degree should last at least three years and "be relevant to the European labour market as an appropriate level of qualification". Such changes of course take time in a diversified European market, and the ministers have wisely given themselves until 2010 to achieve these goals.

The final report can be obtained from Hahn Druckerei GmbH & CO with the address Im Moore 17, D-30167 Hanover, Germany.

ADAPTING ACTIVE LEARNING APPROACHES IN PHYSICS EDUCATION TO LOCAL ASIAN ENVIRONMENTS

O.L. CAMBALIZA, A.P. MAZZOLINI AND M.C. ALARCON
Dept. of Physics, Sch. of Sci. & Eng., Ateneo De Manila University, Loyola Heights, Quezon City, Philippines.

There is now considerable evidence [1] to suggest that "active learning" (AL) approaches are far superior to traditional passive learning modes for improving students' conceptual understanding. Over the last 2 years (after the NSF Chautauqua Workshop [2,3] held in Australia) the Asian Physics Education Network (ASPEN) [4] has been actively introducing these new student-centred learning approaches via national physics education workshops (which to date have been held in Laos, Vietnam, South Korea, Sri Lanka, and the Philippines). Supported by UNESCO, these workshops are designed to expose Asian academics to the rationale behind AL and its implementation in the Asian context. The workshops also train Asian personnel to help organise and run future workshops.
Often AL workshops use sensors interfaced to computers so students can make quantitative measurements (from laboratory experiments or lecture demos) and develop their own understanding about the laws of physics. The ASPEN workshops have tried to adapt these AL techniques to local Asian cultures and the resource limitations that often exist in many developing countries [5]. This involves not only computer-interfaced equipment, but also inexpensive or traditional non-computer-based equipment, computer simulations, and demonstration video clips. The ASPEN workshops have been very successful and the feedback extremely positive. With the continuing support of UNESCO, ASPEN plans to expand its AL workshop actvities, especially in developing Asian countries.

1. The Active Learning Perspective

For nearly two decades, the physics community has witnessed the development of a relatively new area of interest: that of physics education research. Many academics acknowledge the importance of physics education research and its role in exposing students' learning difficulties in introductory physics courses. Richard Hake [1] has suggested that this has resulted in major efforts to improve the way introductory physics is taught, especially after Halloun and Hestenes published a massive and careful study on student learning in both non-calculus and calculus-based introductory physics courses.

Many different studies have shown that students' misconceptions are deeply rooted and develop over a lifetime of incorrectly interpreted personal experiences and observations [6]. Physics education research has shown that because of these deeply-held misconceptions, students' understanding and hence appreciation of very basic physics concepts does not greatly improve based on traditional instruction [2, 6]. However, the quantification and interpretation of student misconceptions has given insights for the development of more effective methods of instruction in introductory physics courses [7]. Thus, over the past few years, a number of curricula have been developed in the United States and elsewhere that are based on the constructivist model of student thinking and learning. Reddish [8] summarized these into four categories: (1) Full Studio Models (e.g. Workshop Physics by P. Laws), (2) Discovery Labs (e.g. Real Time Physics by R. Thornton, D. Sokoloff and P. Laws), (3) Lecture – Based Models (e.g. Concept tests by E. Mazur; Interactive Lecture Demonstrations by R. Thornton and D. Sokoloff), and (4) Recitation-Based Models (e.g. Cooperative Problem Solving by Ken and Pat Heller).

A common factor in all these innovative teaching techniques is that they actively engage students in the learning process. They are commonly called active engagement or active learning techniques [5,8]. Indeed, there is considerable evidence that these active learning or interactive-engagement methods are far superior to traditional teaching methods for improving students' understanding of physics concepts at both secondary and tertiary education levels [1]. However, it is important to note that teaching techniques that make use of demonstrations and computer-aided instruction do not necessarily guarantee an improved conceptual understanding of physics, especially if the techniques themselves are not "student-centered" in the sense that students remain mere passive receivers of information. Hence, lecture demonstrations and computer-aided instruction can only be effective when incorporated into a carefully designed student-centered learning strategy that encourages active participation of the students in the learning process.

The active learning perspective has three underlying assumptions [9]: 1) learning, by its very nature, should be an active process; 2) different people learn in different ways; and 3) learning is only meaningful when learners discover knowledge for themselves and make it their own. Thus, to create an active learning environment inside the classroom, the instructor's facilitating technique should accommodate a diverse range of student learning styles and encourage students to become actively engaged in the learning process. A student-centered, activity-based [10] learning technique should always provide ample opportunities to constantly challenge students' misconceptions about very basic physics principles. Table 1 summarizes the main differences between an active and a passive (traditional) learning environment [11].

Table 1. Passive versus Active Learning Environment

Passive Learning Environment	**Active Learning Environment**
Instructor (and textbook) is the authority – source of all knowledge	Students construct their knowledge from hands-on observations. Real observations of the physical world are the authority.
Students' beliefs are never overtly challenged.	Uses a learning cycle in which students are challenged to compare predictions (based on their beliefs) with observations of real experiments.
Students may never even recognize the difference between their beliefs and what they are told in class.	Changes students' beliefs when students are confronted by differences between their observations and their beliefs.
Instructor's role is as the authority.	Instructor's role is as a guide in the learning process.
Collaboration with peers is often discouraged.	Collaboration and shared learning with peers is encouraged.
Experimental results are presented as facts in a lecture.	Results from real experiments are observed in understandable ways – often in real time with the support of microcomputer – based tools.
Laboratory work, if any, is used to CONFIRM theories "learned" in lecture – often in a RECIPE – TYPE APPROACH.	Laboratory work is used to learn basic concepts – DISCOVERY – APPROACH.

In an active learning environment, the instructor ceases to be the wellspring of all information, but rather assumes the role of a facilitator that guides the students through the learning process. Carefully interpreted quantitative observations now become the foundation on which the physics understanding of the student is developed. These basic principles, which underpin the active learning environment, help discourage the so-called "teacher knows all" syndrome. Active, student-centered learning can be a challenge for instructors (most of whom were probably taught in traditional passive learning environments as students). Hence, an active learning environment creates significant demands on the instructor, as well as the student.

With active learning, the students (who are the focus of concern) cease to be mere receivers of information from the instructor. With the instructor as a guide, they are presented with numerous opportunities to observe, experiment, discuss, and exchange ideas. They are encouraged to work in teams or groups and engage in various activities, such as interactive experiments for laboratory classes, interactive lecture demonstrations for large classes, interactive computer simulations, and interactive problem solving [10]. The activities must be carefully prepared so that they always stimulate the students to think, predict, observe, compare, and reason. This may involve preparing materials with a series of connected sequential questions that are carefully constructed to help students discover answers for themselves. At times, the instructor may have to provide explanations that are kept as short as possible. Often, the instructor summarizes answers elicited from the students. Thus, "right" or "wrong" answers are not determined by the teacher but rather discovered by the students working in groups as they are confronted with activities that challenge their understanding of basic concepts.

2. ASPEN and Active Learning

Over the past two years, ASPEN has been working in close association with the UNESCO Regional Office for Science and Technology in South East Asia. Its mission is to introduce the principle of active learning and various student-centered learning and teaching techniques to Asian member countries through international and national physics education workshops [4]. To date, these workshops have been held in Laos, Vietnam, South Korea, Sri Lanka, and the Philippines. The recent ASPEN workshop in Manila was a follow-up workshop that provided a venue for training physics educators to implement many of the active learning techniques encountered in preceding workshops. Thus, the participants designed, developed, and presented their own activity-based lesson plans making use of computer or non-computer demonstration materials. Figure 1 shows some of the participants training with a simple Bernoulli's principle demonstration that was then used as an aid in one of the active learning presentations. Figure 2 shows a simulated classroom environment where the participants assumed the role of students as two of them presented an active learning lesson plan on the topic of heat. After each presentation, the group constructively discussed the strengths and weaknesses of the presentation, focusing on the development of learning strategies relevant to an Asian context (culture and resources).

Fig. 1. Some participants train with a simple Bernoulli's principle demonstration. (ASPEN Active Learning Trainers' Workshop, Manila, Philippines, Feb 26 – Mar 2, 2001.)

Fig. 2. Simulated classroom environment where participants assume the role of students as two present an active learning lesson plan on the topic of heat. (ASPEN Active Learning Trainers' Workshop, Manila.)

Table 2. Scientists and Technicians in the Pacific Rim

Selected Countries	Year	Total Population (million[1])	Scientists and Engineers (per million population)	Technicians (per million population)
Australia	1994	17.853	3 166	792
China	1995	1 208.841	350	201
Indonesia	1988	194.615	181	-
Japan	1994	124.815	6 309	828
Korea, Rep. of	1994	44.563	2 636	317
Malaysia	1992	19.695	83	84
New Zealand	1995	3.592 [2]	1 699	790
Philippines	1992	66.188	157	22
Singapore	1995	2.821	2 728	353
Thailand	1995	58.183	119	40
Vietnam	1985	72.931	308	-

[1] Population data for 1994.

[2] 1995.

Source: UNESCO Statistical Yearbook 1996; - Statistical Yearbook 1997;

for New Zealand: Ministry of Research, Science and Technology

When introducing the rationale behind active learning, important issues need to be discussed. Active learning techniques often make use of sensors interfaced with computers for real-time measurements that allow instantaneous, quantitative feedback to the students. The foremost concern in introducing the active learning perspective to local Asian environments is its implementation. Different Asian countries are not only characterized by diverse cultures but also by uneven or non-uniform development [5]. For example, the latter issue is clearly evident in Table II, which shows the number of scientists and engineers in the Pacific Rim per million of population as recorded in the 1998 World Science Report [12]. However, it should be noted that this data may have changed significantly over the past few years. As shown in Table II, Japan, Australia, Korea, and Singapore certainly have more than enough qualified teachers (human resources) not to mention the facilities and equipment for immediate implementation. The different rates of development and cultural diversity of Asian countries illustrate that they have different needs and limitations. The ASPEN workshops adapt active learning techniques to present various options for successful implementation, especially in developing countries, and to address specific physics education concerns of member countries.

The ASPEN workshops in Vietnam, Sri Lanka, and the Philippines exposed local participants to some of the various active learning techniques that have been developed. The workshops also demonstrated how these techniques can be adapted and customized to meet local Asian needs. One form of adaptation is the use of non-computer-based active lecture demonstrations, especially for situations and environments where computers are not always readily available. Often, very successful demonstration aids can be fabricated from local, inexpensive, and easy-to-find materials. Although many non-computer-based demonstrations are qualitative in nature, some demonstrations can result in quantitative measurements. Fundamentally, it should be emphasized that the active learning perspective does not depend on the state of the equipment, but rather on the state of mind of the instructor.

In contrast, the ASPEN workshop on Physics Curriculum Design and Content held in February 1999 at the National University of Laos (NUoL) was organized to provide the opportunity for Lao physics instructors to discuss the coherence and relevance of their physics program, the compatibility and comparability of the NUoL physics curriculum with that of other universities in Asia and the Pacific, and to exchange experiences on problems related to teaching approaches and physics research. ASPEN supported and designed the workshop to strengthen the physics curriculum of NUoL and promote the development of active learning techniques suitable to local conditions. The workshop involved paper presentations on physics curricula, teaching approaches, and physics education research, plus discussion groups.

It is also worthwhile noting that the uneven rate of technology-related education development is not only evident on the international level, but also on the national level. In the Philippines, for example, there is a very high density of qualified faculty and scientists (together with technological facilities and other resources) clustered around Metro Manila. Hence, national teacher training and education upgrading programs can be successfully implemented when hosted by one of the well-resourced "Metro Manila" Universities. An example of this is the Summer Instructors' Physics Workshop conducted for the last three years at the Ateneo de Manila University based on a grant from the Commission of Higher Education. Each year, a maximum of 30 instructors from 9 different colleges and universities from different regions throughout the Philippines participate in these activity-based workshops. Physics laboratory and demonstration equipment are fabricated in the university machine shop, while the facilitating faculty prepares the corresponding laboratory manuals. The participants simultaneously learn concepts while being trained to use the equipment through a series of laboratory experiments that are generally problem solving in nature. As an enabling activity, the equipment that the participants use in the workshops is donated to their respective institutions so that it can later be reproduced. Thus, adaptation takes on various forms in order to address diverse needs. Often, limited resources represent significant challenges to university physics teachers. These teachers need to become creative and innovative if they are to implement active learning strategies with their limited resources. The lack of high-technology equipment, such as computers, is a condition that generally calls for the design and fabrication of alternative demonstration materials and experiments.

Examples of such adaptation techniques and practices are documented and shared in the workshops and training sessions conducted by ASPEN. ASPEN has become a very important forum for sharing and discussing the active learning techniques that can be successfully adapted to a cultural-specific environment. Guided by the principle of self-help, ASPEN brings university physics educators together from developed and developing Asian countries to discuss common problems and elicit feasible solutions.

Despite cultural diversity and uneven development among ASPEN member states, the challenge and aspiration to make physics interesting, effective, and relevant for students is paradoxically one concern that knows no boundaries and cultural differences.

References

1. R.R. Hake, "Interactive-engagement vs traditional methods: A six-thousand-student survey of mechanics test data for introductory physics courses", Am. J. Phys, 66, p. 64, 1998.
2. D.R. Sokoloff and R.K. Thornton, "Using Interactive Lecture Demonstrations to Create an Active Learning Environment", The Physics Teacher, 35, p. 340, 1997.
3. P.W. Laws, "Promoting Active Learning Based on Physics Education Research in Introductory Physics Courses", Am. J. Phys., 65, p. 14, 1997.
4. http://www.swin.edu.au/physics/aspen/
5. M.C. Alarcon, "Recent Innovations in University Introductory Physics Teaching", AAPPS Bull., 9(2), p.23, 1999.
6. D. Hestenes, M. Wells, and G. Swackhamer, "Force Concept Inventory", The Physics Teacher, 30, pp. 141 – 151, 1992.
7. L.C. McDermott, "Research on Conceptual Understanding in Mechanics", Physics Today, pp. 24 – 32, July 1984.

8. E. F. Redish, " New Models of Physics Instruction Based on Physics Education Research: Part 1", Proceedings of the Deustchen Physikalischen Gesellschaft Jena Conference, 1996.

9. *Creating an active learning environment*: http//oregonstate.edu/fye/resources/notebook/section1h.htm (21 July 2001).

10. M. C. Alarcon, A Talk on Active Learning, ASPEN Active Learning Trainer's Workshop, Feb 26 – March 2, 2001, Manila, Philippines [unpublished power point presentation].

11. D.R. Sokoloff, A Talk on Improving Conceptual Learning in Lecture with Interactive Lecture Demonstrations, ASPEN Workshop on Physics Demonstrations and Classroom Innovations, Nov 11 – 14, 1999, Hanoi, Vietnam [unpublished power point presentation].

12. T. Turpin, H. Spence, S. Garrett-Jones, and A. Marsh, *Status of World Science (South-East Asia and the Pacific Rim)*, World Science Report 1998, UNESCO Publishing, Elsevier.

A SOUTH EUROPEAN PERSPECTIVE ON PHYSICS EDUCATION

M. VICENTINI
Dept. of Physics - Univ. "La Sapienza" - P.le A.Moro, 2 - 00185 Roma (Italy)
E-mail: matilde.vicentini@roma1.infnf.it

Europe is a union of different national States which, at this moment, share the geographical position (with a somewhat undefined boundary toward Asia), a common market, a European parliament and a working language (english). It will also soon share a common money, the Euro. We must however acknowledge a large diversity of cultural contexts each one characterized by its language, history and social rules. Of course the cultural contexts influence education in general and therefore also physics education. The paper will present a perspective on physics education research and practice from Italy, one of the countries of Southern Europe with a root in the latin culture (Spain, Portugal, France and Italy). The paper will then focus similarities and differences in the approach to education among the latin countries (also with some reference to the latin-american ones), with southern-european countries of different roots (mainly Greece and the Jugoslavian ones) and with northern-european anglosaxon countries. In particular I will discuss the critical role of the choice of one particular language in the communication of the cultural contexts related to education.

Looking at the research literature on Science Education it seems that the line of research on the socio-cultural perspective in teaching and learning is a line of growing interest [1].

In that article some research questions relevant to the line have been proposed as follows:

"1) What is scientific activity/science education as a social institution, and what is its relationship to other institutions?

2) What practices, beliefs, and values constitute the culture of science/science education in a given time and place, and how does this culture change across historical time?

3) How does the subculture of science/science education fit into the overall cultural ecology of a larger community? With what other sub-cultural systems is it allied or in conflict?

4) How is the specialised language of science/science education and its forms of writing and discourse similar to those of other sub-communities and different from them? Why? With what larger-scale social consequences?

5) How do the metaphors and practices of the scientific community influence the kinds of research questions that are asked in particular historical periods, or help determine which kinds of people feel attracted to or excluded from its culture?

6) How is science/science education as a community dependent on economic and political forces outside it, and how does it both resist and accomodate to this dependence?

7) How has science/science education as a culture, including its beliefs about legitimate methods and questions, and its beliefs about its objects of study, been shaped historically by the overrepresentation and under-representation in its ranks of different social categories of people: men and women, Europeans and non-Europeans, wealthier and poorer classes, young and old?

8) How does science/science education as an institution and a culture define the kinds of personal identities it welcomes and supports, and in what respects is science/science education more and less compatible with masculine versus feminine identities, middle-class versus working-class identities, and the global spectrum of national and ethnic cultural identities?"

The author also suggest "Look at each of the ...... questions as a question about science, and as a question about science education. Ask yourself in each case how important the question is, how relevant it is to achieving the goals of science education...". I then tried to do the exercise.

A far as science (and physics in particular) is concerned the questions remind the different perspectives of philosophers and historians of science respect to the scientists for what concerns the "image of science" to be conveyed to the students [2].

In fact the "images of physics" of the three sub-communities are different and more co-operation and debate should take place for solving the problem for education. Anyway the debate can be facilitated by the recognition that Physics is an old science with a corpus of structured organisation of knowledge that has been analyzed from the historical and philosophical points of view. Therefore there is a wealth of information in the literature that we can use to answer the questions at least formulating hypothesis for a constructive debate.

The same is not true if we look at the questions as concerning science-education. In fact the communities of "science" and "science education" are different as in "science" one may recognize the merging of disciplinary subcommunities (biology, physics, chemistry...) that articulate on the specific content knowledge the common methodology of the experimental sciences. Moreover each of the subcommunities is a social institution with an international character and little constraints by the local cultural context.

In "science education" the local constraints are very strong because the results of the research must be meaningful for the specific educational systems.

Here it is then difficult to imagine a general answer to the questions. Indeed, looking at the literature reported in the various international journals it does not seem that we have an integrated community but a set of different social institutions. To frame a complete picture of the research it is not enough to look at the journals that claim an international status by the use of the English language and some attention to the works of non-English authors. One should also consider the national journals sometimes written in other languages (Educacion a la ciencia, Didaskalia, Didaktic der Physik, La Fisica nella Scuola, ... are examples) sometimes using two languages (examples are the greek journal Themes in Education or the Brazilian one Acta Scientiarum).

The relevance of the relation with the local social context brings, as a consequence, that each author priviledges the point of view of his/her country. As an example let's take the questions quoted at the beginning. The article appears in an American international journal as it is revealed, in some way, by the reference in question 7, to "Europeans and non-Europeans". This simplistic subdivision of cultural backgrounds in only two large fields made me feel like being at any one airport of the European Union and having to choose between the two lines of EU and non EU citizens. In the second line we may see Americans (north and south), Asians, Africans... many different cultures. But also in the first line we may see Italians, Spanish, Greeks, Germans...many different cultures!

The analogy is not misleading: for what concerns science education Europe is not a unique sociocultural context. Let me try to present a picture of Europe as a multicultural conglomerate. I am not using the word "Union" notwithstanding the fact of the establishment of the European Union (EU) by the agreement of different national states on some economical and political issues. Recently there is also an agreement (Sorbonne-Bologna) for the establishment of a unified structure of the educational system at the tertiary level (universities). A unified structure within the recognition of our cultural diversities [3].

I will then try to point some of the important cultural diversities. First there is the problem of languages: we have all agreed to use English as the communication language of the Union but this does not mean, of course, that we will not continue to use, and create poetry and literature in general, in our specific languages.

Languages are embedded with culture, therefore we may recognise more cultural similarities among the states which share a common origin of their language. From this point of view one may identify a European-Latin culture uniting the states that have a language that roots back to latin (Italy, France, Spain, Portugal, Romania), a saxon-European culture with a root in celtic language (Germany, Austria, Netherlands...), a slave-European culture, a greek culture... what about the states up in the North?

Last question explicits my ignorance of the roots of all the manifold of European languages and is therefore an instance of the existence of European subcultures. I am not a linguistic expert but I know that there are indications of a connection between the language structure and some thought procedures (f.i. analytical versus synthetical reasoning/alphabetical versus pictorial written languages). The structure of language, together with the historical development, may also relate to specific cultural aspects. Therefore possible similarities, or differences, in the cultural aspects may exist in, or among, the classes identified by the language roots.

An example may be found in the principal religions of the citizens of the different states: latin people are in majority catholic, anglo-saxon christian protestant, slave and greeks christian orthodox or muslems.

I may here recall two instances of educational problems which, while important in the anglo world, do not appear in the latin world. The first instance is the debate creationism/evolutionism. In a catholic country (even in Italy where the Church has acted not only as a religious authority but also as a political one) the problem simply does not exist. The second instance is the gender problem which in the latin country is essentially a social problem not an educational issue: the percentage of students in the scientific disciplines is reasonably high (in Italy the physics students are 30% women since many years) but the number of women in the top level of the Academic ladder is very low.

The language classification points also to possible similarities between Europeans and non Europeans. Latin roots connect Latin Europeans with Latin Americans while anglo-saxon roots give a connection of Saxon Europeans with North Americans and Australians. It should also be mentioned that the need of communication had pushed the European elite class to learn at least a second language. Back in the past latin was a must (we still have latin as a school discipline in some secondary school streams in various countries). In more recent years the choice of the second language was influenced by political issues. In my country (Italy) we had a French period, a German one and an English one only after 2nd world war.

Today not just the elite class but everyone is supposed to learn English as a second language in all countries. This means that all European citizens with the exception of one country (UK) will have in their cultural background the maternal language and a second one. [We continentals do not understand why British students should not be obliged to learn a second language of their choice].

Another important point which states a difference between UK and non-Europeans English speaking countries is given by the amount of editorial translation work which may give the possibility of approaching different cultures by reading, in the maternal language, works (literacy and others) originally produced in another language. The editorial market of the English speaking countries restricts the translation work to few particularly important books. [Again an example from Italy related to science education: many textbooks from UK and the USA are available in an Italian version. No Italian textbook is considered on the English market].

However, notwithstanding the possibility of learning about other Europeans cultures by reading (or TV) and travelling around (but how important it is to be able to speak the language!), our reciprocal knowledge needs to be improved.

A good possibility of improvement may lie in the learning of the historical development. However this is usually country centered: we Italians - where history is a very important discipline of the preuniversity curricula (Historia magistra vitae) - may learn a lot about ancient Egypt, Greece and Roma, the Arab invasion, the French revolution, the two world wars and know very little of the history of the northern countries which had little impact on our historical development.

Historia magistra vitae: here we find a similarity in science education among the latin countries (also non Europeans). In discussing about education the historical and philosophical issues are a must to be afforded just at the beginning. On the other hand we are not very good on the empirical approach. The reverse is true for anglo countries.

In Italy, in Physics, we have a strong community of historian of Physics who share with the researchers in Physics Education the choice of working in Physics Departments. The two communities (education and history) are moreover supposed to interact in the University teaching in particular for teacher training. This brings about another difference with the anglo countries where the researchers in education work in Science Education Departments.

Coming back to the questions raised by Lemke I do not recognize "science education" as one unified community but a set of diversified sub-communities characterized by the cultural context of each country.

Unfortunately the most diffuse international journals (which come from the anglo editorial companies) seem to convey a more unified version. This is probably due to the use of English as the international language on one side and to the partial ignorance of the anglo community of what is produced in other countries but published in the countries language.

Language: When we send articles to an international journal often the first criticism of the referees concerns language problems. Apparently the requirements are for a high level written production with little, if none, consideration of the possible influences of our first language on the English expressions. We even found that it is not enough to have the paper checked by a native English speaker to satisfy some "oxfordian" referee.

Partial ignorance: Our set of references is often criticized if it contains more non-anglo-works which are relevant for the author than anglo references which seem to be anyway a must to be quoted also if they are not considered pertinent by the author.

This brings me to question the values that constitute the culture of the anglo science education community and this is for us, non anglo, a big problem in an era of globalization which sometimes appears to lead to the cancellation of our cultures. I will then conclude by suggesting that an effort should be made for improving the non-anglo representation in the international community.

A final remark: Cultural similarities world wide may exist but not be recognized. I may quote two examples.

The first one is given by the comparison of African with western culture proposed by Horton [4] who recognizes that the principal difference in the understanding of scientific knowledge does not lie on the empirical phenomenology but in the definition of the explaining entities apparently more related to magic in African culture than in the western one: but can one really state that "gravity" or "electric field", if not understood in their relation with phenomena, are not magic entities?

The second example has to do with the pseudo scientific beliefs which may coexist with the scientific ones. A research [5] in some third world countries reported experimental evidence of such a coexistence. We repeated the research by using the same methodological tool (only adapting it to our cultural context) in Italy and Brazil finding an analogous experimental evidence in our western countries [6]. No analogous research was thought to be important for other western developed countries.

References

1. Lemke J.L. - Articulating Communities: sociocultural perspectives on Science Education, J. Research in Sci. Teaching, 38, p.296-316, 2001

2. Vicentini M. - The nature of science: a didactical issue, Acta Scientiarum, 22, p.1285-1295, 2000

3. Ferdinande H., Petit A. (ed.) - Inquiries into European higher education in Physics, 1999

4. Horton R. - Rationality and Relativism, Hollis and Lukes (ed.), Blackwell Oxford, 1982

5. Ogunniyi M.B. et al. - Nature of worldwide presuppositions among science teachers in Botswane, Indonesia, Japan, Nigeria and the Philippines, J. of Res. in Sci. Teaching, 32, p.817-831, 1995

6. Bandiera M., Danhoni Neves M.C., Vicentini M. - Nature of worldwide presuppositions among science teachers in Italy and Brazil, Acta Scientiarum, 21, pp.97-106, 1999

QUANTUM PHYSICS FOR ENGINEERS AND APPLIED PHYSICISTS: APPLIED HOMEWORK ASSIGNMENTS

ZUYUAN WANG*, EDWARD F. REDISH°, SETH ROSENBERG**°
** Southwest Jiaotong University (China), ° University of Maryland (USA)*
***City College of New York (USA)*

Applications of quantum physics have played a major role in technological advances of the 20th century and are certain to play a role in the developments of the 21st. As a result, quantum physics is a topic of increasing interest to scientists and engineers outside of pure physics As it is traditionally taught, quantum physics is not well matched to the needs, backgrounds, and interests of students who are particularly interested in applications, such as electrical engineers. In the project "New Model Course in Applied Quantum Physics," a group of researchers at many universities have researched student backgrounds and learning in a one semester upper division course in quantum physics. the group has developed a variety of materials have been developed for this course. One component, applied homework assignments, is described in detail here.

1. Motivation

Physics departments all over the globe have to deal with the issue of teaching a fairly standard curriculum within a variety of cultural contexts. Students' understanding of physics may be affected by their culturally determined experiences. For example, students raised in rural areas without electrification may have less familiarity with electrical devices than students raised in cities.[1] Namibian students' have natural metaphors concerning time that emphasize its cyclic character rather than its linearity.[2] These cultural influences may change the way these students interpret standard lessons.

Even within the cultural frame of a single college or university, different populations such as physics majors, biologists, and engineers, bring with them different cultures. They have different expectations for what they should be learning and how, they are motivated by different goals, and they are interested by different topics.

A particular problem that is of considerable importance to the further development of modern technologies is the teaching of quantum physics to engineers and applied physicists. Quantum physics as usually taught by physicists often focuses on highly abstract and mathematical presentations with little emphasis on real-world applications.[3] This approach makes good sense for physics students intent on developing understanding of fundamental laws, but it provides little motivation for students whose ultimate goal might be to construct a more cost-effective quantum device. In order to teach these students more effectively, the cultural expectations of the populations must be considered.

In the project "A New Applied Model Course in Applied Quantum Physics," a group of physics education researchers originally at the University of Maryland have studied the population of engineers taking quantum physics as an elective and developed a new instructional approach for teaching them quantum physics more effectively.[4] Most of these students are upper division university students, most engineers. In this paper we give an overview of the project and discuss some details concerning one element that we have developed to improve the cultural match between these students and our course in quantum physics: applied homework assignments.

2. The New Model Course

In the current era, university students in a wide variety of subjects, such as electronics, materials science, biology and computer science, should not only understand classical physics, but also develop an understanding of quantum physics. Quantum physics is an essential component for the development of many of the technical advances of the 20th century. The invention of transistor in 1947 built directly on quantum principles. The laser, positron emission tomography (PET), and superconductors all rely heavily on our understanding of the quantum character of matter. The modern technologies made possible by our understanding of quantum physics are changing our lives at an unprecedented pace. In the 21st century, quantum physics is sure to play a major role in a wide range of technical developments ranging from nanometer engineering to quantum computing. For modern students in applied physics and engineer a knowledge of quantum physics is increasingly necessary.

But quantum physics is a large and difficult subject. Novice students are confronted not only by new mathematics but also by difficult conceptual issues that underlie the physics. For many students, the mathematical treatment normally offered to physicists serves as a barrier that keeps them from choosing to study quantum physics.

For students such as these, especially those for whom quantum physics is optional study, courses must be designed to match the needs, interest, and experience of each particular population of students. The "New Model Course in Applied Quantum Physics" provides a sample of how education research methods can be applied to upper division physics courses in order to provide materials that better match both the needs and interests of particular student populations. For the particular population of engineers and applied physicists the materials have been developed in the context of a one-semester quantum physics course for upper division electrical engineers. By picking and choosing an appropriate and coherent subset of quantum topics, it remains an "honest" quantum course while "impedance matching" to the mathematical strengths of the population.

The focus is on one-dimensional Schrödinger quantum mechanics and relies heavily on the mathematics of ordinary differential equations and Fourier expansions, topics in which the electrical engineers tend to be strong. It suppresses the matrix and state methods, eliminating linear algebra and partial differential equations, topics in which the electrical engineers are often weaker. By eliminating most three-dimensional quantum problems (particularly, angular momentum and related issues) and relativity, time remains within the constraint of a single semester for a serious treatment of tunneling, conductivity, and semiconductors, with a basic introduction to the quantum mechanics underlying such devices as the STM, diode, and transistor.

The materials development for the project is strongly guided by careful educational research and uses techniques that have proven effective at the introductory level. Materials provided use a variety of modern instructional techniques including:

Tutorials — guided-discovery group-learning worksheets

Modern technology — simulations and computer-assisted data acquisition

Applied Homework Assignments — context-rich device-based assignments

Just-in-Time-Teaching — Web-delivered essay questions for strong feedback

Exam questions — problems that stress conceptual learning and understanding

These materials are available on CD and on the web.[5] The CD contains discussions of the research as well as the materials. In the remainder of this paper, we discuss the structure of one of the elements contained in these materials, applied homework assignments.

3. Applied Homework Assignments

To engage the interests and motivation of the particular population of students we are considering, we have developed activities that relate the use of quantum physics to everyday life and to specific devices the students can plausibly expect to work with during their careers. Many end-of-chapter textbook problems concern device applications. However, these problems tend to be too simple and straightforward to have an authentic "real-world" feel. More appropriate for this population are "context rich" problems of the type developed in the US by the Minnesota group for the introductory physics class.[6] Furthermore, at the advanced level at which we are working, a context-rich device-oriented quantum-physics problem can involve an interesting variety of problem solving techniques.

These assignments have two major goals. First, we want the students develop basic physics ideas in their study and apply them in appropriate ways. In addition, we feel that it is vital that the students see that quantum mechanics relates to both the world outside the physics class and their lives. They should be able to make scaling and estimating arguments as well as understand how fundamental quantum principle govern many relevant technologies.

These goals require rather longer and more extended problems than are traditionally presented for homework at the undergraduate level. The group of problems in an Applied Homework Assignment (AHA) can be offered as a special project for a few weeks of outside-of-class work or can be presented in parts throughout the semester. This latter approach could have the advantage of helping students see the course as cumulative and giving students the sense that they have learned something both practical and substantial. To provide ourselves with guidance in constructing the AHAs we propose 7 design principles. Not all of these principles are appropriate to every AHA, but they give guidelines for developing them.

An AHA relates to some fundamental quantum principle or idea

The point of an AHA is not simply to carry out a calculation in order to show that quantum principles allow one to construct devices. Rather, students have severe difficulties in understanding and believing fundamental quantum principles. An AHA should effectively explore the workings of the device in terms of the relevant principles. This should help the students think about fundamental quantum principles and come to grips with their implications for real-world applications. The students will begin to see these principles as important not only in the classroom, but also in contexts that may affect their lives and careers.

An AHA involves some real-world devices that students can imagine using in their professional work.

The device context is important, both to motivate this particular population of students (engineers and applied physicists) and to provide a bridge to the real world. If the students do not see any connection between the context and their own lives, then the "reality" of the quantum principles will not be enhanced. In our teaching, we often adopt extreme idealizations to focus a student's attention on the physics principle. The differences between an ideal model and real-world devices help make a connection between the classroom concept and its realization in everyday life. The applied homework plays a bridging role.

An AHA has a part in which the students describe and explain something qualitatively.

Students often treat quantum physics as a purely mathematical exercise without thinking about its meaning or implications. Expressing the physics in qualitative terms helps them make sense of physical principles by linking to their semantic knowledge.

An AHA has a part in which the students express a relationship in terms of an equation, which they have to derive.

The ability to use equations to solve physics problems is a skill of considerable importance for technical students. In a mathematically-based science such as physics; equations are not simply ways of calculating numbers, but are ways of expressing relationships between measurable quantities and the system's parameters. The mathematical relationship also serves as a way of organizing one's conceptual knowledge and of displaying dependences and correlations. Working with equations appropriately demands continually mapping between the mathematics and the interpretation of the mathematics in the physical system.

An AHA has a part in which the students put in numbers to get an idea of the actual physical scales involved.

Students often treat putting in numbers as a plug-it-in-and-turn-the-crank activity. But one of the critical skills one learns in physics is to pay attention to scales. Deciding how big an effect is and whether it is important or not is a critical component of the physicist's art. Making a connection to numerical scales with which they have experience is particularly valuable.

An AHA contains an estimation problem in which the students calculate something that is actually interesting and relevant to some real problem.

Having to create their own input numbers from information they know helps students make connections between the physics and their real world experience. If the students also find the problem interesting, it can increase their motivation.

An AHA has references to allow students to extend their study to a term paper.

The activity should not be seen as simply a small, closed-ended task but as a part of their ever-growing knowledge relevant to their profession. Giving connections to the literature (including sites on the web) can help students broaden their view of the value of their learning.

4. Examples of AHAs.

We have developed a series of applied homeworks related to quantum phenomena. Some introductory topic covered include: The Phototube, Orientation Interferometer and The Scanning Tunneling Microscope.

We continue to develop AHAs on more advanced topics. A major topic of interest for electrical engineers is conductivity, including band structures and semiconductors. Today, rapid development of semiconductor devices and integrated circuits has been accompanied by an enormous increase of information in the field of semiconductor physics and microelectronics. The materials for the new course includes reading, tutorials, homework problems, and essay questions on the topic of band structures and conduction. Applied homeworks are being developed on the following topics: The *p-n* junction, The Diode, MOSFETs and The Laser. An example of applied homework concerning the photomultiplier is given in the Appendix.

Each problem in an AHA is of a reasonable level of difficulty, comparable to a typical homework problem at the upper division level. But when the students complete the entire related set of problems, they can discover how intimately engineering technologies rely on physics knowledge and on quantum physics knowledge in particular.

Appendix: The phototube AHA

The photomultiplier (PM) tube is an example of how the photoelectric effect finds application in a real-world device. The PM tube consists of a photocathode. When a photon strikes the cathode it emits an electron. This is then attracted by an electric force to a secondary electrode called a dynode. When an electron of energy E strikes a dynode, it emits approximately E/E_0 electrons, where E_0 is the ionization potential of the dynode. (The ratio of emitted to incident electrons at the dynode is called the *emission coefficient*, δ.) These secondary electrons are attracted in turn to the next dynode by an electric force where the number of electrons is again multiplied. A succession of these electron multiplications turns a single electron knocked out by a photon into large numbers of electrons that are detectable by a standard ammeter.

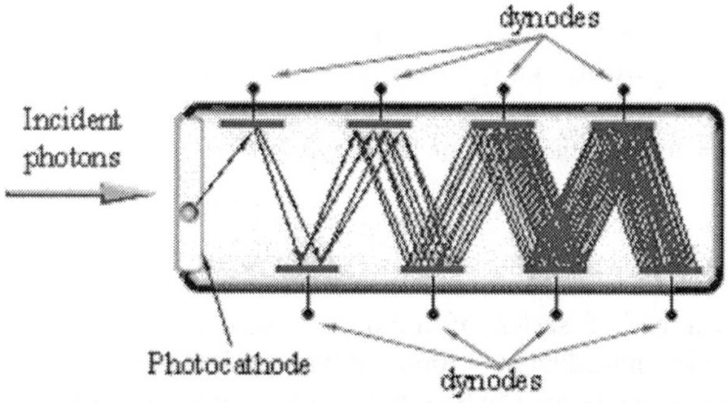

The electric force between the dynodes is produced by increasing their electric potential in succession along the length of the tube.

1) In a PM tube, assume that there are seven dynodes with potentials of 100V, 200V, 300V, ... 700V. The average energy required to free an electron from the dynode surface is 10eV. For each incident electron, how many electrons are freed (a) at the first dynode? (b) at the last dynode ?

2) The ratio of the number of electrons emitted from the PM tube after multiplication to the number of photons incident on the cathode is called the *gain*. From the previous question, we can understand how the overall gain (G) of a photomultiplier tube depends on the number of dynodes in the multiplier section and the emission coefficient or average number of electrons produced per primary electron for each dynode stage. Assume the emission coefficient at the dynode is δ, and there is a chain of n dynodes in the PM tube. What is the relationship between G, δ, and n?

3) The energy of the electron incident on each dynode stage is determined by the potential difference, ΔV, between the dynodes. Then the emission coefficient δ is a function of the potential difference, ΔV, between the dynodes, i. e., $\delta = k\,\Delta V$, where k is an overall constant. (What is it determined by?) Suppose we desire a particular gain, G. Write an expression for the total voltage across the photomultiplier tube (from the cathode to the last dynode) in terms of G, n, and k. We would like to function with as low a voltage as possible. Find the minimum total voltage necessary to produce the gain G, and also find the number of stages necessary as well.

4) The situation is complicated by the fact that not every photon incident on the photocathode causes the emission of a photoelectron. For the photoelectric effect, the *quantum efficiency* (*QE*) is defined as the number of photoelectrons emitted from the photocathode divided by the number of incident photons. If N_0 photons are incident on the cathode, use this fact and your previous results to write an equation to show how many electrons are emitted from the last dynode in terms of the parameters of the tube. (Neglect edge effects, non-linearities in the tube's response, statistical noise etc.)

5) Now that you have explored the workings of the photomultiplier tube, how do you think that the accuracy of the response might break down if the number of incident photons is too high. In other words, what kinds of effects might occur due to a large number of incident photons which will effect the linear relationship between the number if incident photons and the PMT's output current.

6) For real devices, the QE is a function of the wavelength of the incident light. For example, the QE of a potassium-cesium photocathode at λ =589 nm is 16%. Assume a PM tube with an emission coefficient at the dynodes of δ = 10, and n =7 dynode stages.

(a) For a single incident photon, how many electrons come out of the last dynode of this PM tube?

(b) Prototype quantum computers have been made by manipulating the spins of a small number of isolated atoms. One possible method of "trapping" a small number of atoms for use in such a computer is by forming an "optical molasses." (This involves placing the atoms at the center of six laser beams, all pointing at the sample. In this center region, a moving atom will preferentially absorb photons that are coming toward it, and then re-emit the photon in a random direction. Thus, the atom will feel an average net force in the direction opposite its motion.) For Sodium atoms, the light being absorbed and re-emitted by the atoms is 589 nm and each atom scatters a photon approximately once every 30 ns. The trap is located at the center of a 10cm radius spherical chamber, and a phototube is placed behind a 2cm by 2cm viewpoint window (assume that the window's transmission coefficient is 100%) Assuming that you have 10 mA of electronic noise on your signal, what is the smallest number of atoms that you can detect with your PMT?

Readings :

Hamamatsu Photomultiplier Tube Characteristics datasheets :

http://usa.hamamatsu.com/cmp-pdfs/pmts.asp

Photomultiplier Tube Testing, U. if Illinois Urbana-Champain :

http://web.hep.uiuc.edu/atlas/indexPMT.html

Particle Detectors, CERN :

http://training.web.cern.ch/Training/ACAD/Transparencies/Joram300398/pd3/P001.html

Fluorescence photospectrometry, Loyola U. Medical Center :

http://www.meddean.luc.edu/lumen/DeptWebs/physio/pmt.htm

References

1. Grayson, D.. Some Ideas About Electricity Held by Zulu Students, American Association of Physics Teachers meeting, Orono, Maine, August 1992.

2. M. Lemmer, T. Lemmer, and J. Smit, Perceptions of Time Held by First Year Students of the University of Namibia, *SA Journal for Higher Education*, vol. 9(2),pp. 123-128, 1993.

3. R. L. Liboff, *Introductory Quantum Mechanics, 3rd Ed.* Addison Wesley, 1998.

4. Project participants include E. F. Redish (U. of Maryland), R. Steinberg and S. Rosenberg (CCNY), M. Wittmann (U. of Maine), L. Bao (Ohio State U.), Z. Wang (Southwest Jiaotong U.) and Pratibha Jolly (U. of Delhi).

5. To obtain a CD, contact Michael Wittmann via email at Wittmann@umit.maine.edu. The materials can also be found on the web at http://www.physics.umd.edu/ perg/qm/qmcourse/welcome.htm.

6. P. Heller, R. Keith and S. Anderson, Teaching problem solving through cooperative grouping. Part 1: Group versus individual problem solving," *Am. J. Phys*. vol. 60, pp. 627-636, 1992; P. Heller and M. Hollabaugh, Teaching problem solving through cooperative grouping. Part 2: Designing problems and structuring groups, *Am. J. Phys*.vol. 60, pp. 637-644, 1992.

Acknowledgments

This work has been supported in part by grants from the US National Science Foundation (grant DUE-9652877) and the Fund for Improvement of Post-Secondary Education (grants 116B70186 and 116B000300).

ANALYSIS OF KOREAN AND ISRAELI SCIENCE CURRICULA FOR JUNIOR AND SENIOR HIGH SCHOOL

EUN-SUK SEO[a] & MEE OK CHANG[b]
[a] *Susung Girls' Junior High School, Daegu, 706-844, Korea*
[b] *Wolbae Junior High School, Daegu, 704-836 , Korea*

In Israel, the syllabus for Science and Technology in junior and senior high school was developed in the wake of the Harrari report, which recommended combining the study of science and technology. This recommendation reflects the relations and mutual influences between science and technology in modern society. In Korea, the 7th National Curriculum from 2001 is currently taught in junior high school. This study reviews the junior and senior high school science education curricula in Korea and Israel. The analysis concentrates on the characteristics, objectives, time allocations, structures, and contents of the Korean Science subjects in the 7th National Curriculum and Israel's Science and Technology subjects for junior and senior high school. Various training programs and organizations for science teachers are also compared between Israel and Korea.

1. Introduction

The goals of the current study were to identify special features in the Israeli science curriculum for secondary school in comparison with the Korean science curriculum, and determine the most efficient way to train science teachers in Korea based on investigating the advantages of the Israeli training system for science teachers.

Israel is currently one of the strongest countries in science and high technology and invests heavily in education, especially Science and Technology. Since 1968, Israel's educational system has been a 6-3-3 structure, the same as in Korea. Kindergarten and ten years of primary and secondary school(junior and senior high school) are compulsory and free of charge, while two additional years of schooling(grades 11-12) are noncompulsory, yet also free of charge. In Korea, only primary education is free and compulsory.

In the light of the strong link between scientific and technological disciplines in Israel, their mutual interactions, the blurring of the boundaries between them, and the need to teach them in an integrative manner are strongly increased. Therefore, **multi-disciplinary** subjects were introduced into the educational system in 1992, **Science & Technology** in junior high school and **Science & Technology in Society** in senior high school. These subjects are taught on the basis of a curriculum that takes an **interdisciplinary approach**, from the earliest school years through various high-school frameworks intended for students who do not major in scientific or technological fields.

In Korea, the 7th National Curriculum has been in effect since 2000. According to this curriculum, in junior high school, science subjects are not combined with technology, while computer and technology subjects are combined with home economics. In senior high school, science is divided into 4 parts ; Physics, Chemistry, Biology, and Earth science. This is quite different from Israel's case. Therefore, the current study analyzed and compared the science education curricula for junior & senior high school from both countries. A survey was also conducted on the training programs available for science teachers in the two countries.

2. Structure of Educational System

The Israeli society has various cultural characteristics which affect the educational system. The school system is divided into state schools, attended by the majority of pupils, state religious schools, and Arab and Druze schools. The various types of school are shown in Table 1, including any charges and whether or not they are compulsory.

Table 1. Structures of Educational Systems in Two Countries(Jang, 2001)

Schools	Grade	Israel	Korea
Kindergarten And Nursery	Ages 2–5	Noncompulsory	Noncompulsory
	Ages 5-6	Free and compulsory	
Primary School	Grades 1-6	Free and compulsory	Free and compulsory
Junior High School	Grades 7-9	Free and compulsory	Charge and noncompulsory
Senior High School	Grade 10	Free and compulsory	
	Grades 11-12	Free and noncompulsory	

3. Goals of Science Education in Junior High School Curricula

Science & Technology is taught as a single integrated subject in Israeli junior high schools.

3.1 *Goals of Science and Technology in Israel*

The Goals of Science and Technology are very concrete and clear.(3)

1) To acquire knowledge and understanding of facts, concepts, laws, and principles from a variety of scientific and technological disciplines, which are vital for every citizen in the modern world and which constitute a basis for continued studies in the fields of science and technology
2) To learn about phenomena in the world around us.
3) To develop critical and creative-inventive thinking; to understand research and problem-solving methods.
4) To develop the ability to characterize and understand complex systems in science and technology.
5) To understand the mutual relations between the sciences, technology, and society .
6) To become familiar with the historical development of concepts and their implications for the development of science, technology, and society; to understand contemporary theories with an awareness of their limitations in explaining phenomena.
7) To develop a range of capabilities and skills.
8) To understand the unique role and involvement of human begins in nature and the environment.
9) To develop an awareness of intelligent consumerism, fostering decision-making that weighs the choices for a product or system, by means of examination and evaluation.
10) To recognize the value of work and productivity for the individual, the economy, industry, and society.
11) To become acquainted with natural habitats in the country: its regions, landscapes, physical infrastructure, flora, and fauna.
12) To develop willingness to foster values related to nature and to preserve the environment.
13) To foster the interest and desire to expand and deepen one's knowledge in the fields of science and technology.

3.2. *Goals of Science in Korea*

The goals of science in the 7th National Curriculum in Korea are to understand the knowledge systems of science, to have an interest and curiosity in natural phenomena, to be able to use the inquiry method, and to have a positive attitude toward nature.

The specific goals have four parts; cognitive field, inquiry process field, affective field, and Science-Technology-Society field(Table 2).

4. Organization And Time Allocation of Science And Technology in Junior High School

4.1 *Organization of Science Subjects*

The Israeli organization of science subjects is divided into two parts; Required and Elective parts. There are 7 main topics in the required part, which are divided into basic and enriched subjects.

The Korean organization of science subjects is divided into three parts. All students complete the required parts, yet the enriched and supplementary parts are selected depending on the ability of the student(Table 3).

Table 2. Goals of Science in Korea(Korean Ministry of Education, 1999)

Goals	To understand basic concepts of science and apply them to real life through inquiry of nature	
	To foster scientific inquiry ability and apply to real life	
	To have interest and curiosity in natural phenomena and science study and to develop scientific problem-solving abilities	
	To understand that science affects technological development and the advance of society	
Specific Goals	Cognitive field	To understand basic concepts of science and apply them to real life through inquiry of nature
	Inquiry process field	To acquire basic concepts of science through inquiry of natural phenomena
	Affective field	To have an interest in scientific study, scientific inquiry attitude, and scientific problem-solving ability
	Science-Technology-Society field	To understand that science affects technological development and the advance of society

Elective topics in Israel include various projects: Flight and Space, Lasers, The Israeli Electric Power System, Photography, Physical Principles in the Motility Systems of Organisms, Dental Health and New Materials in Dentistry, Weather Forecasts, Microclimate, and so forth.

Table 3. Organization of Science Subjects(Jang, 2001)

	Israel	Korea
Required topics	1. Materials: Structure, Properties and Processes 2. Energy and Interaction 3. Technological Systems and Products 4. Information and Communication 5. Earth and the Universe 6.Organisms: Phenomena, Structures and Processes 7. Ecosystems	1. Energy 2. Materials 3. Life 4.Earth
Elective topics	Systems in Science and Technology	Each unit has enriched and supplementary parts

4.2 ***Time Allocated to Science Subjects***

In Israel, a minimum of 540 hours are allocated to the required parts and about 60 hours to the elective parts. Whereas, only 238 hours are allotted to science in Korea. Therefore, three of the topics included in the Israeli curriculum are not included among the Korean science subjects : Technological Systems and Products, Information and Communication, and Ecosystems. Table 4 presents the time allocated for the main topics in the required and elective parts in Israel. Table 5 presents a comparison of the time allocated for the main topics in Israel and Korea.

4.3 ***Aspects of Topics in Science and Technology***

The close interaction among science, technology, and society constitutes an important component in the syllabus. As such the detailed syllabus, which distinguishes between the scientific, technological, and social aspects of each topic and points to the connections between them, reflects the interdisciplinary nature of the subject and the range of curricular combinations with which it is possible to teach each main topic.

One example of the detailed syllabus is shown in Table 6 in the case of the sub-topic, 'Characterizing and sorting materials' within the main topic of Materials.

5. Goals of Science Education in Senior High School Curriculum

5.1 ***Goals of Science Education in Senior High School in Israel***

The goals in the area of content refer to general aims, with an emphasis on understanding science and its central components within the context of human culture. Meanwhile, the goals regarding the modes of thinking, learning, and skills focus on approaches and patterns of thought vital to science, intelligent consideration of information, and accepted scientific and technological methods.(3)

Table 4. Time Allocation for Main Topics in Required and Elective Parts in Israel (Israeli Ministry of Education Curriculum Center, 2000)

Main Topic	*Teaching Hours*		
	Core Curriculum	Enriched Material	Total
Required			
Materials	75	30	105
Energy and Interaction	60	30	90
Technological Systems And Products	60	30	90
Information and Communication	20	15	30
Earth and the Universe	30	15	45
Organisms	120	30	150
Ecosystems	20	10	30
Total	385	155	**540***
Elective **			
Systems in Science and Technology			60
TOTAL			**600**

*Minimum required for junior high school (540 hours = 18 hours a week, i.e. 6 hours a week for each grade)
**Electives are selected by each school.

5.1.1 Content Goals

1. To promote the recognition of science and technology as part of human culture and develop an awareness of the interaction between science/technology and society.
2. To generate curiosity and interest in scientific – technological topics and issues, particularly those on the public agenda and/or relating to the life of the individual and his / her environment.
3. To provide knowledge and understanding of several scientific and technological fields.
4. To promote awareness of the manner in which scientific and technological knowledge is constructed and understanding of its evolving, non-absolute nature.
5. To develop the ability to relate to information in an intelligent and critical manner.

6.To generate an awareness of inductive reasoning and considerations based on probability in science and technology.
7.To create an awareness that solutions to scientific and technological questions are based on learned knowledge and on logical modes of thought.
8.To provide knowledge and engender an understanding of the reciprocity between human beings and their environment, near and far.
9.To promote positive attitudes toward and appreciation of scientific and technological issues and activities.
10. To foster an awareness that the laws of nature are common to different fields in science and technology.
11.To develop an awareness of and a rational approach toward potential conflicts between science and ethics/religion, which arise from scientific and technological advances, and to promote recognition that scientific knowledge may be of assistance in reaching decisions in the public, private, and social spheres.

Table 5. Time Allocated to Science in Israel and Korea(Jang, 2001)

Main Topic	**Israel**			**Korea**		
	Grades 7, 8	Grade 9	Total	Grade 7	Grades 8, 9	Total
1 Materials	90	15	105	26	34	60
2 Energy and Interaction	45	45	90	26	34	60
3 Technological Systems and Products	60	30	90	-	-	-
4 Information and Communication	30		30	-	-	-
5 Earth and the Universe	30	15	45	24	34	54
6 Organisms: Phenomena, Structure and Processes	90	60	150	26	34	60
7 Ecosystems	15	15	30	-	-	-
Total	360	180	540	102	136	238

Table 6. Example of detailed syllabus for Materials, " Characterizing and sorting materials"

(Israeli Ministry of Education Curriculum Center, 2000)

Topic/Sub-topic	Scientific Aspect	Technological Aspect	Social Value
1.1.1 Basic concepts	-Characterizing objects and materials: distinguishing between objects and materials; mass, volume, surface area -Means of measurement, units of measure		
1.1.2 Sorting materials	Sorting materials by origin, properties, uses. Examples: -organic vs. inorganic materials -pure vs. impure materials -natural vs.man-made materials (e.g, alloys, polymers, complex materials) -sorting by states of matter	-The need to prepare materials from raw materials	Use and production of materials throughout history Stone Age, Bronze Age, Iron Age, "Age of Plastic" The impact of the development of civilization and quality of life
1.1.3 Properties of Materials	-Physical and chemical properties of materials, such as: solubility, activity, thermal conductivity, optics, mechanics(strength, load)	-Use of a raw material in keeping with its properties(at the macro level)	Humans and the environment: -Human use of natural resources; understanding the cycle of materials and the need for recycling, while preserving the quality of life and the environment

12. To expose students to scientific- technological issues of particular importance to the State of Israel.
13. To promote an understanding of the use of scientific knowledge and technology as an aid for finding solutions to problems involving values or religious laws.

5.1.2 Goals Regarding Modes of Thought, Learning, and Skills

1. Students should be able to explain natural phenomena and technological developments by means of scientific knowledge.
2. Students should be aware of different thinking strategies, both general and typical of science or technology, and know how to identify them.

3. Students should be able to read a popular scientific article and explain its contents.
4. Students should be capable of distinguishing between facts and suppositions, between causes and results, and between observations and conclusions.
5. Students should understand scientific experiment and its limitations, and gain experience with guided laboratory work.
6. Students should understand the historical development of scientific ideas and concepts. Students should be able to decipher different formats for presenting information and comprehend the relation between them.
7. Students should become familiar with different types of information sources, be able to locate information in them, edit it, and make use of it.
8. Students should gain experience in team work.
9. Students will be required to demonstrate knowledge and support opinions both in writing and orally.
10. Students should recognize the use of science and technology in solving problems involving values and religious laws.

The Korean goals for senior high school are similar to those for junior high school, as shown in Table 2.

5.2 *Structure of Science Subjects in Senior High School*

In Israel, Science and Technology in Society is a required subject and taught in all high schools. This discipline is intended to serve as an enjoyable way to impart tools and scientific-technological thought processes.

In Israel, the structure of science subjects is divided into two parts; required in grade 10 and elective in grades11-12. The Korean structure is also divided into two parts. Table 7 shows a comparison of the structures of science subjects in senior high school in both countries.

Table 7. Structure of Science Subjects in Senior High School

grades		Israel	Korea
10	Required subjects	STS (Science and Technology in Society)	Science
11~12	Elective subjects	STS (Science and Technology in Society)	Life and Science Physics(Ⅰ,Ⅱ) Chemistry(Ⅰ,Ⅱ) Biology(Ⅰ,Ⅱ) Earth Science(Ⅰ,Ⅱ)

5. 3 *Time Allocated In Senior High School*

In Israel, the curriculum for STS is modular and non-hierarchical in nature. Subjects are comprised of units of 30-45 hours each. Each student is entitled to a total of 240 hours of study in the subject of Science and Technology in Society, over the course of which 6-8 units are taken. These units are offered at least in grades 10 and 11, and it is recommended that they continue through grade 12.

In Korea, science subjects comprise 102 hours in grade 10 and each student is taught 6 units (during 17 weeks). Students in grades 11 and 12 study 'Life and Science' for 68 hours as a general elective then select two more subjects from among Physics(Ⅰ , Ⅱ), Chemistry(Ⅰ , Ⅱ), Biology(Ⅰ , Ⅱ), and Earth Science(Ⅰ , Ⅱ) for 102-68 hours as enriched elective subjects. Table 8 presents the time allocation for science subjects in senior high school in Israel and Korea.

6. Content Comparison in Senior High School

The contents of the STS curriculum are not specified in Israel. The teacher chooses the textbook and arranges the instruction in each regional school. For example, there is no consensus regarding a description of the curricula. Thus, the content map presented in Tables 9 and 10 allows for a flexible design of 'paths' that cross a variety of scientific/technological disciplines and incorporate the ideas or principles they have in common. [4]

Table 8 . Time Allocation for Science Subjects

Grade	Israel		Korea	
10	STS	240	Science	
11~12	STS	240	Life and Science	68
			Physics(Ⅰ , Ⅱ) Chemistry(Ⅰ , Ⅱ) Biology(Ⅰ , Ⅱ) Earth Science(Ⅰ , Ⅱ)	68-102

A topic selected for a given unit will be anchored in one or more disciplines and clarify one or more scientific or technological ideas: As this schematic map illustrates, the discussion of social or philosophical issues can occur in any of the chosen paths. The contents of the Korean senior high school science curriculum are specified in Table 11.

7. Teacher Training And In-Service Education

The proposed model for in-service training allows faculty members to choose to teach topics they are more interested in and closer to their own education and to expand their range of pedagogical and content-related knowledge confidently. Conversely, it enables the training center to develop a more general type of in-service training suited to the nature of STS. The bodies developing these programs are distributed throughout the country and integrated within the existing system of in-service training. In time, they will be responsible for the establishment of a team of outstanding teachers who will become “leaders” in teaching specific STS units and eventually the entire subject in keeping with the curriculum framework

To enable the required work, the Israeli Ministry of Education guaranteed the following hours of in-service training, making them available to the chief inspector:

1998: 1. An experiment encompassing 100 schools. The experiment applied to both the curriculum units and training models and the bodies that designed them.

2. The training and development of 100 teachers specializing in STS instruction, to serve as "leaders" in 1999.

Table 9. Guiding Principles used in Making Content Maps

(Israeli Ministry of Education Curriculum Center, 2000)

Organizing Principles	Material Sciences	Life Sciences	Earth Sciences& the Universe	Technology
Models, Complex Systems and Their Organization	•The atomic model and the molecular structure of material •Powers in nature	•Organization of living organisms as an energy-dependent phenomenon •Adapting structure to function • uniformity of the model and differences in form in the structure and processes of living organisms	•Astronomical and Cosmic systems •Earth systems	•Monitoring and feedback
Evolution	•Natural and man-made materials	•Development of species from previous species •The principle of natural selection	•Origin and development of the universe and the Earth	•Technological developments and innovations
Permanence and Cycles, Preservation and Change	•The principle of conservation •The pace of change •Laws of thermo-dynamics	•Control and regulation, dynamic balance •Growth and development •Continuity	•Transformation of material and energy on Earth •Seasonal change and cycles and changes in geological time	•Processing of raw materials •New materials •Recycling
Information and Communication	•Measurement •Transfer of information	•Interaction at various organizational levels within and between living organisms, dependent on communication	•Measurement and prediction of natural phenomena on Earth	•Mean of collection, Processing, organizing, retrieving and transferring information
Interaction	•Causality •Thermodynamic and kinetic •Considerations in chemical and physical processes	•Living organisms as distinct from their environment, how they affect it and are affected by it • Interactions within the bodies of organisms between species	•Biochemical cycles in Earth systems	•Adapting the environment to its users •How technological means increase the range of relations between human kind and the environment

Table 10. Example of Content Map for " Space"

(Israeli Ministry of Education Curriculum Center, 2000)

Organizing Principles	Material Sciences	Life Sciences	Earth Sciences & the Universe	Technology	Social and Philosophical Aspects
Models, Complex Systems and Their Organization	•Creation and reversal of chemical bonds •The atomic model	•Physical model: lock and key(antigen-antibody, enzyme-substrate) •Mathematical model: ratios of area and volume in organisms •Conceptual model: homeostasis	•structure of the galaxy •solar system •Earth's geosphere	•Transportation systems •production systems (including automation and robotics) •Energy used by humankind and its sources	•Agriculture •Smoking and narcotics •Driving
Evolution	•Materials	•Development of the species, origin of man, origin of life •Adaptation	• The "Big Bang" •Formation of the solar system •Geological development of the earth	•Synthetic materials •New materials (composites, plastics)	•Archeology •Science as developing entity •Development of scientific theories for understanding the earth
Permanence and Cycles, Preservation and Change	•Transformation of materials •Transformation of energy •Heat •Dynamics	•Mechanisms for neural, hormonal, enzymatic feedback •Periodic phenomena such as: flowering, germination, migration, reproductive cycles, biological clocks	•The Earth and its orbit •Water cycle	•Processing of materials •Heating and cooling systems •Food engineering	•Preservation versus development •Quality of the environment •Permanence and periodicity in the Jewish calendar (solar year, lunar year)
Information and Communication	•Electromagnetic radiation •Motion	•Senses, sensitivity and response, camouflage, warning behavior •Molecular communication, information in genetic material, interpersonal communication	•Space research •Research of the Earth from space •Research of the ocean floor •Research of the atmosphere	•Symbols and signs •Non-verbal communication (color and sound) •Information processing (computers) •Advertising and marketing	•Music •The written word •Forms of interpersonal communication
Interaction	•Statistics and their significance	•Regulating enzymatic processes, regulation by environmental means, immunological responses •Prey-victim relations, adaptation to the environment, care for offspring	•Man and the environment • the carbon cycle	•Movement and transportation •Structures and shelter •Health and way of life	•Science and art •Police •Pest control •Medical ethics

Table 11. Content of Science Subjects in Korea

(Korean Ministry of Education, 2000)

Organization Principle	Inquiry	Energy	Materials	Life	Earth	Environment (Technology)
Knowledge	•Inquiry in Science	•Force and Energy •Electric Energy •Wave Energy •Trans-formation Energy	•Electrolyte and Ion •Reaction of an Acid and a base •Reaction Velocity	•Material Metabolism •Stimulation and response •Reproduc-tion	• Change Earth •Atmos-phere and sea • Solar system and galaxy	•Acid rain •Green-house effect • Noise
Inquiry	Inquiry Processes	Simple Inquiry processes : Observation, Classification, Measurement, Prediction, Reasoning Complicated Inquiry processes : Problem Perception, Establishment of Hypothesis, Variable Control, Transformation of Data, Production of Conclusion, Generalization				
	Inquiry Activity	Discussion, Experiment, Survey, Fields Trips, Research				

1999: 1. Teaching of STS in 500 schools, by 1,000 teachers (of biology, chemistry, and physics). Part of the guidance offered to new teachers was given by the 1998 trainees.

2. Continued in-service training for "leaders" in terms of guider instruction.

The in-service training has a modular structure, with three sections in each unit, lasting for 60 hours. This was designed to enable the participants to finish the program within three years.

Training for science teachers in Israel is divided into IT(Information Technology) and STS training. In technological and knowledge education all the teachers use computers directly in instruction. Moreover, it is systematic in the national program and Israeli STS teachers participate in long term training(3-4 years). The methods of teacher training are shown in Fig 2 based on a "Fan Model". One of the central activities of the National Teacher Centers is to prepare educational leadership and support the activities in regional centers. The leader-teachers undergo an intensive training procedure, consisting of a profound elucidation of their educational-didactic work, thereby enriching their subject-matter knowledge, and developing their guiding skills.

Typically, the courses for leader-teachers extend over three academic years, during which the participants spend one day a week at the National Center. The overall program is designed on the basis of the following principles ; (1) Participating teachers need to develop professionally as teachers. (2) leader-teachers need to develop professionally as leaders. (3) The program is designed to help the leader-teachers choose and/or design models for the programs. The courses are run at the regional centers and provide a framework for the initial preparation of the tools necessary for the leader- teachers. Israeli universities work in collaboration with the regional schools, thus the teachers and students are able to maintain an organic relation with each other.

The current Korean training programs for science teachers include experiment training, qualification training, and computer training. The common science and science teacher qualification training is conducted in teacher training centers at regional universities, while the computer training is conducted at the regional Education Office. The qualification training takes 315-hours and the unit experiment training takes 60-hours. Accordingly, various training programs need to be developed to improve the quality of science teachers.

8. Summary

Table 12 provides a summary of the main contents in this report. In Israel, the instruction and learning of Science and Technology emphasizes team work, group activity, scientific methods, cooperative learning, discussion, and demonstration. However, in Korea, the current programs emphasize self-oriented individual learning and theoretical learning.

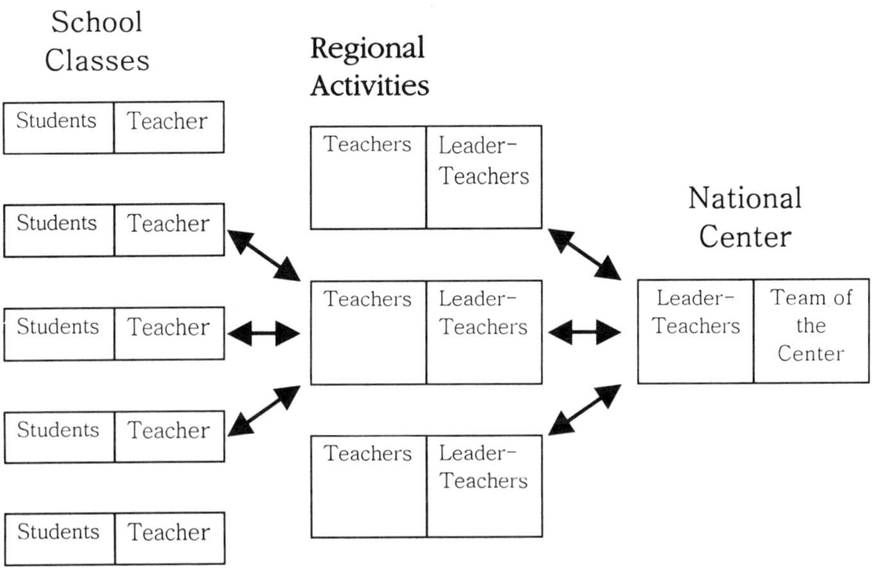

Figure 1. Structure of Teacher training: Fan Model. (Ganiel, 2000)

For grades 7-9 the Israeli curriculum has 7 parts, while the Korean curriculum only has 4. The three topics not included among the Korean Science subjects are: Technological Systems and Products, Information and Communication, and Ecosystems. In Israel, each topic is taught from three aspects: scientific, technological, and social value, thereby relating the content to real life.

The time allocated to science education in Israel is a minimum of 540 hours over three years(grades 7-9) and 240 hours in grade 10. In contrast, the time allocated in Korea is only 238 hours in grades 7-9 and 102 hours in grade 10. Accordingly, the time allocated to science education in Korea needs to be increased, plus students need to develop scientific and technological problem-solving abilities and skills related to real life. In Israel, Science and Technology is combined into one integrated subject called **Science & Technology** in junior high school. **Science & Technology in Society(STS)** is also a new subject in the required curriculum for senior high school students who are not majoring in natural science or technological-engineering subjects.

Table 12. Comparison of Science Education Curriculum between Israel and Korea

	Israel	Korea
Goals and Contents	Related to real life Emphasis on fostering scientific capabilities and skills, Interest and curiosity	Related to theoretical study Emphasis on understanding scientific concepts
Subjects	Interdisciplinary approach Integrated subjects	Separated subjects Integrated subjects Technology and Home-Economics
Methods of Learning	Team work Inquiry and Discovery, Lecture, Discussions, Demonstrations, Laboratory Activities, Fields trips, Activities, Project simulation	Self-oriented individual learning Lectures, Inquiry and discovery, Laboratory activities
Attitudes	Positive attitude Understanding of organic relation between science and technology Fostering scientific knowledge and problem-solving ability	Creativity, Openness, Objectivity, Rationality, Collaboration
Application of Computer	Combined learning in science and technology	Separated learning in other subjects
Structures of Studies	Required and elective parts	Required, enriched, and supplementary parts
Aspects of Topic	Three aspects: Scientific, technological, and social value	Two aspects: Knowledge and inquiry

This interdisciplinary approach exposes pupils to a scientific and technological content and emphasizes the relation to the social context, both in terms of the application of principles and in terms of defining human and social needs and solving problems, with the aim of improving the quality of life. The teaching of Science & Technology gives precedence both to the needs of the learner as a citizen of the future and the human impact on the environment and society.

When comparing the Israeli and Korean science curricula, it is clear that Korea needs to increase the time allocated to science subjects in senior high school. Furthermore, Korean science education needs to develop problem-solving abilities in students so as to meet the increasing social challenges related to the interaction between science, technology, and society.

The Korean teacher training system also needs revision. Science teachers should not just receive simple information during their training programs. Rather, teacher leaders need to be selected and trained to help develop more training programs for all science teachers. Plus, a communication and support network needs to be built among science teachers.

The most urgent change needed in Korean science education is for science teachers to join the debate, participate in training programs, and initiate new developments for improving Korean science education.

References

1. J.J., Jang, The Junior High School Science Curricula Comparison in Korea and Israel, Kyungpook National University, Korea, 2001.
2. Yaacov Iram and Mirjam Schmida, The Educational System of Israel Contributions to the Study of Education ; No. 70., 1998.
3. Ministry of Education Curriculum Center in Israel, Curriculum for Science and Technology Studies, Jerusalem, 2000.
4. Ministry of Education, Analysis of Education Curriculum in Junior and Senior High School (Ⅰ), (Ⅲ), Korea, 1999.
5. Ministry of Education, Culture and Sports In Israel, Tomorrow 98 : Report of The Superior Committee on Science, Mathematics and Technology Education in Israel, 1992.
6. Dalia Sprinzak, Yedidia Segev, Ehud Bar, Daniel Levi-Mazloum & Drorit Piterman, Facts and Figures about Education and Culture in Israel, Jerusalem, 1998.
7. Uri Ganiel, Fostering Change In Science Education: the Israeli Experience, Department of Science Teaching, the Weizmann Institute of Science, Israel, 2000.

ANALYSIS OF CURRENT SCIENCE EDUCATION RESEARCH TRENDS IN KOREA WITH A FOCUS ON CONSTRUCTIVISM

MANHEE KIM, HYUNJU JEONG & BEOMKI KIM
Dept. of Physics Education, Korea National University of Education, Cheongwon, 363-791, Korea.
http://phys.knue.ac.kr/~kimbk/ (Kimmh2297@hanmail.net)

The purpose of this study was to analyze the current science education research trends in Korea and to discuss their significance, with a focus on constructivism. While traditional science education has been grounded on philosophical objectivism, a new paradigm of science education emerged in the late 20C converged on a constructivist theory, reflecting the short comings of science education to meet the needs of the post-industrial society. Constructivism is rooted in philosophical epistemology, which basically has two major traditions based in the psychological and sociological disciplines. However, the constructivist movement in current science education is more heterogeneous. A recent review identified various interpretations of constructivism (R. Matthews, 1994), and Fetherston (1997) has argued that the constructivism as applied to science teaching is undertheorised. Therefore, this lack of common meaning for most constructivist approaches can not contribute significantly to theory development. Furthermore a firm theory could lead to practical reform in science education. However, the Korean situation has not yet been investigated with regard to this problem. Accordingly, the current study analyzed research papers that appeared in the *Journal of the Korean Association for Research in Science Education* since 1978, which is the most popular and influential Journal published by the Korean Science Education Society. As a result, some hidden meanings concerning constructivism are discussed, the trends of science education research in terms of current themes are identified, and finally some implications and recommendations are suggested with regard to the basis of science education.

1. Introduction

The theoretical understanding of education gives rise to practical methodologies. However, recent science education research has fallen short in meeting the needs of our post-industrial society, and lacked effort to gain a theoretical understanding, if not pragmatic solutions. Traditional science education has been grounded on philosophical objectivism, partial understanding, and analytic methodologies. In contrast, the new paradigm of science education that emerged in the late twentieth century has converged on constructivist theories and STS strategies.

Constructivism is rooted in philosophical epistemology. Yet, what is the perspective of epistemology throughout philosophical history? It may be how conceptual change is possible, the so-called Menon dilemma. It also embraces scientific to religious awakening. Constructivism can only partially answer the mystery of conceptual change, which is so heterogeneous and undertheorised (Treagust, 1991 ; R. Matthews, 1994 ; Fetherston, 1997). However, constructivist approaches in science education continue to increase, thereby reflecting the lack of theoretical background in science education. As such, this research trend can not contribute significantly to the development of science education.

Accordingly, the purpose of the current study was to analyze the science education research trends in Korea and discuss their meanings and implications, with a focus on constructivism.

2. Research Method

The current study analyzed the research papers that appeared in the "*Journal of the Korean Association for Research in Science Education*" from 1978 to 2000, which is the most popular and influential Journal produced by the Korean Science Education Society since 1978. The total number of volumes was 51 and total number of papers 546.

3. Epistemology And Constructivism

3.1 *What Is The Meno Dilemma?*

"You argue that a man cannot enquire either about that which he knows or about that which he does not know; for if he knows, he has no need to enquire; and if not, he cannot; for he does not know the very subject about which he is to enquire." (Plato, Meno, 80E)

This passage is one of the few instances in the Platonic dialogues in which Socrates' "yes man" poses what seems to the reader a real problem for Socrates. This problem is still with us today, and the mechanism of conceptual change may be the answer.

3.2 *Traditional Epistemology*

Archimedean perspectives have been suggested by some to be the foundation of inquiry in traditional epistemology, and include reason, experience, or *a priori* inborn ability. These trends are called foundationalism and consider the individual as the origin of knowledge and premise knowledge acquisition on the static state. However, this has limits for practical education, as it can not explain the 'generative' characteristics as distinct to the 'discovery' characteristics of knowledge. (Hamlyn D.W. , 1978)

3.3 *Areas of Constructivism*

R. Matthews(1994) studied constructivism from various perspectives and classified it into specific areas. Constructivism is rooted in philosophical epistemology, which basically has two major traditions based on the psychological and sociological disciplines. Psychological constructivism originates with Jean Piaget's account of children's learning as a process of personal, individual, and intellectual construction arising from their activity in the world. This tradition bifurcates into, on the one hand, the more personal and subjective tradition of Piaget that can be seen in von Glasersfeld's work, and on the other hand, into the social constructivism of the Russian Vygotsky and his followers, who stress the importance of language communities for the cognitive construction of individuals, as can be seen in the work of Duckworth (1987) and Lave (1988).

The second major tradition is sociological constructivism, originating from Emile Durkheim and augmented by sociologists of culture, such as Peter Berger and, more recently, by sociologists of science in the Edinburgh School, such as Barry Barnes, David Bloor, Harry Collins, and Bruno Latour. This sociological tradition maintains that scientific knowledge is socially constructed and vindicated, and investigates the circumstances and dynamics of science's construction. In contrast to Piaget and Vygotsky, it ignores the individual psychological mechanisms of belief construction, and focuses upon the extraindividual social circumstances, which, it claims, determine the beliefs of individuals; the individual becomes a sort of "black box" for the theory. Extreme forms of sociological constructivism claim that science is nothing but a form of human cognitive construction comparable to artistic or literary construction, and having no particular claim to truth. The claims of sociological constructivism, and its contentious and revolutionary implications for science education, were examined by Slezak (1994).

4. Analysis of Research Papers

Over the past several years, educational research undertaken from a constructivist perspective has encompassed a wide range of phenomena, from probing the actual content of student conceptions or mental stages to the study of learning environments and philosophical background of curricula (Steffe, L. et al., 1995). Research related to constructivism also reflects this broad scope of interest.

The four themes underlying recent research seem to be:

(l) Surveying conception (SC): attempts to understand the pre-instructional conception of students or teachers.

(2) Conceptual change (CC): studies of instructional models and teaching techniques for conceptual change/ the benefits of constructivist epistemologies.

(3) Expansion of constructivism (EC): studies on the expansion of constructivist epistemology, theory, practice, curricula, teacher preparation, texts, evaluation, stages of Piagetian cognitive development, and education policies.

(4) STS: studies on STS reflecting constructivism.

The percentage of constructivist papers among the total number was 36%, which were classified according to the four themes: CC 14%, SC 9%, EC 8%, and STS 5%. The total number of constructivist papers for each category is shown in Figure 2. The remaining papers (64%) were not included in the current analysis.

The first paper reflecting constructivism was a "Study on the Piagetian child's conception of movement and speed" by H.J. Kim in 1978. However, the first paper introducing a "constructivist theory" was actually "A study of the philosophical basis of preconceptions and the relationship between misconceptions and science education" by H.H.Cho in 1984. Actual research on conceptual change started in 1988. J.S. Kwon reported on "A comparative discussion on the instructional procedures and strategies in learning scientific concepts" in 1992, which encompassed almost all previous learning models. J.S. Kwon argued that cognitive conflict between cognitive structure and environment is important. The first paper introducing "STS" was "Problems of discipline-centered science education and methods of utilizing everyday materials in science education" by J.S. Kwon in 1991. Thereafter, more and more papers dealing with "STS-reflecting constructivist terminology" started to appear. T.H. Noh was the first to introduce "Vygotskian social constructivism" in 1998 in his paper "The effects of grouping in cooperative learning strategy". Subsequently, Piagetian constructivist research changed to Vygotskian.

The relative rates of each category differed over time. The SC rate was initially higher than the CC rate until 1990, then the CC rate surpassed the SC rate. However, the overall number and rate of constructivist papers gradually increased, as shown in Figure 1.

5. Implications

Various implications were identified based on the current analysis:

The constructivist approach is widely considered as the most fruitful research agenda. However, advances in terms of being able to explain the learning of science as a personal construction are still relatively slow.

There is a weakness in conducting a sociological analysis of the microculture established by a classroom community compared with worldly trends.

Constructivism is a topic that seems destined to remain in science education literature for some time, with fruitful results. There is certainly much that remains to be elucidated about constructivism, both as an epistemology and as a theory of teaching and learning.

Constructivism is related to epistemological relativism without specific foundation. However school education should transmit public knowledge as a worldly heritage with authority. Should the use of contradiction in education be reconsidered; is it compatible?

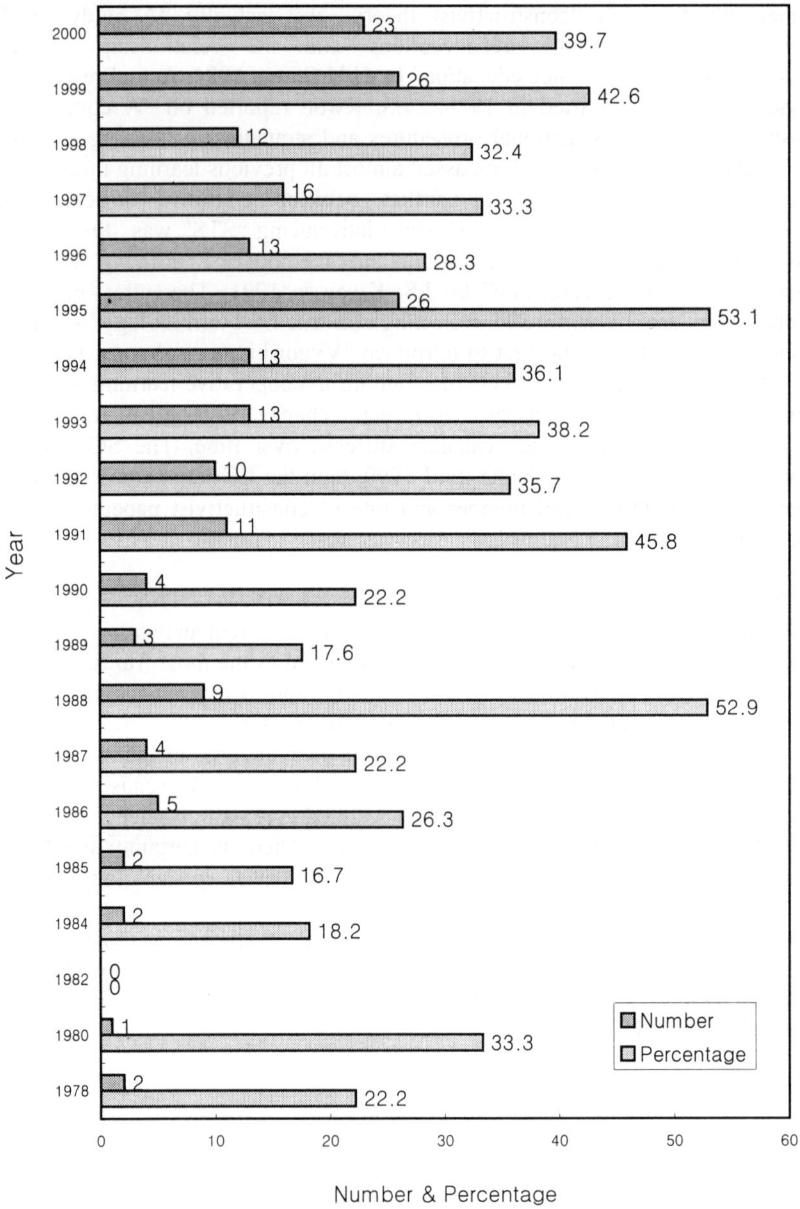

Figure 1. Number & percentage of constructivist papers over time

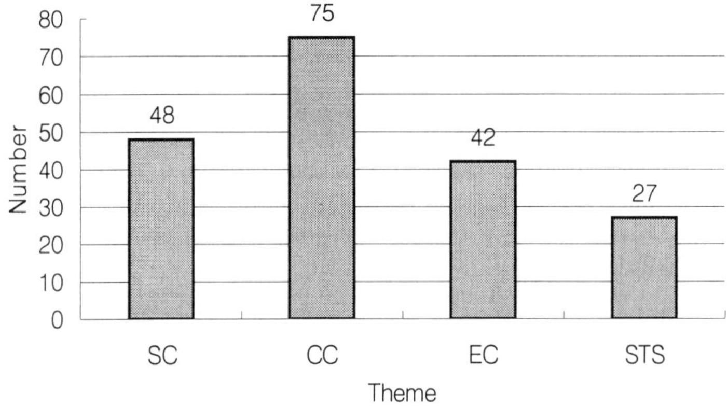

Figure 2. Total number of papers according to each theme

6. Conclusions and Recommendations

The current study attempted to analyze the science education research trends in Korea and identify their implications, with a focus on constructivism. Probing conceptions and Piagetian approaches were mainly conducted until 1990. However, studies on conceptual change increased after 1991. In particular, Vygotskian social constructivism was introduced in 1998.

There is less theoretical research than class application research as regards constructivism. Constructivist research has gradually converged into studies on conceptual change. Various recommendations are suggested for advanced science education research: Theoretical research efforts are necessary to organize different areas of constructivism, i.e., as an epistemological or pedagogical belief. The relevance of constructivism should be extensively examined in relation to science traditions and education.

References

DeBoer, G. (1991). A history of ideas in science education. Teachers College Press, New York.

Duckworth, E. (1987). The having of wonderful ideas. Teachers College Press, New York.

Fetherston, T. (1997). The derivation of learning approach based on personal construct psychology. International Journal of Science Education, 19(7). 801-819.

Hamlyn, D.W. (1978). Experience and the growth of understanding. R.K.P., London.

Lave, J. (1988). Cognition in practice : Mind, mathematics and culture in everyday life. Cambridge University Press, New York.

Matthews, M. (1994). Science Teaching. Routledge, New York.
Petrie, H.G. (1981). The dilemma of enquiry and learning. Chicago University Press.
The Korean Association for Research in Science Education. (1978-2000). Journal of the Korean Association for Research in Science Education.
Slezak, P. (1994). Sociology of scientific knowledge and science education. Science & Education, 3(3).
Steffe, L. and Gale, J. (Eds.). (1995). Constructivism in education. Lawrence Erlbaum, Hillsdale, N. J.

Acknowledgement
This work was supported by the Brain Korea 21 Project in 2001. *

JOHN TYNDALL(1820-1894), WHO BROUGHT PHYSICS AND THE PUBLIC TOGETHER

JINWOONG SONG *& SOOK KYOUNG CHO°
*Seoul National University, Seoul, Korea, °Korea Science Foundation, Seoul, Korea

The developments of science education until the middle of the 20th century were often driven by personal ideas and achievements of some influential individual scientists (e.g. T. H. Huxley, H. E. Armstrong, L. Hogben, J. Conant), while that of the 2nd half of the 20th century can be characterized as collective efforts through various research groups of science educators (e.g. PSSC, HPP, Nuffield, SATIS). In this respect, John Tyndall(1820-1894), a physicist of the Victorian England best known as Tyndall's Effect, can be considered as one of the great scientists who had a big influence on science teaching and the popularization of science before science secured its place in school curricula. Tyndall worked as a research scientist at the Royal Institution of London, where various lectures and demonstrations of physical sciences were regularly performed for general public, and he was particularly famous for his fascinating physics demonstrations. In this study, we will summarize his activities and achievements as a teacher as well as a popularizer of physics, illustrate some of his famous demonstrations and his ideas concerning physics teaching and discuss implications to today's physics education.

1. Introduction

The history of school science education can be largely divided into two phases of development. The second phase began around 1960s, when science education has been established as an autonomous discipline for itself. After this 'period of the innovation of science education' as it was called, there appeared many characteristic features in science education as seen today. A new group of professional science educators emerged, new research methods had been developed, and various theories for science education have been proposed. Science educators collectively worked for developing contents of science education, for finding out cognitive processes of science learning, and for investigating science teaching strategies and

John Tyndall(1820-1894)

materials.[15]

Before the 1960s, however, individual and influential scientists spoke for science and the science education. They raised the necessity of introducing science as school curriculum and expressed their views on what would be the proper content for science education. They also investigated effective and ideal methods for science teaching and addressed the importance of science education to other intellectuals. Since there was not a proper sense of science educators yet, individual scientists' experiences, insights, and views of science strongly influenced the practice of science education in general. It was the period that was driven by personal ideas and achievements of individual scientists rather than the collective efforts of various research groups of science educators. Among those scientists, M. Faraday, J. Tyndall, T. H. Huxley, H. Spencer, J. von Liebig, J. Hooker, H. Armstrong, L. Hogben, J. Conant were notable figures.[9][15]

In this respect, we will discuss John Tyndall. A Victorian physicist, Tyndall worked as a research scientist at the Royal Institution of London, where various lectures and demonstrations of physical sciences were regularly performed for the general public. He was famous for his fascinating physics demonstrations, especially. In this study, we will discuss him as a teacher as well as a popularizer of physics not as a physicist, by showing his activities (performing physics demonstrations and writing books on physics and science) and his ideas for science education. This will be done within the context of the 19th century in Britain, when science in general and science education in particular were not organised by the government. Then, we will discuss some implications of Tyndall to physics education today.

2. Victorian Science: Popularization of Science

There were many significant achievements in science during Victorian period in Britain. The famous book of *On the Origin of the Species by Means of Natural Selection, or the Preservation of Favoured Races in the Struggle for Life*, written by C. Darwin in 1859 clearly proclaimed that the 19th century was the age of biology, following the great advances in geology during 18th century. Darwin's theory of evolution, with the implication that humans had originated from apes, aroused great embarrassment to theologians. After all, theologians and scientists conflicted at the famous Oxford meeting for the British Association for the Advancement of Science in 1860, which has been thought of as the origin of the war between science and religion.[5]

However, this was the age of physics also. The independent topics of physics such as 'heat', 'work', 'electricity', 'magnetism' were united by a newly emerged concept of 'energy'. Together with the formalization of the conceptual foundation of electrodynamics by R. Clausius in the 1850s, the concept of 'energy' and 'entropy' finally opened a new area of thermodynamics. In addition, the concept of 'field' which M. Faraday had invented to describe electrical and magnetic forces, was developed and formalized into Maxwell's electromagnetic equations. All those developments contributed to establishing a modern sense of physics as a secure discipline.[11]

The rapid progress during this period was not confined only to the contents of science, such as new scientific discoveries, new concepts and the formalization to represent them. There was a series of arguments and movements asking for organization of science and science education, after the Great Exhibition of 1851 held in Hyde Park of London, and some practical results began to appear. Various societies for every branch of science were founded one after another, old universities, such as Oxbridge, were reformed and new Red Brick universities like the University of London were opened for the general people. This was the period when the professionalization of science gained momentum.[6],[19]

But, the most important development occurring during this period was that science came in contact with people for the first time in history. Science, which had been monopolized by a small group of gentlemen amateurs, could be opened to the general public owing to some influential scientists' endeavor. G. Birkbeck, H. Brougham, H. Davy, M. Faraday, L. Playfair, H. Roscoe, A. Strange, and T. H. Huxley tried to reach to people through various formal and informal channels such as science lectures and demonstrations, Mechanics' Institutes, science schools, journals and newspapers at a low price, and science exhibitions or museums.[8]

3. The Royal Institution And Science Lectures

One of the main channels for science to meet wider audience was various science lectures. The Literary & Philosophical Societies, founded from late 18th century in industrial cities such as Manchester and Birmingham, began to arrange science lectures once or twice a month for the local public. Gradually, science lectures spreaded to other local cities and became popular. In London, although the Royal Society was already existed, it was mainly a home for gentlemen scientists and its original aim for spreading useful knowledge faded. This led the foundation of the Royal Institution.

The Royal Institution, founded by Count Rumford (Benjamin Thomson), aimed for « bettering the condition and increasing the comforts of the poor ». The president of the Royal society, Sir J. Banks invited many influential individuals to his house, 32 Soho Square, London in 1799. After he explained his grand idea for founding a new science institution, he asked financial support for it. With the kind help of 132 donators, the Royal Institution opened its door to the public, fully equipped with a new laboratory and a theater classroom in 1800.

Royal Institution of London

Tyndall's Lecture at the Royal Institution

In spite of its founding aim, the early period of the Royal Institution oriented for scientific investigations and experiments by its own not for spreading useful knowledge. For example, T. Young, followed by Dr. Garnett and Webster, discovered a theory of interference and published it in the *Journal of Proceedings*, but he was not a good conveyor of science at all.[7] That had to wait for H. Davy, who moved the work of the Royal Institution to science lectures for the public. Davy, appointed as the professor of the Institution in his 20s, attracted 'the educated class' as it was called, by his gifted talent of speaking and of demonstrating some fascinating experiments.[18]

While science lectures delivered by Davy were more for the upper class people, those of M. Faraday were more for middle class people. After returning from Europe in 1815, Faraday began his research on his own and became the superintendent for experiments in 1821, and finally appointed as the Professor of Natural Philosophy following Faraday. He, then, organised the Friday Evening Lectures and the famous Christmas Lectures, which have continued until today.[18] His lectures were about electricity, magnetism and chemistry all of which were accompanied by vivid experiments and demonstrations. One of notable features of his lecture was that children and ladies were invited as audience. Faraday, by expanding the audience, made the Royal Institution a center for spreading scientific knowledge and the best place for actualization of the popularization of science.

This was the time when John Tyndall joined to the Institution as professor. Educated in Germany, Tyndall enhanced the experimental aspects of science and science teaching. For the next 20 years, he also actively worked for transforming the Royal Institution into an institution for science education for the general people.

4. X-Club Scientists

If the Royal Institution was one of the formal channels for the popularization of science, the X-club organised in 1864 was an informal but most effective channel for connecting science to people.[8]. All the nine members of this group - T. H. Huxley, T. A. Hirst, E. Frankland, J. Tyndall, Rubbock, G. Busk, H. Spencer, J. D. Hooker, W. Spottiswoode - had known each other through BAAS meetings since 1830s and had actively worked for the Royal College of Mines and new Department of Science and Arts from 1850.[3]

The X-club scientists met once a month in an informal way and had many things in common compared with other contemporary scientists. Firstly, most of them studied science by themselves without proper formal education. Secondly, few of them entered Oxbridge and three of them (i.e. Tyndall, Frankland, Hirst) had some experience studying in Germany. Thirdly, they were not traditional amateurs but professional scientists doing original research, whom R. Macleod called 'researchers'.[8] Moreover, they were the real reformers who realized that a proper foundation for science education could only be made by the government and who actively worked for the organization of science. Also, they published a weekly magazine for science, *Nature*, for spreading their voices.

Tyndall was one of this X-club scientists who were involved in many Royal Commissions related to science in one way or another from 1860s to 1890s and « played enormously important role to create a new images of science ».[5],[12]

5. Life of John Tyndall

John Tyndall was born in Ireland in 1820 and educated at a National School and Preston Mechanics' Institution. In 1847, he was appointed as a teacher of mathematics and engineering, surveying and physics at the recently established Queenwood College. In this College, a wide range of pure and applied sciences covering especially topics such as the steam engine, railways were taught. One of X-club scientists, Frankland was already there.

Tyndall was an excellent science teacher and lecturer, delivering his words by beautiful illustrations. In 1853, he gave an invited lecture at the Royal Institution and succeeded as Professor of Natural Philosophy. There he devoted himself for the next 20 years, just as his predecessor had done.

J. Tyndall and his wife

Tyndall's main contribution to science was the study of diffusion of light by large molecules and dust, which is known as Tyndall Effect. In January 1870, he gave a Friday Evening Lectures on Dust and Disease where the Prime Minister was present. Here, he combated the doctrine of spontaneous generation of life, which brought him much criticism and even ridicule.

Avalanche at the Alps

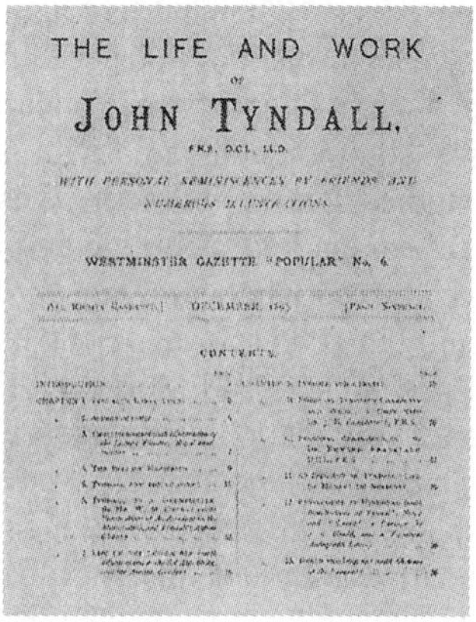

THE LIFE AND WORK

JOHN TYNDALL,

WESTMINSTER GAZETTE "POPULAR" No. 6

CONTENTS.

1st Biography of J. Tyndall

A physicist Tyndall was also a well known climber. He began mountaineering in the early days, when most of the great peaks were still unclimbed and the Alps were partially explored. His first visit to Switzerland was made in 1849 and his second visit with T. H. Huxley was made in 1851. However, he and his companions met the great avalanchean Avalanche at the Alps in 1864 on the Peak of Morteratch. Luckily, no one was hurt seriously except a few bruises and cuts, but Tyndall lost his precious watch. Naturally, this led him to write some books on Glaciers such as *The Glaciers of the Alps* (1860)[21], *The Forms of Water in Clouds and Rivers, Ice and Glaciers* (1872).

In 1887, Tyndall resigned his chair at the Royal Institution because of increasing illness. He died in 1893 not because of his illness but because his wife inadvertently gave him chloral for his indigestion instead of milk of magnesia.[4] His sudden and unexpected death was a great shock to Victorian intellectual societies, especially for X-club scientists. This implied the end of Victorian science and of the influence of X-club scientists and the emergence of new 20th century science.

The papers such as *The Westerminster Gazette*[25] and *the Daily Telegraph* mourned his death through the articles remembering his life and works. In fact, his wife Louisa Tyndall published a book titled *Life and Work of John Tyndall*[10], where his works were carefully recorded. A biography of Tyndall, including his selective writings, was also published as *John Tyndall: Essays on a Natural Philosopher*(Macmillan & Mollan, 1981). In 1945, *The Illustrated London News* memorized Tyndall in the special corner of « A Full-Length Biography of a Great Scientist ». Recently, Barton interpreted him as a pantheist in a paper « John Tyndall, Pantheist: A rereading of the Belfast Address » in 1987.[2]

6. Activities of The Popularization of Science

During the 19th century, it was fashionable to attend science lectures accompanied by demonstrations when people wanted to know about new scientific discoveries and some scientific information. The Royal Institution, where Davy, Faraday and Tyndall were once the professors, was the first place for that purpose. They were, in fact, very good at scientific demonstrations, from whom many audiences including science teachers learned science. Science teachers, especially, had the experience of seeing the demonstrations, hearing the way of delivering lectures, and seeing with the mind's eye.[16],[17]

Tyndall was an excellent experimenter attracting audience by fascinating demonstrations. He, especially, was very good at trying many different experimentations to see other unseen aspects of existing experiments. For example, he would focus unseen radiant heat tried to ignite a cigar at the focal point, mould ice by pressure into almost every shape, display large line spectra of the chemical elements, and even produce 'a singing flame' which responded to notes from his voice.[17] These experiments seemed to be spectacular but not just spectacles themselves. They were part of lectures and demonstrations for helping audience's understanding of his lectures. The following are examples of the experiments he made during his lecture demonstrations;

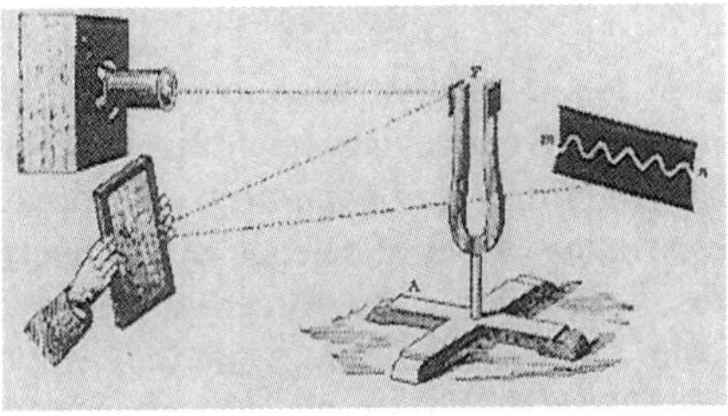

from *Light* (1882)

from *Magic of Science* (1860)

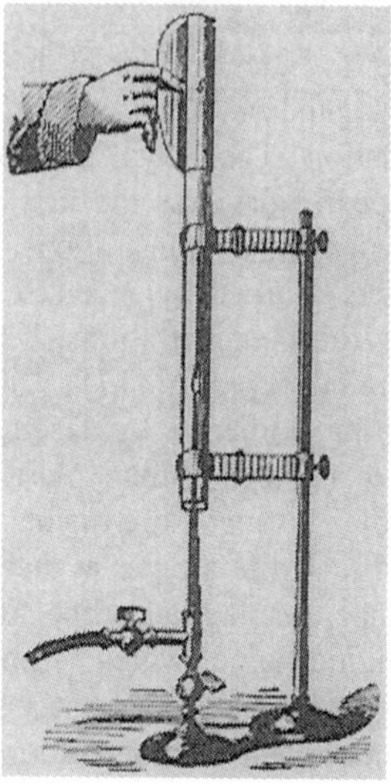

from *Sound* (1875)

from *Heat: a Mode of Motion* (1898)

Tyndall was also a wonderful lecturer. His lectures were enjoyed with great fun owing to his easy and accessible way of conveying scientific ideas and discoveries (The Standard, 1877). In fact, *the Illustrated London News* evaluated him as « the person to make audiences swallow a pill of science by themselves by transforming it to a sweet, contrast to Huxley, who forced audiences to swallow a pill of science ». He knew who the audience were, what they wanted to know, and how he could access them.[8] H. E. Armstrong, who attended his marvelously illustrated lectures on Physics remembered him as follows:

... he was no mathematician but as a teacher he was worth all the Professors of the Royal College of science and the Royal School of Mines put together and no other actor on the Royal Institution stage has surpassed him as a popular lecturer ... ([1], p.204).

In addition, Tyndall himself published many books. He was one of the first people to provide notes with details of demonstrations and to publish books after the lectures. All the processes and results of experiments were so well explained that readers could follow his processes without great difficulties. Considering the fact that publication was another important channel for popularization of science during the 19th century, he was one of the scientists who fully used that. The Following are examples of his publications :

- *The Glaciers of the Alps*

- *Heat Considered as a Mode of Motion*
- *Fragments of Science for Unscientific People*
- *Lessons in Electricity*
- *Essays on the Floating of the Air in Relation to Putrefaction*
- *The Forms of Water in Clouds and Rivers, Ice and Glaciers*
- *Mountaineering in 1861: a vacation tour*
- *The Floating Matter of the Air*
- *Hours of Exercise in the Alps* (New edition)
- *Natural Philosophy in Easy Lessons: A School Text*
- *New Fragments of Science*

Among his books, the most important one, seen from science education perspective, is *Fragments of Science for Unscientific People*.[23] As the title shows, he would liked teaching science for the general people who had no scientific background knowledge at all. The Following are the titles and the contents of the books, where we can see his wide range of ideas relating to science and his special respect for Faraday.

The Constitution of Nature
Thoughts on Prayers and Natural Law
Miracles and Special Providence
Matter and Force
An Address to Students
Scope and Limit of Scientific Materialism
Scientific Use of the Imagination
On Radiation
On Radiant Heat in Relation to the Color and Chemical
Constitution of Bodies
On Chemical Rays and the Structure and Light of the Sky
Dust and Disease
Addition to "Dust and Disease"
Life and Letters of Faraday
An Elementary Lecture on Magnetism
Shorter Articles :
    Slates
    Death by Lightening
    Science and Spirits
    Vitality
    Additional Remarks on Miracles

In this book, Tyndall especially emphasized the importance of scientific imagination in science education. He said that « we are gifted with the power of imagination and by this power we can lighten the darkness which surrounds the world of senses. ... imagination becomes the mightiest instrument of the physical discover. Newton's passage from a falling apple to a falling moon was, at the outset, a leap of the imagination. When William Thomson tried to place the ultimate particles of matter between his compass-points, and to apply to them a scale of millimetres, he is powerfully aided by this faculty... ».([23], p.132). Then, he illustrated a few simple instances which scientists had already made such as observing the falling of heavy rain-drops into a tranquil pond. Take another example, audiences had to go with him in imagination to the level of water molecules in the ice crystals, to share his vision of what might be going on there, and then upwards to the level of the Alpine range - the great condenser of Europe as he called it -, in order to appreciate his science of ice.[16]

Tyndall's approach to physics was intellectual as well as emotional. He argued that although « the ultimate problem of physics is to reduce matter by analysis to its lowest condition of divisibility, and force to its simplest manifestations, and then by synthesis to construct from these elements the world as it stands., the solution comes ... more of spiritual insight than of actual observation ».([23], p.100). He, strongly believed "men feels as well as thinks", thought that it was vain to attempt to separate moral and emotional nature from intellectual nature. For him, the ultimate aim of science and science education was to make people feel the beauty of nature. He wrote as follows:

The sound of the village bell which comes mellowed from the valley to the traveller upon the hill, has a value beyond its acoustical one. The setting sun when it mantles with the bloom of roses the alpine snows, has a value beyond its optical one. The starry heavens, as you know, had for Immanuel Kant a value beyond their astronomical one ... ([23], p.103)

7. John Tyndall And Science Education

It was mid 19th century when school science education began to be introduced in a limited way in Britain. Owing to the arguments and suggestions by statesmen of science such as Strange and Huxley, the Department of Science and Art was formed in 1853. The DSA, then, encouraged the introduction of science classes in various schools by adapting the science examination and subsiding teachers and students who got high marks. However, the majority of children still had not much chance of being educated, and if they could, it was limited to a elementary level. It was very difficult for school science teachers to have practical science education, because of the deficiency of facilities for scientific education.[14]

As mentioned earlier, Tyndall filled the gap by making full use of his position as the professor of the Royal Institution. He especially used the channels of science lectures and publications in order to connect science with people. His fascinating science demonstrations and well written books played crucial roles in the popularization of Victorian science, especially popularization of physics.

Tyndall had his own view on the modernization argument in science education. During the 19th century, there arose a great debate between a group of scientists who tried to introduce science as school curriculum and the humanists emphasizing traditional classical studies. Tyndall's view on this was that there should be harmony between traditional studies centered on the humanities and new developments of human civilization through science.[9]

Furthermore, Tyndall was one of the examiners of science examination as T. H. Huxley.[15] As discussed in an earlier paper, this was a very important position to influence science and science education, directly and indirectly, in Victorian Britain.

So far, we briefly summarized the life and activities of a Victorian physicist Tyndall. After 150 years, what do they tell us about science education? This question, of course, will be answered more clearly when more historical and educational materials are critically analysed. However, it can give some useful implications and new insights for school physics education today.

First of all, Tyndall's experiment and demonstrations were full of fun and imagination. He showed the delicateness and wonders of physical science through the experiments and the audience had a chance to experience a vivid science education. His lectures were alive themsclvcs. In this sense, if we find the lists of all his experiments and revise them within our context, and then retry it in our own classrooms, it will help science classes to be more lively. Indeed, this will be the actualization of science education through imagination, as Tyndall had dreamed.

Secondly, Tyndall's original researches on physics are still important teaching topics today. For example, Tyndall's effect of light, heat as a mode of motion and the production of ice and glacier themselves would make good contents for physics education.

Thirdly, Tyndall's way of conveying his scientific discoveries and ideas is a good example for science teachers and the people who would like to popularize science. Facing the worldwide current situation of physics education, it might be very effective for making science teachers be familiar with his methods of conveying and communicating science to students.

Finally, his experiments were always connected to everyday life. This is an important strategy to attract people who have not enough scientific knowledge. This way of approach will be effective because it will lessen tensions and stresses which the audience or student might have.

References

1. H. E. Armstrong, *The Teaching of Scientific Method and Other Papers on Education*. Macmillan and Co., Ltd., London, 1925
2. R. Barton, "John Tyndall, Pantheist: A Rereading of the Belfast Address", *Osiris*, 2nd series, 3: pp.111-134, 1987.
3. R. Barton, "An influential set of chaps : The X-club and Royal Society politics 1864-85", *The British Journal of The History of Science*, vol.23, pp.53-81.
4. G. Bishop, *Eight Hundred Years of Physics Teaching*. Fisher Miller Publishing, Hampshire, 1994.

5. P. J. Bowler, *Evolution: The History of an Idea.* California University Press, California, 1987.
6. D. S. L. Cardwell, *The Organization of Science in England.* Hinermann, London, 1972.
7. G. Caroe, *The Royal Institution.* John Murray Ltd, London, 1985.
8. S. K. Cho, *The Special Loan Collection of Scientific Apparatus, 1876 : The Beginning of the Science Museum of London and the Popularization of Physical Science.* Ph.D. thesis, Seoul National University, Seoul, 2001.
9. G. E. DeBoer, *A History of Ideas in Science Education*, Teachers College Press, New York, 1991.
10. A. S. Eve, and C. H. Creasey, *Life and Work of John Tyndall.* Macmillan & Co. Ltd, London, 1945.
11. P. M. Harman, *Energy, Force, and Matter.* Cambridge University Press, Cambridge, 1988.
12. T. W. Heyck, *The Transformation of Intellectual Life in Victorian England.* Croom Helm, London, 1992.
13. *The Illustrated London News*, "A Full-length Biography of a Great Scientists". July 14, 1945
14. J. Song, "The process of the quickening and development of science-technology-society education in the United Kingdom(1): between the beginning of the 19th century and the middle of the 20th century". *Journal of the Korean Association for Research in Science Education*, vol.19, pp.409-427, 1999.
15. J. Song and S. K. Cho, "T. H. Huxley as a pioneer of British school science education : focused on his life and activities". *Journal of the Korean Association for Research in Science Education*, vol.21, pp.38-58, 2001.
16. C. Sutton, C. "A hundred and one science teachers: they still live on in us", *School Science Review*, vol. 79, pp.21-26, 1998.
17. C. Sutton, "Science as Conversation", In J. Wellington (Ed.) *Practical Work in School Science: Which way now?* Routledge, London, pp.174-191, 1998.
18. C. Taylor, *The Art and Science of Lecture Demonstration.* Adam Hilger, Bristol, 1988.
19. D. M. Turner, *History of Science Teaching in England.* New York: Arno Press, New York, 1927.
20. J. Tyndall, *Lectures on Sound 1854-82* (manuscript notes in the archives of the Royal institution).
21. J. Tyndall, *The Glaciers of the Alps.* J. Murray, London, 1860.
22. J. Tyndall, "On the importance of the study of physics". In E. L. Youmans (Ed.), *The Culture Demanded by Modern Life*, New York: Appleton, pp.57-85, 1867.
23. J. Tyndall, *Fragments of Science for Unscientific People : a Series of Detached Essays, Lectures, and Reviews.* D. Appleton and Company, New York, 1874.

24. J. Tyndall, *Lessons in Electricity.* Longmans, Green & Co., London, 1876.
25. Westminster Gazette, *The Life and Work of John Tyndall.* "Popular" No. 6, 1893.

ANALYSIS OF FACTORS RELATED TO CAREER CHOICE IN SCIENCE

JIN YOON[a] & SUNG-JAE PAK[b]

[a] *Kanghyeon Middle School, Seoul, Korea*

[b] *Department of Physics Education, Seoul National University, Seoul, Korea*

The purpose of this study was to investigate students' career choices related to science. After identifying the important factors related to choosing a career, the differences in the student responses were compared with their career choice, grade, and gender. 'Perceptions about science-related careers', 'participation in science-related activities', and a 'preference for learning science' were identified as the three main factors related to a career choice in science. The relative importance of these three main factors influencing a career choice related to science and the causal relationships between the career choice and these three factors were also investigated. The results showed that the relative importance of the three main factors was 'participation in science-related activities', a 'preference for learning science', and 'perceptions about careers related to science' in an order of magnitude of ß, yet differed with grade and gender. The choice of a science-related career was both directly and indirectly influenced by the 'perceptions of careers related to science', while a 'preference for learning science' and 'participation in science-related activities' only had an indirect effect.

1. Introduction

Choosing a career is one of the most important things in life. This decision is often based on various experiences both in and out of school over many years. Many countries have reported a decline in the number of students wishing to continue in the field of science, especially physics [1][2]. Although not all students need to become scientists or engineers, it is still necessary to be scientifically literate in today's highly technologically developed society. Yet, with the rapid changes in the modern world, the demand for scientists and engineers is steadily rising. As such, the importance of making a good career choice cannot be overemphasized. Although science teachers and educators want their students to enjoy, appreciate, and benefit from their science education so that a sufficient number of able students can ultimately decide to continue into scientific and technological careers, it is not so easy to motivate students to choose a career in science and technology. The factors that affect student attitudes to school science and careers in science are varied and depend on the individual student [3]. If the process of choosing a career related to science were better understood, it may be possible to improve career education in science education.

The first thing in science career education is to survey students' career choices related to science. An analysis of the factors influencing student career choices related to science is also important for constructing good science career education materials. Various previous studies have already investigated these factors[4], [5], [6], and[7]. However, the classification of such factors and their selection vary from researcher to researcher according to their background, purpose, research methods, and so on. In reality, it is impossible to determine all the factors influencing a science career choice. Out-of-school factors, such as the home background, job opportunities, and job status of careers in science and technology are difficult to change through school science education, yet they are probably as influential or even more influential than in-school factors. Accordingly, the current study paid more attention to the in-school factors influencing a science-related career choice, as these are the areas where teachers and science educators can have more potential impact.

2. Purpose

The purpose of this study was to survey the factors related to a student choosing a career in science. From examination of previous research and the results of a pilot study, three important factors related to a career choice were identified: 'Perceptions about careers related to science', a ' preference for learning science' and 'participation in science-related activities'. The differences in the student responses were compared according to their career choice, grade, and gender. The current study also attempted to identify the relative importance of the three main factors influencing the choice of a science-related career and the causal relationship between the career choice and these three factors. To determine the complementary reasons for a career choice, the reasons for choosing, changing, and not choosing a science-related career, were also analyzed.

The research questions were as follows.

What is the ratio of students who have science- related career plans ?

What are the differences in the student responses to the factors influencing their choice of a science-related career in relation to their career choice, grade, and gender?

What is the relative importance of the three main factors related to choosing a career in science for each grade and gender ?

What are the causal relationships between a science career choice and these three factors ?

What are the reasons students choose a science- related career ?

3. Methods

3.1 *Subjects*

The sample selected for the study consisted of 947 students in grades 6, 9, and 11 in a middle class area of Seoul. The numbers of boys and girls in each grade was the almost same.

3.2 *Questionnaire*

A questionnaire was developed to identify the factors influencing students in making a science-related career choice after preliminary research and a literature survey. The questionnaire consisted of two parts: the first was to identify a career choice and the reasons for choosing it, while the second was to identify the factors influencing a science-related career choice. The first part of the questionnaire asked the students to select a career choice from 9 categories, then indicate when this decision was originally formulated and the reasons why. In the case of choosing a science-related career, the students were also asked to write down a more concrete plan for their science-related career, and choose a specific field, for example, a pure scientist, engineer, mathematician, computer scientist, medical doctor, and so on. The 48 items were then selected from the second part of the questionnaire after a factor analysis.

The three main factors related to a career choice in science were 'perceptions about careers related to science', a 'preference for learning science' and 'participation in science-related activities.' The 'perceptions about careers related to science' included 'advantages of entry to higher education', 'social evaluation of science-related careers', and 'job prospects, salary, and possible leisure time'. The 'preference for learning science' included 'a preference for science lessons', 'preference for laboratory activities', 'usefulness of science', and 'teacher's influence'. The 'participation in science-related activities' included 'hobbies and activities through the media' and 'club activities, competitions, and science festivals'. For each factor, the students were asked to respond to a sentence based on a five-point Likert scale.

3.3 *Procedures*

The students were asked to respond the questionnaire by their homeroom teacher or science teacher during class in October 2000. The expected response time was about 30 minutes, however, sufficient time was given to allow a full response. The students indicated their grade and gender and chose a future career from 9 categories. They were also asked to indicate the time when they originally decided on their career choice and explain the reasons. Thee students were encouraged to respond as frankly as possible without any notion of being assessed. The student responses were analyzed after discarding any insincere ones.

3.4 *Statistical Analysis*

SPSS 8.0 was adopted for the statistical analysis. The student responses to the three main factors were compared between groups using ANOVA. A regression analysis was conducted to determine the relative importance of the three main factors. AMOS 4.0 was used to identify any causal relationships between the career choice and the three main factors.

4. Results

4.1 *Ratio of Students with Science-related Career Plan*

Of the students surveyed, 26% indicated a science-related career choice. There was significant difference in this percentage between the various grades. However, more boys(31%) chose science-related careers than girls(20%). Those who chose a science-related career wanted to be a computer scientist(22%), medical doctor(20%), pure scientist(17%), and engineer(16%). The number of students who wanted to be a pure scientist decreased with age.

4.2 *Differences Between Groups*

The differences in the student responses to the factors influencing a science-related career choice between the groups were compared using ANOVA (See Table 1). Those who indicated a science-related career choice exhiited a higher means as regards their 'perceptions about careers related to science', 'participation in science related activities' and 'preference for learning science'($p<0.001$). The means of the 'preference for learning science' decreased with the grade($p<0.05$), the means of the 'perceptions about careers related to science' increased($p<0.05$), and the means of the 'participation in science-related activities' did not differ ($p>0.05$). The boys exhibited a higher means than the girls in both their 'participation in science-related activities' and 'preference for learning science'($p<0.001$), yet the means of the 'perceptions about careers related to science' did not show any significant gender difference($p>0.05$).

Table 1. Differences in student responses between groups

	Perception	Preference	Participation
	Mean(Standard Deviation)		
Science(N=239)	3,22(0.52)***	3.34(0.55)***	3.09(0.67)***
Non-science(N=692)	2.95(0.47)	2.89(0.54)	2.55(0.60)
Boy(N=466)	3.04(0.53)	3.07(0.61)***	2.91(0.64)***
Girl(N=458)	3.00(0.46)	2.92(0.52)	2.47(0.61)
Grade 6(N=312)	2.94(0.51)*	3.11(0.57)*	2.76(0.70)
Grade 9(N=328)	3.06(0.48)	2.97(0.56)	2.67(0.65)
Grade 11(N=287)	3.06(0.49)	2.91(0.56)	2.66(0.64)

[a] N represents the number of respondents.
[b] * indicates the significance level.(*** P<.001, * P<.05)

4.3 *Relative Importance of Factors*

According to a stepwise regression analysis, the relative importance of the three main factors was 'participation in science-related activities'(β=0.33), a 'preference for learning science'(β=0.27), and 'perceptions about careers related to science'(β=0.18) in an order of magnitude of a standardized regression coefficient β. These three factors explained 40% of the variance in career choice. This order was the same in grades 6 and 9, yet in grade 11, the order of importance was a 'preference for learning science'(β=0.37), 'participation in science-related activities'(β=0.29), and 'perceptions about careers related to science'(β=0.17)(See Figure 1). For the boys, the order of importance was 'participation in science-related activities'(β=0.34)', 'perceptions about careers related to science'(β=0.22), and a 'preference for learning science'(β=0.21). For the girls, the order of importance was a 'preference for learning science'(β=0.36), 'participation in science-related activities'(β=0.23), and 'perceptions about careers related to science'(β=0.15)(See Figure 2). From a comparison of the standardized regression coefficient and amount of explanation, it was found that 'participation in science-related activities' and a 'preference for learning science' both influenced the career choice, whereas the 'perceptions about careers related to science' were influenced by the career choice.

4.4 *Causal Relationships between Career Choice And Three Main Factors*

According to structural equation modelling with AMOS 4.0, the most fitting model was where 'perceptions about careers related to science' and a 'preference for learning science ' influence each other, a 'preference for learning science ' influences 'participation in science-related activities', 'participation in science-related activities' influences 'perceptions about careers related to science', and 'perceptions about careers related to science' and a science-related career choice influence each other.

The effect of 'perceptions about careers related to science', a 'preference for learning science', and 'participation in science-related activities' on a science-related career choice had a standardized total effect coefficient of 1.41(direct effect 0.83, indirect effect 0.57), 0.86, and 1.03, respectively.

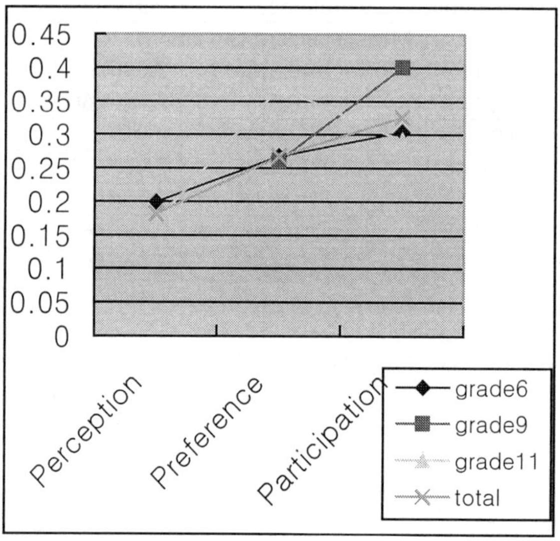

Figure 1. Comparison of β for each factor in three grades.
[a]numbers on y-axis are standardized regression coefficient β

4.5 *Reasons for Career Choice*

The most frequently mentioned reason of a career choice related to science was "an interest in science and learning about science". Computers, the Internet, and science hobbies were also frequently mentioned, along with the influence of mass media, reading science books, and biographies of famous scientists. The reasons for changing a career choice from a science-related area to another field included "the difficulty of science", "not fitting my aptitude", "losing interest", and "having a new career hope". The reasons for not choosing a science-related career included "no interest in science", "a different career plan", "no concern about science", "science is not my strength", and "no confidence in science".

5. Conclusions And Implications

In conclusion, the ratio of students with a science-related career choice was not high (26%). Student responses in relation to the the three main factors and their relative importance differed with the grade and gender. The choice of a science-related career was directly and indirectly influenced by 'perceptions about careers related to science', whereas a 'preference for learning science' and 'participation in science-related activities' were only found to have an indirect effect.

Accordingly, the current results imply that helping students to prefer learning science and giving them more opportunities for science-related activities is more important than engendering positive perceptions of careers related to science. Therefore, materials are required for science career education that consider the different factors influencing student career choices, their age, and gender, as highlighted by the current results.

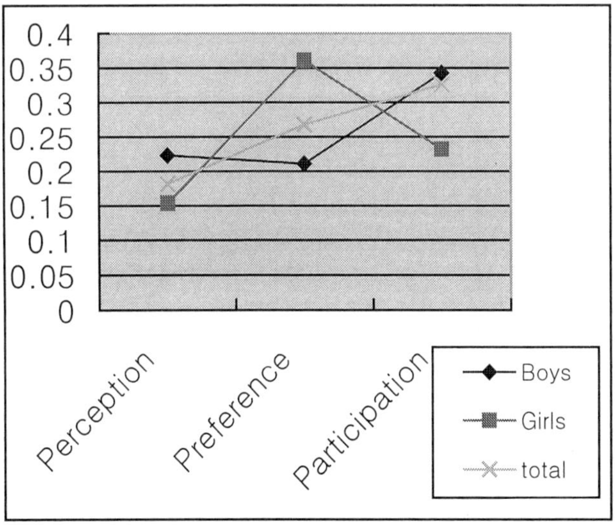

Figure 2. Comparison of β for each factor between boys and girls

References

1. B. E. Woolnough, Why Students choose Physics or Reject it, *Physics Education,* vol. 29, pp. 368-374, 1994.
2. EC, Research on causes of and solutions to the decrease in the number of students in engineering and science education in Europe (EU) during the last 10 years. Working paper of DG XXII at EC, Brussels, 1996.
3. B. E. Woolnough, Changing pupils' attitudes to careers in science, *Physics Education,* vol. 31, n5, pp. 301-308, 1996.
4. B. E. Woolnough, Factors Affecting Students' Choice of Science and Engineering, *International Journal of Science Education,* vol. 16, 659-676, 1994.
5. J. O. Myeong, Predicting and Understanding Korean High School Students' Science Tack Choice: Testing the Theory of Reasoned Action by Structural Equation Modelling, Ph.D. thesis, The University of Texas at Austin, Texas, 1990.
6. B. E. Woolnough, et al, Factors Affecting Students' Choice of Science and Engineering: Parallel Studies in Australia, Canada, China, England, Japan and Portugal, *Research in Science & Technological Education,* vol. 15(1), 105-121, 1997.
7. B. T. Eiduson & L. Beckman(ed.), *Science as a Career Choice. Theoretical and Empirical Studies.* Russell Sage Foundation, New York, 1973.

ANALYSIS OF FACTORS AFFECTING THE PROCEDURE OF BECOMING A SCIENTIST

KYOUNG-AE JANG*

Seoul National University, san 56-1, Shinrim-Dong, Kwanak-gu, Seoul 151-742, Korea

The current paper investigates the factors affecting the procedure of becoming a scientist in Korea. Sixty-minute individual interviews with scientists were used to assess the main factors involved in personal career choices. The investigation was based on a framework composed of personal ability, home environment, school environment, and sociocultural environment. The personal ability included personal interest, aptitude, and achievement; the home environment included parents, brothers and sisters, relatives, and hobbies; the school environment included teachers, friends, and club activities; the socio-cultural environment included public events, mass media, social recognition, and social prospects.

Thirty scientists that had received science awards in Korea during 1990-2001 were sampled. Nineteen scientists (63.3%) stated that they had a specific dream before entering high school, yet only 5 scientists (17%) dreamed of becoming a scientist. Twenty-one scientists (70%) said that they decided on their future career in 12th grade or as a freshman at university, which is related to the Korean educational system. However, this decision time is relatively late.

The results were analyzed based on the interview transcripts. According to the career choice procedures, personal ability, home environment, school environment, and socio-cultural environment were found to be almost equally important. In particular, personal interest and scientific aptitude, paternal influence, teachers (especially science teachers and professors), and social prospects were all emphasized. The younger scientists were more influenced by their personal aptitude and social prospects, while chemists and engineers were more influenced by their parents and social prospects. The qualitative research results are presented in detail.

1. Introduction

Since the 1980s, ‘science for all’ has been a slogan of school science education and the concept of ‘scientific literacy’ has been regarded as a counter-concept against the discipline-oriented movement of science education(Song, 1999). It is true that one of the main aims of science education is to develop a scientifically literate public. However, other important aims of science education include fostering competent students who are interested in studying science and encouraging science-related careers. In fact many countries are concerned about the number of students attracted to become a scientist or engineer.

Various researchers have already attempted to identify the main factors that affect a student's decision to pursue higher education in science and engineering and science-related careers (Woolnough, 1991, 1993, 1994, 1995, Young et al., 1997). However, there has been little research on what makes students become scientists. Although, it is difficult to determine precise career choice procedures due to the time lapse involved, it is still possible to identify the main factors affecting career choices. Accordingly, the current study attempted to analyze the key factors affecting the procedure of becoming a scientist.

2. Method And Procedure

Interviews were used to gain a deeper understanding of individual career choice procedures. As such, 60-minute individual interviews were conducted with scientists in their offices, and all interviews were tape-recorded.

A total of 34 scientists who had received some kind of science award in Korea during 1990 - 2001 were sampled. All were men. Ultimately, interviews were conducted with 30 scientists, as three scientists from the original group refused due to personal reasons, and one more could not be contacted.

The scientists were divided into 4 groups according to their age and major, respectively. The number of scientists in their 30s, 40s, 50s, and 60s was 7, 7, 12, and 4, respectively, while the number of physicists, chemists, biologists, and engineers was 6, 7, 8, and 9, respectively.

The investigation was based on a framework composed of the following four categories: personal ability, home environment, school environment, and socio-cultural environment. Plus, each category consisted of several factors as follows: personal ability included the meta-ability to figure out aptitude, personal ability, and achievement; the home environment included parents, brothers and sisters, relatives, and hobbies; the school environment included teachers, friends, and club activities; the socio-cultural environment included public events, mass media, social recognition, and social prospects.

3. Results And Discussion

3.1 *Personal Ability*

The meta-ability for figuring out aptitude and personal ability was highlighted in 25 scientists. These scientists regarded their personal interest and aptitude as the key reason they became interested in science. The scientists in their 30s' and the physicists and chemists indicated that their personal interest and aptitude were relatively important. This would seem to confirm Woolnough's research, Woolnough (1993), where a love for physics and chemistry was given as the reason why students chose to study physics and chemistry at university.

Furthermore, most of the scientists (26 scientists) were always in the top 10% of their school. In particular, they did well in mathematics and science. 19

scientists mentioned that they enjoyed and loved mathematics. The younger scientists and physicists also appeared to have relatively good grades.

3.2 *Home Environment*

Parental influence, particularly from the father, was identified as most important as regards the home environment. Thirteen scientists stated that their father strongly advised them to choose a career with good social prospects. In this case, most of the scientists were chemists or engineers. The other scientists were more indirectly influenced by their fathers, yet the mother's influence was hard to find.

The scientists in their 40s and 50s, and the chemists and engineers indicated a relatively higher influence from their home environment, possible because chemists and engineers are considered to have better career prospects by parents.

In the case of brothers and sisters, brothers were found to be an influence on 7 scientists. Plus, it was interesting that some of this influence was negative. For example, 2 scientists said that they decided not to become a doctor because their elder brother was majoring in medical science. In addition, relatives (5 scientists), hobbies (3 scientists), and one miscellaneous influence (1 scientist) were identified as other factors.

3.3 *School Environment*

Instruction by science teaches and professors was emphasized as regards the school environment (21 scientists). The scientists in their 40s and 50s related that their science teachers were excellent. This was possibly because their respective schools had a very good reputation in Korea at that time. Good lectures by university professors and the chance to participate in scientific research were also mentioned as having an impact on the decision to become a scientist. In addition to science teachers, mathematics teachers and the advice of other teachers also had an influence. Friends (4 scientists) and club activities (4 scientists) were identified as other factors.

3.4 *Sociocultural Environment*

Social prospects were emphasized in the sociocultural environment (12 scientists). This was particularly evident among the engineers and chemists, yet absent among the physicists. This area was also connected to parental approval. Other issues included social recognition (6 scientists), public events (6 scientists), mass media (2 scientists), other (8 scientists). Some of the physicists also confessed that the social image of genius being related to physicists had some influence. Science camps, lectures at science expos, and visiting science museums were identified as important events, while TV science documentaries, scientific publications for students, famous scientists, and religion were identified as other factors.

4. Conclusion

In conclusion, most of the scientists interviewed made their respective career decisions when they were in 12th grade or freshmen at university, based on consideration of intrinsic factors, such as personal interest and aptitude, and extrinsic factors, such as social prospects. Furthermore, the intrinsic factors were related to the influence of science teachers and professors, while the extrinsic factors were related to paternal advice.

Based on the current results, the main areas identified as promoting science as a career choice are recognition of personal aptitude and ability, positive reinforcement from parents about science and science-related careers, outstanding teachers and professors, and positive social recognition and prospects for science and science-related careers.

References

P. F. Brandwein & A. H. Passow (ed.) (1988). *Gifted Young in Science*, Washington, D. C.: National Science Teachers Association.

D. Hodson (1998). *Teaching and learning science: Towards a personalized approach*. Buckingham, Philadelphia: Open University Press.

R. M. Monhardt, J. W. Tillotson & P. D. Veronesi (1999). Same destination, different journeys: a comparison of male and female views on becoming and being a scientist, *International Journal of Science Education, 21*(5), 533-551.

D. A. Roberts (1988). What counts as science education? In P. Fensham (ed.) *Development and Dilemmas in Science Education*. Lewes: Falmer Press.

J. Song & K. Kim (1999). How Korean students see scientists: the image of the scientist. *International Journal of Science Education, 21*(9), 957-977.

B. E. Woolnough (ed.) (1990). *Making Choices*: An Enquiry into the Attitudes of Sixth-formers towards Choice of Science and Technology Courses in Higher Education. Oxford: Oxford University Department of Educational Studies.

B. E. Woolnough (1991a). *The Making of Engineers and Scientists*. Oxford: Oxford University Department of Educational Studies.

B. E. Woolnough (1991b). *The making of engineers and scientist: Factors affecting schools success in producing engineers and scientists*. Oxford: Oxford University Department of Educational Studies.

B. E. Woolnough (1993). Teachers' perception of reasons students choose for, or against, science and engineering. *School Science Review, 75*, 270.

B. E. Woolnough (1994a). Factors affecting students' choice of science and engineering. *International Journal of Science Education, 16*(6), 659-676.

D. J. Young, B. J. Fraser & B. E. Woolnough (1997). Factors Affecting Student Career Choice in Science: An Australian Study of Rural and Urban Schools. *Research in Science Education, 27*(2), 195-21

Part II

Conceptual Change

HOW DO STUDENTS UNDERSTAND ENVIRONMENTAL ISSUES IN RELATION TO PHYSICS?

INOUE TOKUYA, GAKU YAMAMOTO, SHIMOJO TAKASHI
Dept. of Science Education, Faculty of Education, Tokyo Gakugei University, Koganei, Tokyo, Japan.

In current educational programs, environmental issues are not discussed from the viewpoint of physics. Plus, physics education does not demonstrate that the laws of physics can be usefully applied to understand various environmental phenomena. The current study revealed that most high school and undergraduate students are unaware that environmental phenomena can be explained by the laws of physics that they have already studied. Accordingly, a program of physics education is proposed that emphasizes the relation between the laws of physics and environmental phenomena. In particular, the program focuses on issues that can be understood based on the Second Law of Thermodynamics, such as an urban heat island.

1. Introduction

Today's society is faced with environmental issues and energy crises on a global scale, as demonstrated by the strengthened efforts by many countries to solve this international dilemma. For example, in Japan, recent restructuring resulted in the Department of Environmental Issues being promoted to the Ministry of Environmental Issues. It is therefore extremely important that everyone has a chance to understand the significance of these issues and participate in helping to protect our planet.

Currently, there are a variety of successful environmentally related programs within the educational curriculum. For example, programs dealing with ecosystems, enzyme disturbances, and environmentally hazardous materials have been highly effective. However, due to the pressing importance of environmental issues, such programs need to be improved and expanded. Normally, the successful programs listed above are only perceived from a biology or chemistry standpoint, and physics is rarely involved. Generally, physics is a science that deals with universal truths and works from a more fundamental view point than other scientific areas, as such, it seems hard to relate physics to solving ordinary problems in daily life. However, when examining the issues concerned with the irreversible diffusion of hazardous materials, physics is the ideal science to work with. Also, understanding the mechanics of energy- dissipation in internal combustion engines and power generation involves the basic principles of thermodynamics. Therefore, it is only practical that a basic education in thermodynamics and statistical mechanics from an early age (high-school level), including a basic knowledge of entropy and energy, is crucial in the long term for solving the environmental problems that plague us today. Perhaps a new generation that can view today's problems from a broader scientific perspective will be better equipped to protect our future and thus a worthy investment for humanity.

The current study surveyed a range of high school and undergraduate students regarding their concerns about environmental issues. Thereafter, a high school physics program, Physics IB, was analyzed and it was found that many undergraduate students did not realize the importance of thermodynamics in environmental issues. Finally, the presentation of irreversibility in the Second Law of Thermodynamics in several textbooks was examined, resulting in a new interest among high school seniors in problems concerning our environment.

2. Current Student Cognition

2.1 *Student Cognition of Environmental Issues in Relation to Physics*

First, 148 students majoring in science or engineering at Meiji University were surveyed regarding their knowledge of environmental issues and the relation to physics. All students were first-year undergraduate students that had taken the high school Physics IB course. The details of the survey are outlined in Table 1.

Table 1. Student Cognition Survey Questions

Question 1
Please list 4 environmental issues that you believe to be the most important.

Question 2
Which science do you think is the most appropriate in dealing with environmental issues? Please pick two of the following.

1: Physics 2: Chemistry 3: Biology 4: Earth Science

The results of Question 1 are displayed in Fig. 1. Although many students knew about global warming and the destruction of the ozone layer, only a small percentage (4%) showed a concern about energy problems. As such, it would appear that most of the undergraduate students that participated in the survey were unaware that energy problems are environmental issues.

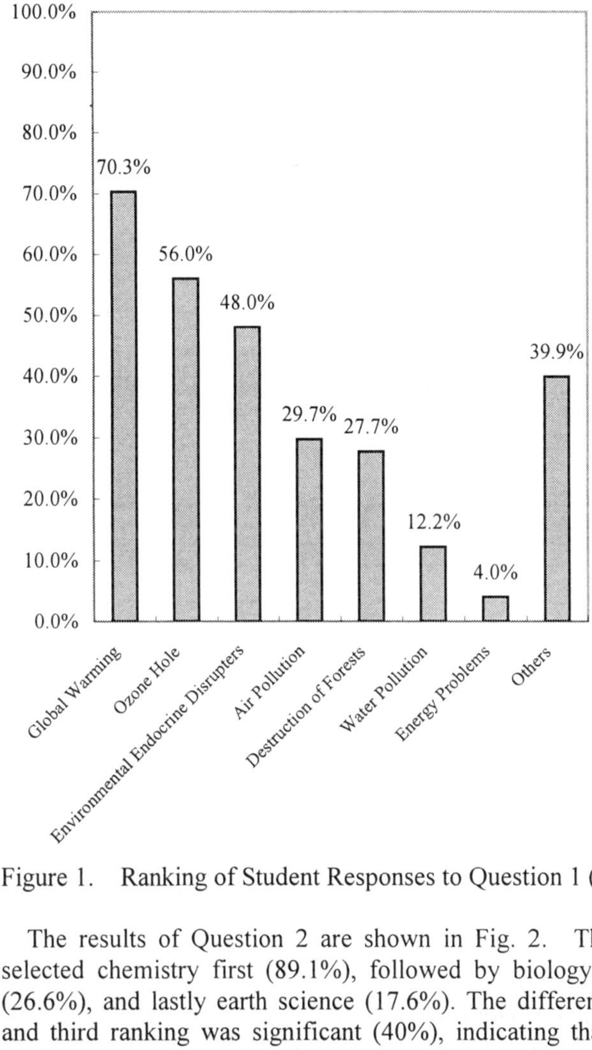

Figure 1. Ranking of Student Responses to Question 1 (N=148)

The results of Question 2 are shown in Fig. 2. The majority of students selected chemistry first (89.1%), followed by biology (66.9%), then physics (26.6%), and lastly earth science (17.6%). The difference between the second and third ranking was significant (40%), indicating that most of the students believed that chemistry and biology were far more appropriate than physics and earth science when dealing with environmental issues. This is even more evident in the results of Table 2, as the overwhelming combination was chemistry and biology.

Table 2. Students' Ideas of Combinations of Science Subjects Contributing to Solving Environmental Issues (N=148)

Physics & Chemistry	18.9% (28)
Physics & Biology	5.4% (8)
Physics & Earth Science	2.0% (3)
Chemistry & Biology	58.1% (86)
Chemistry & Earth Science	12.2% (18)
Biology & Earth Science	3.4% (5)

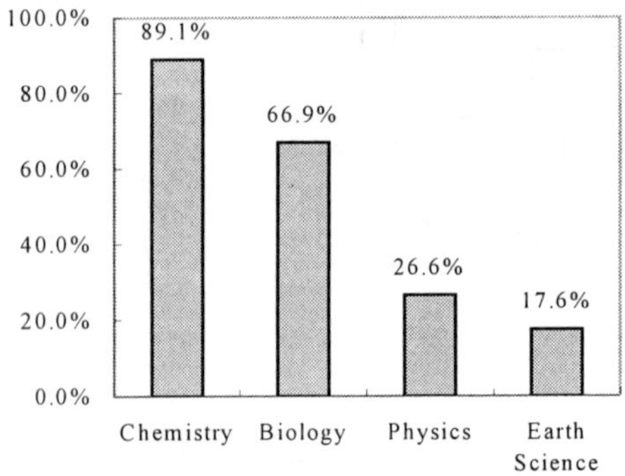

Figure 2. Students' Ideas of Science Subjects Contributing to Solving Environmental Issues (N=148)

2.2 *Student Understanding of Thermodynamics*

Next, a class of 117 second-year high school students at High School A was surveyed to see if they could understand the basics of thermodynamics. Although all students were taking the Physics IB course, they had not taken any classes on thermodynamics. The survey included a total of four problems (listed in Table 4), each dealing with a different subject (listed in Table 3). The same undergraduate class used in the first survey was also surveyed to compare the results.

Table 3. Problem Categories in Students' Understanding Survey

Problem 1	Energy Conversion
Problem 2	Perpetual Motion
Problem 3	Diffusion with Thermal Motion
Problem 4	Efficiency of Heat Engine

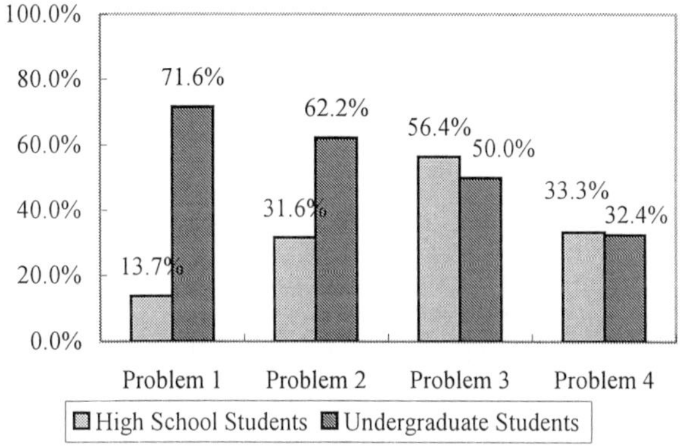

Figure 3. Percentage of Correct Answers (High School Students: N=117, Undergraduate Students: N=148)

The percentage of correct answers to this second survey is shown in Fig. 3. Obviously, the high school students that had not been educated in thermodynamics showed poor results overall, however, the majority of undergraduate students who had completed the course were only able to answer problems 1 and 2 correctly. Therefore, the survey suggests that the current high school program of thermodynamics is insufficient.

2.3 *Contents of Japanese High School Physics Textbooks*

Currently, there are 7 publishers that produce the Physics IB textbooks. Therefore, a comparison was made of the definitions of the Second Law of Thermodynamics among these textbooks. The details are listed in Table 5.

It is evident that these definitions have been modified and simplified for high school students. These textbooks also reflect different approaches to teaching the Second Law of Thermodynamics.

Table 4. Problems in Student Understanding of Thermodynamics Survey

Problem 1:
All moving objects have energy. Suppose a speeding car puts on its brakes and comes to a complete stop. What happens to the car's energy? Explain, in your own words.

Problem 2:
It is possible to use a motor as a generator by turning it. If there are two connected motors, where one is used as a generator, while the other motor is kept running, is it possible to keep the motor going forever?

1. Yes, it is possible 2. No, it is not possible

Problem 3:
There is a cup of water and a sugar cube is placed on the bottom. What will happen to the sugar over a long period of time? Select the answer that best fits your idea of what will happen. Assume that the sugar cube is small enough for it to be fully dissolved, and ignore the effects of evaporation and the warming of the water.

1. Does not dissolve 2. Partly dissolves
3. Completely dissolves

Problem 4:
During the summer, most homes use air conditioners to lower room temperature. Does the efficiency of the air conditioner change depending on the outdoor temperature? Circle one.

1. More efficient on cooler days
2. More efficient on warmer days
3. Efficiency does not change

This analysis was followed by determining if the problems from the four-problem survey could be successfully answered using the textbooks. The results are listed in Table 6. The concept of energy exchange (Problem 1) was found in all seven textbooks. For Problem 2, dealing the impossibility of a perpetual motion machine, only 4 out of the 7 included information on this area. For Problem 3, although the concept of the irreversibility of the process was included in 4 textbooks, only 1 properly explained the phenomenon using thermal motion. Finally, none of the textbooks provided any information that the heat phenomenon depends on the heat source and temperature difference, although all the textbooks mentioned that the effect was dependent on temperature. As such, these results were almost a precise reflection of the performance of the students in the survey, and emphasize the importance of high school physics classes as an opportunity to teach about the relationships between physical phenomena and environmental issues.

Table 5. Descriptions of Second Law of Thermodynamics in Physics IB Textbooks

Publisher	Description
A	All phenomena concerning heat are irreversible, and the following make up the Second Law of Thermodynamics. Heat naturally and at all times transfers from warmer substances to cooler substances. It is impossible for all heat to be used for work.
B	In an irreversible process, although the change is precisely defined, it will only work toward equilibrium. This direction of an irreversible process is defined in the Second Law of Thermodynamics. There are a number of ways to phrase the Second Law, one of which is "It is impossible for all heat from a certain source to be used for work".
C	The phenomenon of heat transfer is an irreversible process, and the direction of transfer cannot be defined only from the First Law. Therefore, the law that defines the natural direction of heat transfer is known as the Second Law of Thermodynamics. This can be further defined as "Heat will never transfer from cooler substances to warmer substances without external operations." Further, "Heat cannot, without external operations, be completely converted into mechanical energy."
D	If two objects at different temperatures are placed in contact there will be a heat transfer, yet, since it is impossible for heat to transfer from cooler objects to warmer objects, this is an irreversible process. Therefore, in a situation where there is only a transfer in heat energy, the process is irreversible.
E	A perfectly reversible phenomenon, much like the world without friction, is only an idealized environment and thus quite impossible. The law stating the irreversibility of real world phenomenon is known as the Second Law of Thermodynamics.

F	One naturally occurring phenomenon is the heat transfer from warmer substances to cooler substances. If a heat transfer occurs from a cooler substance to a warmer substance, although this does not violate the Law of Conservation of Energy, this in fact is impossible since heat transfer has a direction. (Abbreviation). In all instances, ordered states can be just as easily turned to chaos, whereas the opposite is impossible. Many phenomena, including heat loss in friction, energy loss as heat in an energy transfer, dissipation, and diffusion, are based on the random movement of atoms.
G	Heat-based engines always dissipate some heat before returning back to their original state, therefore, the efficiency (e) is always less than 1. Taken further, it is impossible to create an engine that can completely convert all heat energy. This phenomenon is known as the Second Law of Thermodynamics. Of course, through the use of different forms of energy there is no energy loss due to the Law of Conservation of Energy; however all energy used is eventually bound to become heat.

Table 6. Inclusion of Description of Problems in Students' Understanding Survey in High School Physics IB Textbooks

Publisher	A	B	C	D	E	F	G
Problem 1	*	*	*	*	*	*	*
Problem 2	*	*			*	*	
Problem 3						*	
Problem 4							

2.4 *Conclusion and Discussion of Surveys*

From the first survey, it was discovered that although most undergraduate students were aware of various environmental issues, they did not realize that many of the energy problems had a deep connection to thermodynamics. Furthermore, most of the undergraduate students believed that chemistry or biology was much more appropriate for dealing with environmental issues compared to physics. Since physics deals with larger scale universal truths, physics was not considered to be appropriate for solving small everyday problems. However, the survey also revealed that most undergraduate students did not see any connection between their high school physics education and their everyday problems, which also widened the gap between the two subjects.

The results from the second survey showed that the amount of information taken in about thermodynamics, specifically the Second Law of Thermodynamics, was based on the amount of information in the textbooks. The diffusion of the dissolving pollutant in Problem 3 and heat efficiency for saving energy in Problem 4 both represented environmental/energy problems related to thermodynamics. Therefore, the lack of difference in the performance between the undergraduates and high school students on these problems highlighted the importance of a good high school education in thermodynamics for dealing with environmental issues.

3. Environmental Education in High School Physics

3.1 *Inclusion of Environmental Issues in High School Physics IA Textbooks*

Currently, Physics IA is mostly taught in vocational high schools (excluding technical vocational high schools), as such, Physics IB students do not usually have a chance to be exposed to IA. The Physics IA course is designed to be less theoretical and more applicable to everyday life. Environmental issues are no exception to this approach. Out of the 7 publishers that publish the Physics IB course, 5 also publish the textbooks for IA. The chapter titles concerning environmental issues and energy problems are listed below in Table 7. These sections are not found in the Physics IB textbooks, and this may be one important reason why undergraduate students do not realize that energy problems are major environmental problems.

Table 7. Environmental Issues and Energy Problems Included as Chapter Titles in Physics IA Textbooks

Publisher	Included in List of Contents
A	Energy Resources, Usage of Natural Energy
B	The Energy We Use Everyday
C	(No Publication)
D	Solar Energy and Nuclear Power
E	Usage of Energy
F	(No Publication)
G	Solar Energy and Nuclear Power

3.2 *Importance of Teaching Second Law of Thermodynamics*

As shown in Table 5, the definition of the Second Law of Thermodynamics was found to vary between different textbooks. This is most likely due to the attempt to simplify the description for new learners of thermodynamics, and changes, such as the absence of the term entropy, occur so that high school students can more easily grasp the new concepts. However, without entropy, it is difficult to explain various phenomena, such as the irreversibility of the conversion between heat and work, the irreversibility of dissipation, and the efficiency of combustion engines. Thus, the omissions would seem to lead to further confusion among students.

Yet, most importantly, when tackling environmental issues using physics, such as the problems of diffusing pollutants and energy dissipation, then the Second Law becomes vital. Learning thermodynamics is a "must" to arouse the interest of students in environmental issues, therefore, more time and text must be added if this is to happen.

3.3 *Teaching Plan for Thermodynamics that Emphasizes Understanding of Environmental Issues*

Based on the collected information, a teaching plan was developed that emphasizes understanding environmental issues. Parts have been incorporated that were even missing in the Physics IA program, along with more in-depth information on the Second Law of Thermodynamics. The proposed program was implemented with 3 classes of third-year students at High School B (total 104 students). The plan is listed on the following page in Table 8.

3.4 *Evaluation of Lectures, Based on Teaching Plan*

To evaluate the classes based on the proposed teaching plan outlined in Table 8, a survey was conducted to check the level of understanding in the high school students that participated in the classes. The questions on the survey are listed below in Table 9.

Problem A is mainly concerned with the Second Law of Thermodynamics; the irreversible process of the conversion between heat and energy and the dissipation of energy. The ideal response should be "Because the conversion between heat and energy is irreversible, all forms of energy must be, finally, converted into heat. This is one of the causes of global warming or a heat island". A comparatively high percentage (74.0% or 77 students) gave a correct answer. This also suggests that the plan was successful in teaching students about the Second Law and developed a sense of connection between energy problems and environmental issues.

Problem B required students to use their knowledge of thermodynamics in solving a realistic energy problem. The vast majority of answers were correct (94.2% or 98 students) and most students were able to give their own opinions on what they thought best for slowing the depletion of resources. The individual answers were basically categorized into three groups: energy saving (53.8%, 56), substitution of energy resources (32.7%, 34), and improved efficiency in the usage of energy (13.5%, 14). The typical answers to Problem B are listed in Table 10.

Table 8. Teaching Plan for Thermodynamics that Emphasizes Understanding Environmental Issues

Period of Observation	From: April, 2001 To: July, 2001
Classes	3-D, 3-E, 3-F
Number of Students	3-D: 30, 3-E: 34, 3-F: 40
Textbook	Physics IB, Publisher E
Name of Subject	Heat and Energy
Total Time Spent	14 teaching hours
Teaching Contents	**Heat and Temperature: (3)** Heat and Temperature Three States of Matter Thermal Motion Thermal Equilibrium Specific Heat and Heat Capacity **Nature of Gas: (3)** Pressure of Gas Boyle's Low Charles' Low Absolute Temperature Absolute Zero Point Boyle-Charles Low **Work and Heat: (3)** Mechanical Equivalent of Heat Internal Energy of Gas Energy Conversion First Law of Thermodynamics Adiabatic Change Efficiency of Heat Engine (A)
Teaching Contents	**Irreversible Process: (3)** Reversible Process Irreversible Process Diffusion with Thermal Motion Work to Heat Heat to Work Heat as Useless Energy Perpetual Motion Second Law of Thermodynamics Efficiency of Heat Engine (B) **Energy and Our Life: (2)** Past, Current, and Future Energy Increasing Energy Consumption Limited Energy Resources Substitute Energy Resources for Oil Various Environmental Issues Global Warming Heat Islands

Numbers in parenthesis = number of classes spent on subject.

Table 9. Questions on Student Understanding Evaluation

Problem A
Humanity has been using energy resources, such as oil, in internal and external combustion to transform one energy resource into another form of energy. To accomplish this the resources are turned into heat, then converted to other forms of energy. If humanity is to continue converting resources into heat, what do you suppose will happen?

Problem B
Today, there is a steady depletion of energy resources. What measures do you think are the best to slow down this process?

Independent to the survey, the students were asked to prepare an optional report on energy and environmental issues. In the report, the students were instructed to investigate an issue they felt strongly about and finish with their own opinions. Surprisingly, even though the reports were assigned near to the end of the term, a high number of students submitted reports (74.0% or 77 students). Furthermore, the topics of the reports varied widely, thus it is believed that the proposed teaching plan for thermodynamics had a powerful impact on the students. The report topics are listed below in Table 11.

3.5 *Students' Impression*

Table 12 lists some representative reports from the many that were submitted. The students' reports view their subject from a number of different angles and show that the students themselves have honest opinions. It was also observed that the students developed a wide range of interest in environmental issues based on the proposed teaching plan. However, it is still unclear how the new teaching plan affected their understanding of the Second Law of Thermodynamics. Therefore, a method for evaluating the resulting understanding is needed.

Table 10. Students' Ideas for Measures against Energy Problems

Energy Saving	**Student 1** Even if new resources are used, the process of converting that resource into work requires time to build the appropriate facilities. Therefore, oil will probably still be used for some time. **Student 2** I can only do so much, which is saving energy. Yet, since I can do that much to solve this crisis, it must be done. **Student 3** I believe that it is important for everyone to use our resources as best as they can. For example, to switch off unneeded electrical appliances in our homes is one thing we can easily do. If everyone works to save some energy, then it might be possible to save a great deal of energy as a whole.
Substitu -tion of Energy Resource	**Student 4** Since heat dissipates and the process is irreversible, it seems best that we use earth-friendly methods, such as wind and solar power to produce energy. **Student 5** The taxes on oil and coal should be raised. People might be angered at first, but if we use the money raised from those taxes to develop better methods of power generation, such as solar energy, people will gradually accept it and oil will be used less.
Improved Efficiency of Energy Use	**Student 6** We live in a world today that requires the use of energy. We should become more efficient and find ways to reuse exhausted heat as an energy resource. **Student 7** Increase the efficiency of energy use and reuse heat.

Table 11. Report Topics

Topics	Number
Energy (Energy Resource) Problem	19
Global Warming	18
General Environmental Issues	13
Environmental Endocrine Disturbance	7
Air Pollution	6
Heat Islands	5
Acid Rain	7
Ozone Holes	2
Population Explosion	1

Table 12. Student's Impressions

Student A
Theme: Energy and Environmental Issues
Before, I didn't take energy problems seriously. For example, the idea of separating PET bottles at school seemed ridiculous. However, I now truly acknowledge that energy problems are a serious issue. Fossil fuels should be preserved and people need to recycle. And I've heard that PET bottles can even be turned into clothing. Previously, I thought that energy problems could be solved just by energy saving, but now I realize that isn't enough. To use new resources also takes time to build the appropriate facilities, and so we must spend years and months in attempting to solve our global crisis.

Student B
Theme: Global Warming.
After researching about global warming, I realize that only if everyone around the world cooperates can we deal with this most important issue. Every each person must work to help halt global warming. As a first step, people should only buy what they need, use what they have to their full extent, separate recyclables from the trash, conserve water by taking showers, and, in the summer, use air conditioners more efficiently by closing curtains. Such small efforts in saving energy can help save our planet.

Student C
Theme: Recycling Technology
Although more people have started recycling, the passing of laws, such as those that enforce the recycling of electrical appliances, prove that we do not recycle enough. These laws should be common sense, and if everyone added a little effort, we would be able to significantly reduce that amount of trash that we produce.

Student D
Theme: Global Warming
Right now, people should do as much as they can to stop global warming. Some examples include; better care of appliances, only using air conditioners as needed, and other small things that can be easily found around us. That way, if we can save a significant amount of energy in total, it will help slow global warming. It is important that everyone helps save the earth.

Student E
Theme: Heat Islands and Rain in Urban Areas
As the saying goes, "There are good things and there are bad things in life", if we live an easy life then there are sure to be consequences. Humanity has been lazy, and now we must pay the heavy price. Everyone dislikes work and humans have been geniuses at making life easier. Therefore, I believe that we will be able to find a way to make saving a planet a piece of cake.
Yet, time is short, and destiny is drawing closer. We must find a solution now, or it will be too late. Strangely enough, the plagues of today were sown by our own hands. Perhaps later generations will deal with our mistakes. But that is what we must do as humans.

4. Conclusion and Discussion

To a class of undergraduates who had taken Physics IB, physics is not the ideal science to tackle environmental problems. However, it was found that their education in thermodynamics is inadequate, especially in the Second Law of Thermodynamics, due to their high school textbooks, thereby explaining their lack of concern and understanding of energy problems. In addition, since high school students are not taught well about irreversible processes, this also explains their lack of understanding of environmental issues. The confusion generated by inadequate textbooks creates a dislike of physics among students, therefore, immediate modifications are required.

A new teaching plan was developed that allows a fuller understanding of realistic environmental issues based on physics, and its success and effect on students were demonstrated (refer Section 3). The proposed program should be adapted to fit the needs of different schools and classes, however, this is surely a worthy investment.

Acknowledgement

The authors would like to thank the following people for their help in the current study: Mr. Maki Tachikawa, Ms. Tokiwa Miura, Ms. Setsuko Matsumoto, Department of Physics, School of Science and Technology. We would also like to thank Mr. Dygo Tosa for translating this paper into English from Japanese.

References

1. Report on Energy-Environment Education in Physics Class, *Journal of the Physics Education Society of Japan*, Vol.49, No.1, pp.15-20, 2001.
2. An Attempt Environment Education through Teaching the Second Law of Thermodynamics, *Journal of the Physics Education Society of Japan*, Vol.48, No.3, pp.214-219, 2000.
3. Ernest L. McFarland, James L. Hunt, J. L. Campbell, Ruth H. Howes, Energy, Physics and the Environment, *American Journal of Physics*, Vol.63, No.4, pp. 382-383, 1995.
4. John Wheatley, T. Hofler, G. W. Swift, and A. Migliori, Understanding some simple phenomena in thermoacoustics with applications to acoustical heat engines, *American Journal of Physics*, Vol.53, No.2, pp. 147-162, 1985.
5. Egbert Boeker, Rienk van Grondelle, Environmental Physics Second Edition, John Wiley & Sons, LTD, 1995.

STUDY OF STUDENTS' UNDERSTANDING OF BASIC CONCEPTS IN PHYSICS

OANSUP CHOI*, YOUNGMEE LEE° & JAESOOL KWON*

**Dept of Physics Education, Korea National University of Education, Chungbuk, Korea*
°Dept of Science Education, Ewha Women's University, Seoul, Korea

Traditional physics instruction, as embodied in traditional physics textbooks, presents a physicist's understanding of the world. In this situation, students are passive learners and their conceptual difficulties are rarely addressed, thereby contributing to the development of new preconceptions. In many cases, preconceptions are often misconceptions or alternative conceptions, which impede the learning of appropriate physics concepts. Consequently, student preconceptions of scientific phenomena have recently become a focal point in physics education research. The current paper focuses on three problems: identifying students' understanding of basic physics concepts, establishing the students' degree of certainty, and distinguishing between a lack of knowledge and preconceptions for students across several grade levels. The investigation was conducted with female high school students in Incheon.

1. Introduction

Students are often able to use algorithms to solve numerical problems without completely understanding the underlying scientific concepts. Thus, teachers are invariably surprised to learn that, despite their best efforts, students do not always grasp the fundamental ideas covered in class. Even some of the best students give the right answers, yet are only using correctly memorized words[1]. For example, the Harvard physics professor Eric Mazur stumbled onto this fact in 1990[2] when he came across a standardized conceptual physics test that had been developed by David Hestenes and Ibrahim Halloun at Arizona State University and first published in 1985. As conceptual questions are elementary and easy, Mazur assumed that, since his students performed well with standard physics problems, they would also do well on a conceptual physics test. However, he was shocked to find that the performance in the simple conceptual physics test was comparable to that with seemingly much more difficult standard questions. Thus, investigations of the difficulties students encounter with physics have grown in number and sophistication. Yet, as researchers gain deeper insights into students' understanding of the material taught in the classroom, they are able to create curriculum materials that are more effective in improving a student's actual understanding. At the same time, the growing understanding of student reasoning provides the opportunity to find more systematic descriptions of how students come to make sense of physics.

In general, student knowledge on a particular concept can be divided into three categories: scientific conceptions (scientifically correct knowledge), preconceptions (scientifically incorrect knowledge), and lack of knowledge[3].

Depending on the nature of the particular measurement being conducted, these three categories can produce different outcomes, both independently and in combination. One of the major research findings on science education is that before instruction students develop preconceptions about phenomena. These preconceptions usually differ from the taught concepts, referred to as scientific conceptions. Preconceptions have a strong cognitive structure that differs from accepted understanding, are formed at a young age, and create a serious barrier that interferes with learning and the acquisition of new knowledge. Therefore, student preconceptions are extremely significant for teachers in any area of science. Without specific knowledge of preconceptions, teachers are ignorant of the interpretative framework through which students view the material that the teacher is trying to convey. Therefore, what teachers think they are communicating may not be what the students are actually learning. Hence many previous studies have paid particular attention to preconceptions and mental schemes held by students in specific disciplines. Yet, most research does not distinguish between preconceptions and lack of knowledge. Although there are many goals when investigating student understanding of physics, most researchers seek to recognize and "diagnose" specific difficulties, while also identifying the most common difficulties related to a specific topic[4], thereby helping individual students more effectively, while also creating a curriculum that can help the highest number of students overcome the most common difficulties.

2. Purpose of Current Study

As regards research, learning is an active construction process on the side of the learner based on pre-existing knowledge. In the above sense, conceptual change, whereby students learn pathways from certain preconceptions or lack of knowledge towards scientific conceptions, has proven to include both rational (logical) and emotional issues. However, in class, it has been found that students memorize equations and problem-solving skills and only retain a very small fraction of the information they are taught on basic physics concepts. As a result, they perform poorly on tests of conceptual understanding [5-7]. Numerous studies have focused on preconceptions[8-12] and highlighted the importance of distinguishing between a lack of knowledge and a preconception. Although both are relevant to understanding and the assimilation of subject matter, the instructional method for eliminating preconceptions and remedying a lack of knowledge is a fertile area in education research. Accordingly, the current paper hopes to answer certain questions concerning the basic concepts in physics:

What is the conceptual understanding held by groups of students?

What is the degree of certainty held by groups of students?

Do students have preconceptions or lack of knowledge as regards basic physics concepts?

3. Procedure

3.1 *Participants*

A total of 405 female high school students and biology education students participated as volunteers. Among the participants, 132 were 10th-grade students, 132 11th-grade students, and 141 12th-grade students.

3.2 *Test*

The conceptual physics questions were designed to elicit the students' conceptual understanding. The conceptual physics questions usually involved little or no symbolic manipulation or formulas and were taken from three different sources[5,13,14]. The questions were not reviewed in class during the semester, and dealt with different aspects of physics concepts.

Table 1. Conceptual questions.

Question No	Topic
1-3	Electric Currents
4	Negative velocity in Graphs
5	Gravitational force and Normal force
6	Direction of object in Circular motion
7	Newton's third law
8	Surface waves

3.3 Certainty

Each answer was assigned a mark of 1 if correct or 0 if wrong, and the total number of students who answered a question correctly was obtained by adding the marks then dividing this sum by the total number of the students' CRI (from 0 to 4).

4: absolutely certain it is true
3: sure it is true
2: no idea whether it is true or not
1: sure it is a guess
0: absolutely certain it is a guess

4. Result

(1) Question 1 (Electric Currents)

38% of the 10th-graders, 50% of the 11th-graders, and 56% of the 12th-graders answered correctly. Yet many students held the view that a current is energy and energy is used as light in a bulb. Hence, less current flows back to the battery. Therefore, the idea of the consumption of a current was not eliminated by formal instruction. The degree of certainty for the correct answers was 1.23, 1.06, and 1.01 and for wrong answers was 0.79, 0.80, and 0.98 for the 10th-graders, 11th-graders, and 12th-graders, respectively.

(2) Question 2 (Electric currents)

45% of the 10th-graders, 34% of the 11th-graders, and 27% of the 12th-graders answered correctly. These percentages were lower than those for 10th-graders in five European countries - 50%(Shipstone et al, 1988). Plus many students still held the view that a current is consumed in the case of resistance, thus the current flow is lower with resistance. The degree of certainty for the correct answers was 0.57, 0.58, and 1.14 and for the wrong answers was 0.55, 0.50, and 0.60 for the 10th-graders, 11th-graders, and 12th-graders, respectively.

(3) Question 3 (Electric currents)

39% of the 10th-graders, 30% of the 11th-graders, and 27% of the 12th-graders answered correctly. Most students held the view that wire will glow first according to the direction of the current. The degree of certainty for the correct answers was 0.29, 0.42, and 0.47 and for the wrong answers was 0.56, 0.37, and 0.58 for the 10th-graders, 11th-graders, and 12th-graders, respectively.

(4) Question 4 (Negative velocity in graphs)

25% of the 10th-graders, 38% of the 11th-graders, and 36% of the 12th-graders answered correctly. Most students held the view that a decrease in the negative velocity means that the object speeds up. The degree of certainty for correct answers was 1.47, 1.63, and 1.34 and for wrong answers was 1.36, 1.14, and 1.28 for the 10th-graders, 11th-graders, and 12th-graders, respectively.

(5) Question 5 (Gravitational force and normal force)

31% of the 10th-graders, 40% of the 11th-graders, and 39% of the 12th-graders answered correctly. Most students did not hold a conceptual understanding, as they thought that if the normal force is smaller than mg, the ball will not rebound from the ground. The degree of certainty for the correct answers was 0.49, 0.83, and 0.92 and for the wrong answers was 0.66, 0.78, and 1.05 for the 10th-graders, 11th-graders, and 12th-graders, respectively.

(6) Question (Direction of object in circular motion)

23% of the 10th-graders, 21% of the 11th-graders, and 35% of the 12th-graders answered correctly. The degree of certainty for the correct answers was 1.10, 1.37, and 1.82 and for the wrong answers was 0.95, 1.12, and 1.39 for the 10th-graders, 11th-graders, and 12th-graders, respectively. Even though this problem is well known as a misconception, the degree of certainty was sure it is a guess.

(7) Question (Newton's third law)

10% of the 10th-graders, 20% of the 11th-graders, and 26% of the 12th-graders answered correctly. The degree of certainty for the correct answers was 1.00, 1.59, and 1.68 and for the wrong answers was 0.56, 0.87, and 0.78 for the 10th-graders, 11th-graders, and 12th-graders, respectively.

(8) Question (Surface waves)

27% of the 10th-graders, 22% of the 11th-graders, and 71% of the 12th-graders answered correctly. In the case of the 12th-graders, they studied surface waves one week prior to the test. However, even though their correct response rate was significantly higher than for the other grades, their degree of certainty was not as high as the correct response rate. The degree of certainty for the correct answers was 1.03, 1.21, and 1.45 and for the wrong answers was 0.88, 0.66, and 0.74 for the 10th-graders, 11th-graders, and 12th-graders, respectively.

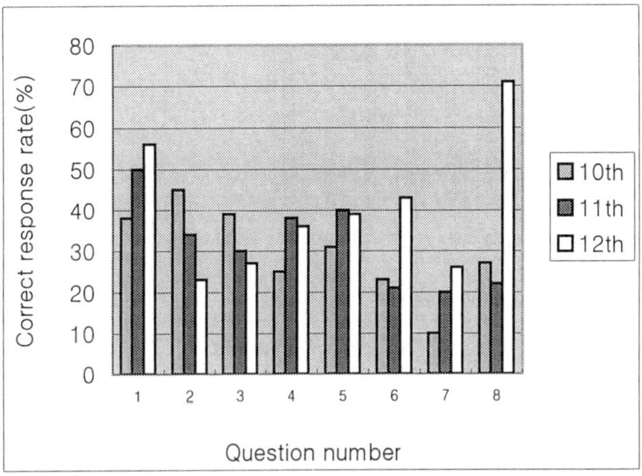

Figure 1. Correct response rate for questions.

(9) Overall the correct answer rate was 30%, 32%, 40% for the 10th-graders, 11th-graders, and 12th-graders, respectively.

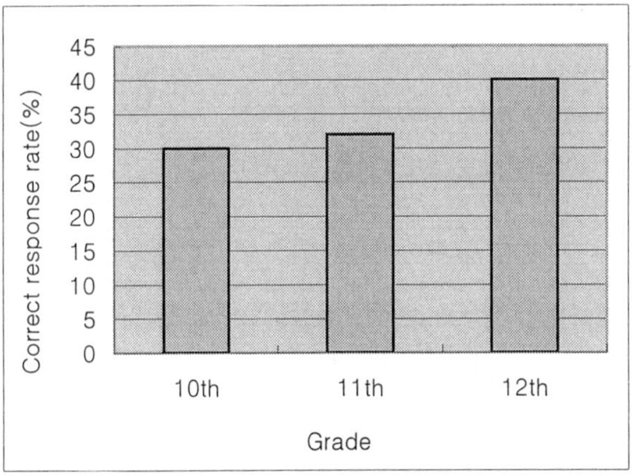

Figure 2. Overall correct response rate for questions.

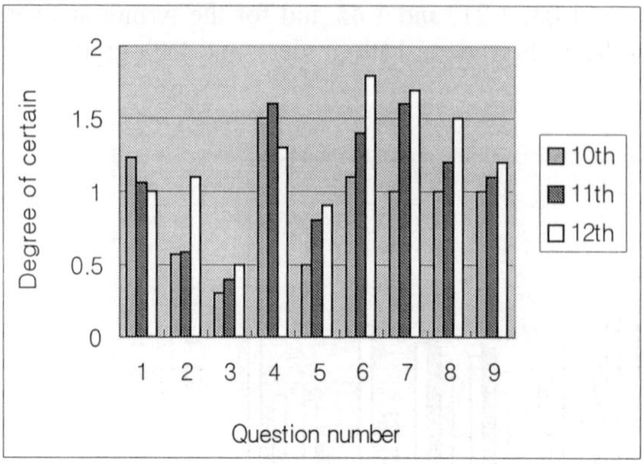

Figure 3. Degree of certainty for correct answers.

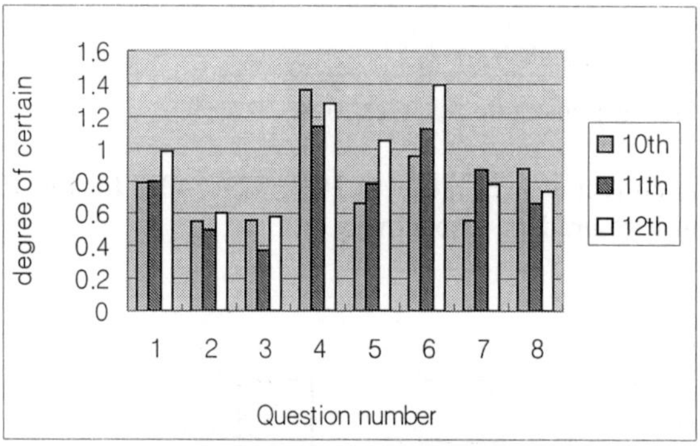

Figure 4. Degree of certainty for wrong answers.

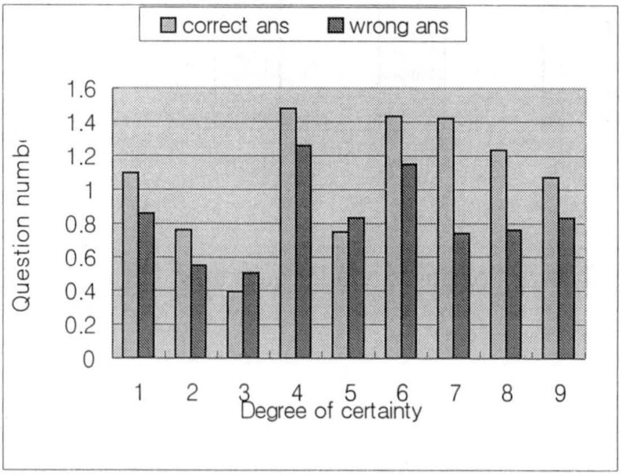

Figure 5. Degree of certainty for correct and wrong answers.

(10) The overall degree of certainty for correct answers was 0.90, 1.09, and 1.23 and for the wrong answers was 0.79, 0.78, and 0.93 for the 10th-graders, 11th-graders, and 12th-graders, respectively.

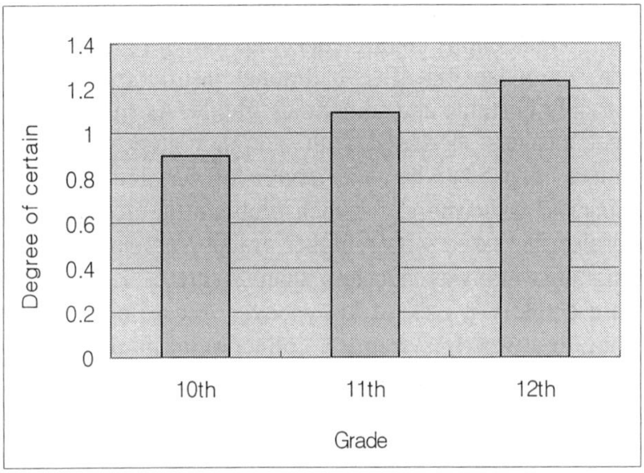

Figure 6. Overall degree of certainty for correct answers.

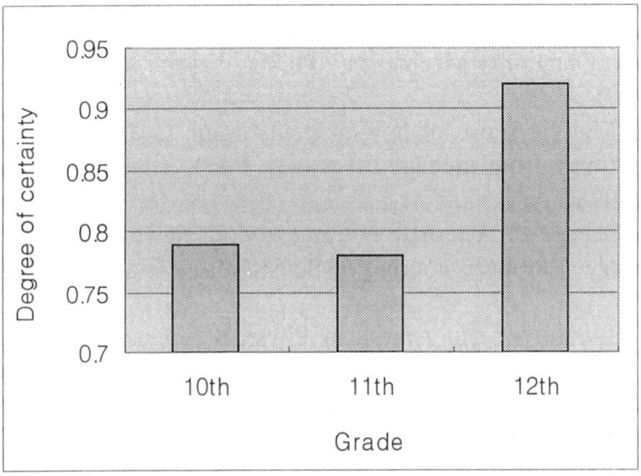

Figure 7. Overall degree of certainty for wrong answers.

5. Conclusion

The current study showed that, when students were first learning new material, their responses exhibited a strong contextual dependence and a lack of confidence as to the conditions under which the rules they had learned applied. As such, the normal understanding of physics concepts increased with a higher grade, yet the rate of increase was not more than 10% of what they were taught. Listening to lectures, reading textbooks, solving quantitative problems, watching demonstrations, and performing experiments seemingly had little effect on the students' learning. Therefore, traditional forms of instruction would appear to be inadequate for helping most students develop a functional understanding of basic physics concepts. Even excellent students memorized equations and problem-solving skills, thus their results as regards physics concepts were not as expected. In addition, connections among concepts, formal representations, and the real world were often lacking after traditional instruction, thus students' concepts were invariably unstable and sometimes fragile. As such, the degree of certainty for correct answers increased with a higher grade, yet the rate of increase was no more than 5%(when the degree of certainty was calibrated based on a percentage). Furthermore, for each problem the degree of certainty for wrong answers also increased with a higher grade. Yet, since the degree of certainty for wrong answers was not less than 1(sure it is a guess), their preconceptions were actually a lack of knowledge. Accordingly, the students reasoned by referring to a weakly systematic collection of intuitions rather than referring to a structured framework of knowledge.

References

1. Klemeš V. Teaching and propaganda. Physics Today, 10 Mar. (2000).
2. Mazur E. Peer Instruction: A User's Manual. Upper Saddle River, NJ: Prentice Hall. (1997).
3. Bao, L. Dynamics of Student Modeling: A Theory, Algorithms, and Application to Quantum Mechanics, Ph.D. dissertation, University of Maryland. (1999).
4. Wittmann, M. Making sense of how students come to an understanding of physics: An example from mechanical waves, Ph.D. dissertation, University of Maryland. (1998)
5. Colin T and George J. Alternative frameworks: Newton's third law and conceptual change. European Journal of Science Education, vol. 8, 291-298, (1986).
6. Heuvelen A. Bulletin of the American Physical Society, 36 1359, April, (1991).
7. Redish F. The implications of cognitive studies for teaching physics. American Journal of Physics 62(6), 796-803,(1994).
8. Sharon K and Thomas A. Gender, prior knowledge, interest and experience in electricity and conceptual change text manipulations in learning about direct current. Journal of Research in Science Teaching 34(2), 107-123, (1997).
9. Zeilik M, Schau C and Mattern N. Misconceptions and their change in university level astronomy courses. The Physics Teacher 36, 104-107, (1998).
10. Hasan S, Bagayoko D and Kelley E. Misconceptions and the certainty of response index. Physics Education . 34(5), 294-250, (2000).
11. Trumper R. University students' conception of basic astronomy conception. Physics Education . 35(1), 9-15, (2000).
12. Ireson G. The quantum understanding of pre-university physics students. Physics Education . 35(1), 15-22, (2000).

13. Shipstone D M, Rhöneck C. von, Jung W, Karrqvist C, Joshua S & Licht P. A study of secondary students' understanding of electricity in five European countries. International Journal of Science Education *10,* 303-316, (1988).
14. University of Montana Physics Department, misconception.

STUDENTS' CONCEPTIONS ABOUT POLARIZED ELECTROMAGNETIC WAVES

HYUN DEOK PARK, KYOUNGDAE KIM, MIRANG KWON, CHUNGHI LEE, SUKYOUNG KIM & JUNG BOG KIM

Department of Physics Education, Korea National University of Education, Chung-Buk, Korea

A survey was conducted on students' conceptions about polarized electromagnetic waves. Three research methods were used, including written responses, drawings, and interviews based on a questionnaire. As a result, students were found to have alternative conceptions of a plane wave related to the basic concepts of linear polarization. There was also a tendency to link an invisible EM wave with a visible water wave or rope wave. Furthermore, although most students were able to calculate the intensities of waves transmitted through two linear polarizers with an arbitrary angle between the transmission axes, they did not distinguish the mechanical structures from the polarization axes. As such, a slit was equated with a polarizer due to its similar geometrical description. Accordingly, the current results suggest that instruction on polarization should be based on the interaction between matter and light.

1. Introduction

Among the various properties of light, the human eye is able to detect brightness and color, yet not polarization. The concept of polarized light is naturally derived from the cognition that light is a transverse electromagnetic wave. The polarized state of light can be measured and changed arbitrarily using a polarizer or retarder etc. Electro-optical switches and polaroid sunglasses both originated from research on polarized light and demonstrate its useful application. In addition, a polarizer can be used to create interference patterns so as to realize various colors and three-dimensional vision. One cannot help being surprised at the beautiful colors revealed when a thin mineral is observed through a polarization microscope. As such, it is interesting and important to understand the complicated process that occurs as a result of the interaction between light and material.

The properties of light in geometrical optics and physical optics have already been the subject of a lot of research in the education field. The results have shown that students have more difficulty in understanding optics compared with other physics fields. Recently, Michal et al.[1] reported on the ability of university students to understand mechanical waves. They found that most students had a tendency to consider mechanical waves as particles. Meanwhile, Bradley et al. [2, 3] stated that the serious difficulties usually associated with physical optics stemmed from a lack of understanding of light as an electromagnetic wave, and that students often credited the slit with the characteristics of a polarizer. Y. Jeon[4] also highlighted several problems related to the way polarization is explained in current Korean textbooks, however, he did not investigate student responses. Therefore, there has been relatively little research on polarization from an educational perspective.

Learning concepts is the main learning activity and thus the core activity of a class [5]. As such, it is important to investigate student conceptions. In Korea, the topic of polarization is included in the high school curriculum. Accordingly, it is important to determine whether students learn about the concept properly and what type of concepts students form in class Consequently, the current study developed a questionnaire to consider these issues and investigate students' conceptions.

2. Research Methods

To investigate students' conceptions about polarized electromagnetic waves, four basic questions were asked. Fourteen college sophomores majoring in physics education were included in the study. After responding to the questionnaire, the students were personally interviewed. All the students had studied about polarization in the previous semester and at high school, except for one student. The students were also observed while taking a class on 'vibrations and waves in physics'. In addition, 17 seniors majoring in physics education at the same university and 19 seniors majoring in physics at another university were also included in the study as a comparison and to verify the efficacy of the current study.

Although it is difficult to generalize the ideas of university students about polarization, the results were categorized into several groups. Not all students who responded to the questionnaire were interviewed. Students whose ideas were singular and difficult to understand were interviewed. The interviews were all recorded.

3. Results

3.1 *Question #1*

The questionnaire consisted of 4 questions. Each question required an answer and explanation. Figure 1 was used with question #1 to establish the general conceptions held by the students of an EM plane wave. The students were asked to rank the magnitude of the electric fields at various points in each figure. Since it was not easy to classify the different answers, the data was categorized according to resembling types, as shown in tables 1 and 2. Other responses, except for the correct responses, were represented in terms used by the students. University B students were found to mix up the vibrating plane conception with the central axis conception. Plus, many students failed to correctly answer the question about a plane wave due to the reasons previously stated by Bradley et al.[3] Most students responded that the magnitude of the electric field changes according to the distance from the x axis, because the position of the sine curve expresses the intensity of the electric field. Although the sine curve expresses the function value at each point on the axis, students did not think of it as a graph. In fact, many students linked the electromagnetic wave to a visible phenomenon. Of most concern was that some students regarded the sine curve as a water wave. In this case, when considering only a vibrating plane, this conception led to the wrong answer because the students thought that the electric field only existed on the plane involving the sine wave, which was a vibrating plane.

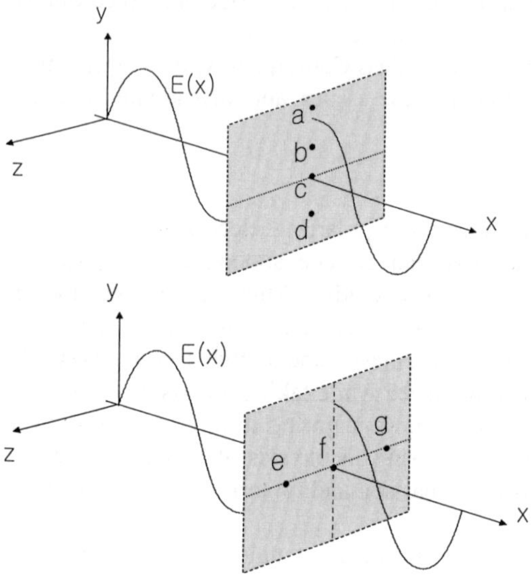

Figure 1. Sine curves describing a plane wave.

Table 1. Response types and possession percentage of results for question #1.

<table>
<tr><th rowspan="2">Response types</th><th colspan="2">University A</th><th>University B</th></tr>
<tr><th>Sophomore (N=14)</th><th>Senior (N=17)</th><th>Senior (N=19)</th></tr>
<tr><td>*Correct responses</td><td>21%</td><td>59%</td><td>0%</td></tr>
<tr><td>Water waves</td><td>21%</td><td>6%</td><td>5%</td></tr>
<tr><td>Vibrating plane</td><td>29%</td><td>24%</td><td rowspan="2">42%</td></tr>
<tr><td>Central axis</td><td>21%</td><td>0%</td></tr>
<tr><td>Others</td><td>7%</td><td>12%</td><td>26%</td></tr>
<tr><td>Blank or incomplete</td><td>0%</td><td>0%</td><td>26%</td></tr>
</table>

*The function values at all points on the wave plane are equal.

Table 2. Responses for sine curve representing a plane wave

Water waves	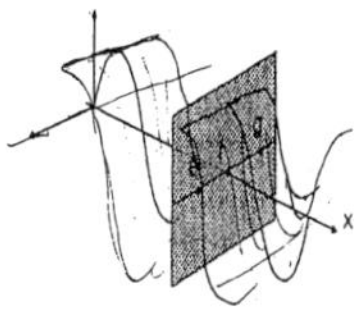 Conception remembering water waves and thinking of the phase as the height of the water wave, that is, **e=f=g**. Since the amplitude represents the intensity of the electric field (E.F.), **a>b>c>d** or **a>b=d>c**
Vibrating plane	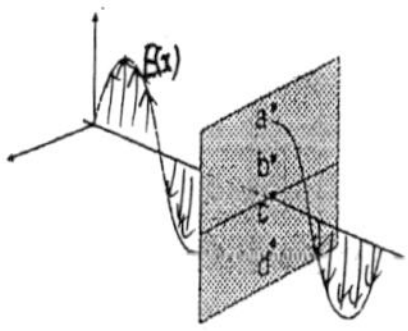 E.F. only exists on the plane involving the sine wave, which is a vibrating plane. The energy of the magnetic field exits outside the plane depicted by the E.F. (There is no E.F. or the E.F. is inversely proportional to the distance from the vibrating plane) That is, **f>e=g**
Central axis	As regarding propagating the x-axis (or sine curve) of the electromagnetic wave as a charge source generating E. F., the intensity of the E.F. is inversely proportional to the distance from the x-axis (or sine curve) That is, **f>e=g and a<b=d<c or a<b<c, c>d (or a>b>c>d)**
Region	An electric field only exists inside the sine curve, except for the intensity of E. F. from **a** or **d**

3.2 *Question #2*

The students were asked to answer freely what would happen if polarizer (3) entered into the space between two polarizers making no transparence. Question # 2 is a formal question given in class exams to investigate how to determine polarized light. Some students thought of the polarizers as retarders and wrote the equation " $1/\sqrt{2} + 1/\sqrt{2} = 2/\sqrt{2} = \sqrt{2}$ " to explain why the oblique polarized light was more intense than the non-oblique light.

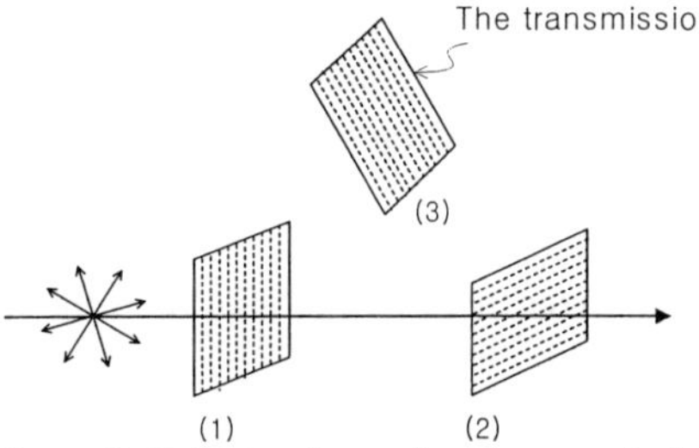

Figure 2. Polarization using three polarizers.

3.3 *Question #3*

Question # 3 examined how students perceived a slit and a polarizer. Fig. 3 is one of the figures used to ask what would happen when rotating the polarizer in front of the slit. Table 3 shows the results. Some students confused the polarizer with the slit, and the polarization axis of the polarizer with the long side of the slit. These students seemed to equate the EM wave with a mechanical rope wave and regard the transmission axis expressing the polarized direction as a visible structure. As such, they believed that the polarized light parallel to the long direction of the slit could pass through the slit.

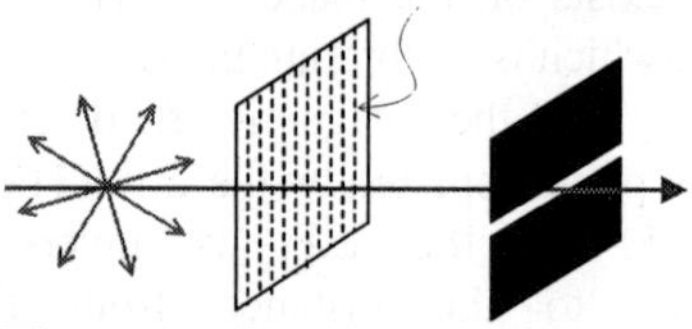

Figure 3. Slit and polarizer

Table 3. Percentage of results for question #3.

Responses	University A		University B
	Sophomore (N=14)	Senior (N=17)	Senior (N=19)
*Correct responses	21%	18%	16%
Slit equated with polarizer	50%	65%	37%
Other	28%	18%	27%
Blank or Incomplete	0%	0%	21%

* Intensity distributions are independent of the angle of the polarization axis because the diffraction by the slit is not dependent on the light polarization.

3.4 *Question #4*

For question # 4, an antenna was used to examine the students' conceptions about the response of an electron and polarized EM wave in material. Using Fig. 4, the students were asked when the antenna would accept the most micro radio waves and what would happen if a wire grating entered into the space between the antenna and the radio transmitter. The results are presented in Tables 4 and 5. Many students thought that the transmission axis was parallel to the wire line, thereby explaining why the ↕polarized light, which had the same direction as the wire, could pass through the wires.

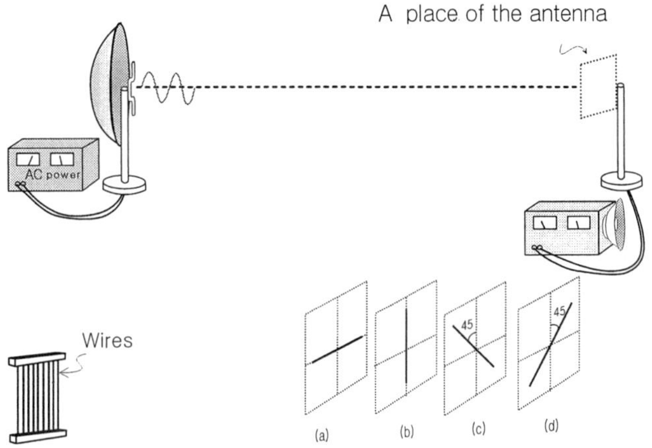

Figure 4. Antenna and polarization

Table 4. Percentage of results for question #4.

Responses	University A		University B
	Sophomore (N=14)	Senior (N=17)	Senior (N=19)
*Correct responses	14%	47%	16%
Transmission axis is parallel to wires	50%	0%	37%
Transmission axis is perpendicular to wires	14%	29%	11%
Other	14%	6%	5%
Blank or incomplete	7%	18%	32%

* The signal at b is the largest, and will disappear after placing the wire grating parallel to antenna b, because the EM wave will be absorbed by the wire grating. The polarization axis of the wire grating is perpendicular to the wire direction.

4. Conclusions and Implications

The results of the current study indicate that students tend to link an invisible EM wave with a visible water wave or rope wave. Students often do not identify a graph function. A slit is often equated with a polarizer and not much distinction is made between polarization and the material structure. Students were only found to remember pieces of knowledge. Therefore, classes connecting theory with the real world are required. Further research is needed for more effective teaching of the abstract concepts related to polarized light.

Table 5. Points in incorrect responses to question #4.

Transmission axis is parallel to wires	Although it will receive well, if (b), it will not receive at all as the wire grid will be perpendicular to the vibrating direction of the EM wave
Transmission axis is perpendicular to wires	·Since the wire is perpendicular to the transmission axis, (a) will receive well. · Because TV antennas have so many transverse elements, (a) will receive well.

References

1. Michael C. Wittmann, Richard N. Steinberg, and Edward F. Redish, Making Sense of How Students Make Sense of Mechanical Waves, *The Physics Teacher, Vol. 37*, pp. 15-21, 1999.
2. Bradley S. Ambrose, Paula R.L. Heron, Stamatis Vokos, and Lillian C. McDermott. Student understanding of light as an electromagnetic wave : Relating the formalism to physical phenomena, *Am. J. Phys. 67*, pp. 891-898, 1990
3. Bradley S. Ambrose, Peter S. Shafter, Richard N. Steinberg, and Lillian C. McDermott, An investigation of student understanding of single-slit diffraction and double-slit interference, *Am. J. Phys. 67*, pp.146-155, 1999.
4. Youngseok Jeon, Seoul National University, a Master's thesis 1997.
5. Nanju Kwon. Korea National University of Education, a Master's thesis, 1994.

STUDY OF STUDENT'S COGNITIVE PROCESS FOR LINE GRAPHS

T. KIM, E. KIM & B. KIM

Dept. of Physics Education, Korea National University of Education, Cheongwon, 363-791, Korea

Graphs, especially line graphs, are one of the most important vehicles in science, as they provide information at a glance within a small amount of space and can highlight relationships between variables. However, there are some important cognitive and educational questions that have been raised by the current use of graphs in many textbooks and articles. In addition, although line graphs are frequently used to communicate all kinds of data in science classroom activities, little has been reported on the student cognitive process based on empirical research, except for Jones et al. (1999).

Accordingly, the current study was intended to empirically investigate this cognitive process. First, the forms and contents of line graphs were compiled from many kinds of secondary school textbooks, especially physics related. Second, a computer program was created to determine the order in which the readers viewed the various components (Title, X-scale, X-label, Y-scale, Y-label, Data region, Legend) of a line graph and the time they examined each component. Third, the viewing order and times were examined in terms of each component using the SPSSWIN statistic program. Consequently, the results explicity identify the student cognitive process for a line graph. Finally, various educational and theoretical implications are presented with regards to the student cognitive process of a line graph.

1. Introduction

The use of graphs is commonplace in many various fields. Graphs are often included in books and newspapers to help communicate social and political information. Especially in the sphere of science, graphs are utilized to interpret information about experimental measurements and convey information clearly. As such, line graphs, in particular, represent a preferred form of communication for scientists, engineers, and mathematicians. In addition, line graphs can communicate trends and comparisons that would otherwise be difficult to convey.

Science education is often concerned with predicting relationships between variables, which is exactly what line graph can do, as line graphs show the relationships that exist between two variables.

Consequently, the cognitive process for a line graph is important in science instruction. Mokros & Tinker (1987, p.370) [10] state, quoting Howard Garder's (1983, p.302) assertion, that "The mastering of symbolic systems ... might even be regarded as the principal mission of modern educational systems."

Since many researchers have sensed that line graphs are important in the science classroom, there has been an increased interest in graphs, including investigations of misconceptions and skills in graphing (Berg & Phillips, 1994 [1]; Berg & Smith, 1994 [2]; Leinhardt, Zaslavsky, & Stein, 1990 [6]; Brasell, 1990 [7]; Brasell & Rowe, 1993 [9]; McDermott et al., 1987 [11]; Wavering et al., 1986 [13]; Rosenquist & McDermott, 1987 [15]; McKenzie & Padilla, 1984 [24]), the study of communication with graphs in the context of MBL (Microcomputer-Based Laboratory) (Berg & Smith, 1994 [2]; Jackson et al., 1993 [3]; Brasell, 1987 [8]; Mokros & Tinker, 1987 [10]; Linn et al., 1987 [12]; Wavering, 1989 [14]; Beichner, 1990 [17]; W-M. Roth et al., 1997 [19]); Adams & Shrum, 1988 [23]), and theories of graph comprehension (Lohse, 1993 [5]; Reynolds & Baker, 1987 [16]; Bertin, 1983 [20]; Moon, 1999 [21]; Pinker, 1983 [22]; Fisher, 1992 [26]; Maichle, 1994 [29]; Winn, 1994 [30]). Yet, there have been few reports concerning the student cognitive process for a line graph based empirical research, except for Jones et al. (1999) [27]. Also, in a study on graph skills, Brasell & Rowe (1993, p.64) [9] found that students thought graphs were useful, but were neutral in rating them as interesting or enjoyable.

Furthermore, even though typical science instruction uses graphs frequently, graphs can be treated as a superfluous adornment (Brasell, 1993, p.69) [9]. Therefore, if line graphs are to become cognitively accessible to students, first, empirical data is needed on the strategies people use to read line graphs. Then the insights and implications ascertained from understanding how students read line graphs can be used as communication information in many courses and contexts.

Accordingly, the current study used the following question to identify clues about the student cognitive process for a line graph:

Is there a statistically significant order and statistically significant difference between the viewing times of each component as regards the cognitive process for a line graph?

Two other assumptions were also made:

(1) A line graph schema is distinct from a general graph schema.

(2) The viewing order, time spent on each graph component, amount of verbalization, and level of subject complexity are all used to interpret line graphs.

As such, the current paper presents empirical research on the way graphs communicate information to a reader from the aspects of the viewing order and time spent viewing each component.

The sub-goals of the current study were as follows:

(1) Is there a statistically significant order in which readers view the components (Title, X-scale, Y-scale, X-label, Y-label, Data region, Legend) of a line graph?

(2) Is there a statistically significant difference between the times spent examining each component?

2. Methods

As a start, the shapes and numbers of graphs included in various kinds of science textbooks were investigated.

2.1 *Analysis of High School Science Textbooks*

Three of the following kinds of textbook were selected, based on an analysis of various kinds of high school science-related textbooks: Common Science (7), Physics I (8), and Physics II (8).

The numbers in parentheses represent the number of relevant textbooks analyzed. The number of graphs was counted using the notion of "plate", according to the characteristics identified by Roth, Bowen, & McGinn (1999, p.984) [18].

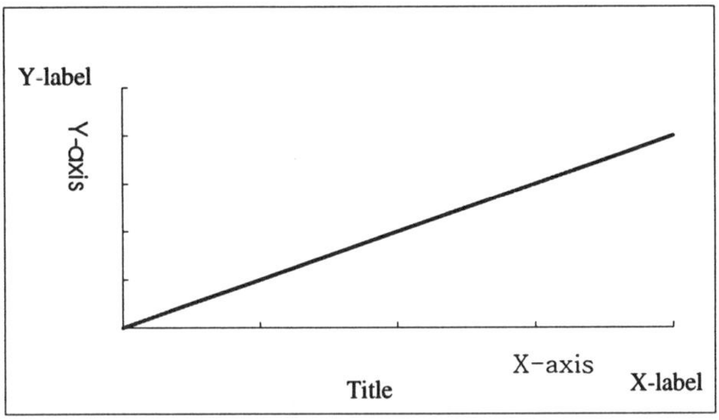

Figure 1. Shape of line graph in typical Korean textbook

Typical shape of line graphs. It was found that almost all graphs were line graphs in the form of Figure 1.

Number of graphs in Common Science textbooks. Table 1 presents the number of graphs found in the Common Science textbooks surveyed. The graphs were then classified into four domains, ie, Physics, Chemistry, Biology, and Earth Science, to determine which domain had the most graphs, especially line graphs. As a result, Physics comprised 37.35 % of the total, Chemistry 10.49 %, Biology 16.67 %, and Earth Science 35.49 %. In addition, almost all the graphs in the Physics domain were line graphs. Accordingly, it was decided to further investigate Physics-related textbooks, such as Physics I and Physics II.

Table 2. Number of graphs in Common Science textbooks

		Textbooks								
		A	B	C	D	E	F	G	Total	%
Domain	P[1]	13	25	14	10	20	19	20	121	37.3
	C[2]	6	3	6	4	2	7	6	34	10.5
	B[3]	10	1	7	3	8	16	9	54	16.7
	E[4]	21	11	15	11	16	19	22	115	35.5
Total		50	40	42	28	46	61	57	324	

[1] Physics, [2] Chemistry, [3] Biology, [4] Earth Science

Number of graphs in Physics I textbooks. Table 2 shows the frequency with which graphs were used as a presentation resource in Physics I textbooks. Among three units, the unit of motion & energy clearly had the most graphs (66.94 %).

Table 3. Number of graphs from Physics I textbooks

		Textbooks									
		I	II	III	IV	V	VI	VII	VIII	Total	%
Unit	M[1]	22	29	9	22	24	17	19	22	164	66.9
	E[2]	5	5	6	3	4	7	5	7	42	17.2
	W[3]	7	3	4	5	5	2	7	6	39	15.9
Total		34	37	19	30	33	26	31	35	245	

[1] Motion & Energy, [2] Electromagnetism, [3] Waves & Particles

Number of graphs in Physics II textbooks. Table 3 shows the frequency with which graphs were used as a presentation resource in Physics II textbooks. When compared with Table 2, Table 3 reveals a higher frequency. Plus, the units of Force & Motion (42.51 %) and Energy & Heat (23.91 %) included more graphs than the others.

Table 4. Number of graphs in Physics II textbooks

		Textbooks									
		a	b	c	d	e	f	g	h	Total	%
Unit	F[1]	24	23	17	29	23	17	21	22	176	42.5
	H[2]	14	15	13	10	24	10	7	6	99	23.9
	E[3]	12	5	13	9	10	11	9	5	74	17.9
	W[4]	9	6	10	7	16	5	3	9	65	15.7
Total		59	49	53	55	73	43	40	42	414	

[1] Force & Motion, [2] Energy & Heat, [3] Electromagnetism,
[4] Waves & Particles

2.2 *Development of Computer Program*

After investigating the shapes and contents of the line graphs, a computer program was developed using Visual Basic version 6.0 to run in a Windows 98 environment.

Issues related to development of software tool for analyzing line graphs. Based on the analysis of graphs in science-related textbooks, the following parameters were established:

(1) Select the main content within Common Science: Physics-related domains.

(2) Divide the graphs into units within Physics I and Physics II: Force & Motion (5 graphs), Energy & Heat (2 graphs), Electromagnetism (2 graphs) and Waves & Particles (1 graph).

(3) Determine the type of graph based on investigating Common Science, Physics, and Physics II: line graph.

One of the graphs used with the test subjects is shown in Figure 2. The main concern was not the subject's answers, but rather their cognitive process (viewing order and time spent viewing each component).

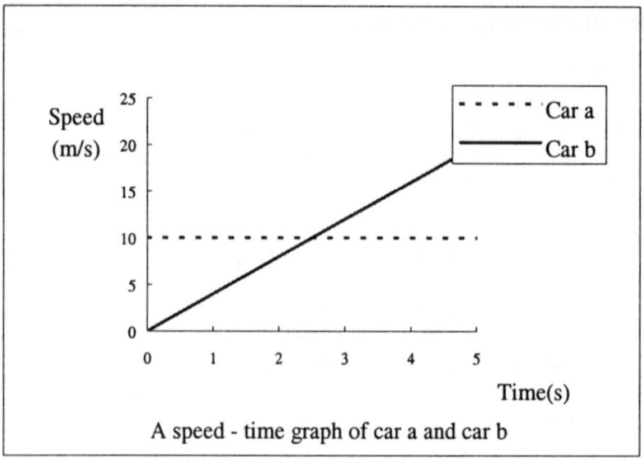

Item 3. How far do the cars go after 4 s?

Figure 2. Example of graph used with test subjects

Program design. The computer program was written using Visual Basic version 6.0 to run in a Windows 98 environment and displayed on a computer VDU (Visual Display Unit). To prepare the subjects, all the questions were displayed on the VDU before showing the graphs.

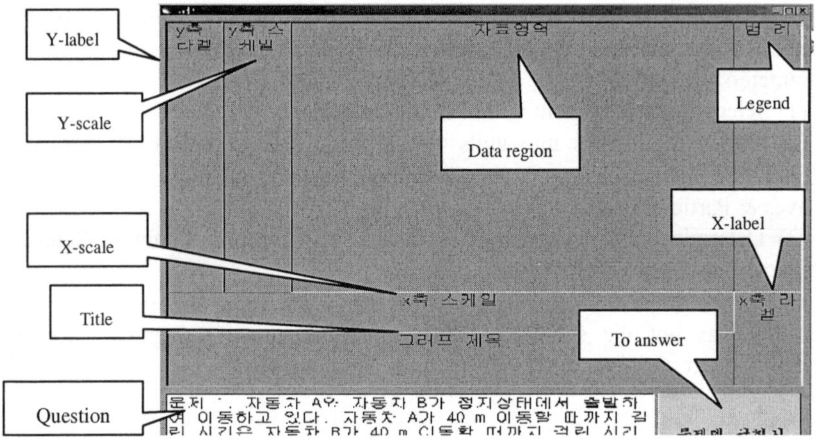

Figure 3. Sample graph covered by jacket

When a graph was initially displayed, it was covered by seven buttons. Then each button would open whenever the subject clicked on it. The subjects were encouraged to open any button they wanted in any order they preferred in response to the accompanying questions. To avoid confusion, the program was explained to the subjects based on a trial example. During the actual test, the program monitored the order in which the subject selected the components and the time spent viewing each component. Figure 3 shows the appearance of the VDU when all seven components (Title, X-scale, Y-scale, X-label, Y-label, Data region, and Legend) are covered by a jacket, while figure 4 shows the appearance of the VDU after the subject has opened two components (Y-scale and Data region).

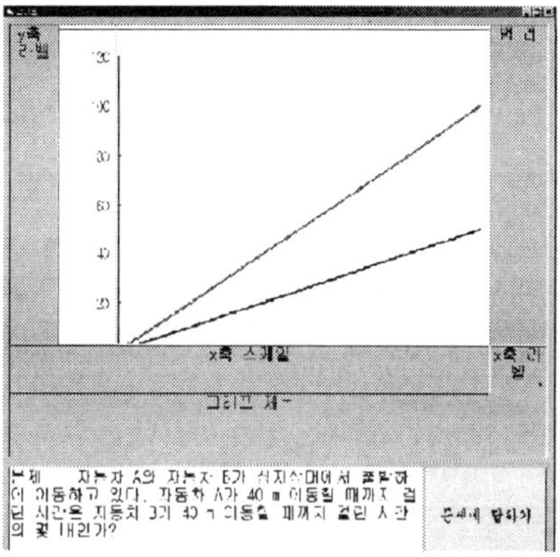

Figure 4. Sample graph with two open components (Y-scale and Data region)

2.3 *Test Subjects*

The test was conducted on 39 university students, mainly freshmen, see Table 4. All the test subjects were taking general Physics courses.

Table 5. Test Subjects

	Subjects		
Grade	Male	Female	Total
Freshmen	19	5	24
Sophomores	3	·	3
Juniors	2	2	4
Seniors	3	5	8
Total	27	12	39

3. Results and Discussion

TOGS (Test of Graphing in Science, McKenzie & Padilla, 1986) [4] and the Visual Basic program were used as the foundation for analyzing the results of the viewing order and time spent viewing each component of a line graph.

3.1 *Graph-related Abilities of Subjects*

All the subjects were asked to complete TOGS as a basis for reading line graphs. TOGS is a 26-part multiple choice test and requires approximately 40 minutes to administer. TOGS scores range from 0 to 26 and the test measures graph-related skills, such as scaling of axes, constructing best line interpolation, and describing the relationships displayed by graphs. The TOGS developers previously reported a reliability estimate of .83 (Kuder Richardson-20, .68 for graph construction, .73 for graph interpretation). Based on the scores from the current study, the reliability estimate for TOGS (Cronbach's α) was .85 (.72 for construction, .76 for graph interpretation).

Figure 5 shows the percentages of correct answers based on the Construction and Interpretation subtests.

3.2 *Order of Component Selection*

Table 5 shows the order in which subjects choose to read the seven graphing components. Table 5 represents the frequencies and percentages of the viewing order according to each component. The total number of possible occasions was 390, based on 10 line graphs x 39 subjects.

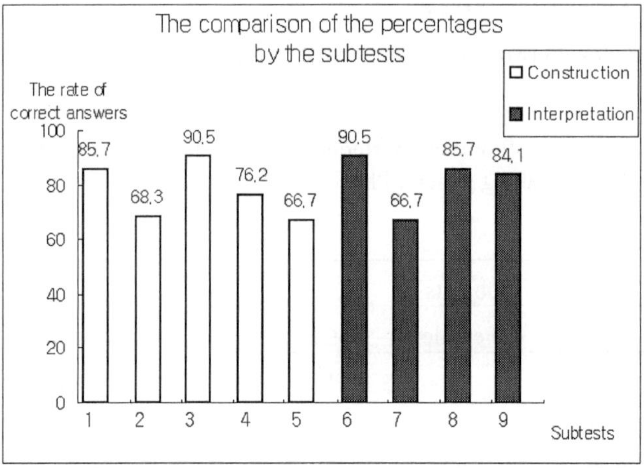

Figure 5. Graph-related abilities of subjects based on construction and interpretation subtests

1 selecting appropriately scaled set of axes, 2 selecting set of coordinates,
3 appropriately assigning data into graph, 4 selecting graph with best fit line,
5 selecting graph that correctly displays data,
6 selecting corresponding value for Y(or X), 7 interpolating/extrapolating,
8 selecting appropriate description of relationship,
9 interrelating data displayed in two graphs.

To verify whether or not there was any difference in the viewing order of each component of a line graph, the results were analyzed based on Chi square analyses. χ^2analyses revealed that the graph components were selected in a statistically significant order [χ^2=841.37, p<.05].

The data region was clicked first in the majority of tests (87.9%), while the other components were rarely clicked first. The legend (36.4%) tended to be selected second. The following order was the Y-scale (31.0%), X-scale (22.1%), and so on. The third choice exhibited a bimodal selection pattern between the Y-scale (31.0%) and X-scale (30.8%). There was also a tendency for the Y-scale (24.9%) to be selected fourth, the Y-label (39.0%) fifth, the X-label (42.3%) sixth, and the Title (63.3%) seventh.

Table 6. Number of occasions according to order in which subjects chose to view seven graphing components

	Viewing order						
	1st	2nd	3rd	4th	5th	6th	7th
	N (%)	N (%)	N (%)	N (%)	N (%)	N (%)	N (%)
Title	10 (2.6)	7(1.8)	20 (5.1)	10 (2.6)	38 (9.7)	58 (14.9)	247 (63.3)
Y-label	9 (2.3)	9 (2.3)	57(14.6)	88(22.6)	152(39)	53(13.6)	23(5.9)
X-label	5(1.3)	4(1.0)	15(3.8)	40(10.3)	58(14.9)	165(42.3)	102(26.2)
Y-scale	4(1.0)	121(31)	121(31)	97(24.9)	35(9.0)	11(2.8)	1(0.3)
X-scale	14(3.6)	86(22.1)	120(30.8)	82(21)	64(16.4)	23(5.9)	1(0.3)
Data region	343(87.9)	21(5.4)	13(3.3)	5(1.3)	4(1.0)	4(1.0)	0(0.0)
Legend	5(1.3)	142(36.4)	44(11.3)	68(17.4)	39(10)	76(19.5)	16(4.1)
Total	390	390	390	390	390	390	390

Based on the above results, a typical viewing route for a Korean reader was formulated, as displayed in Figure 6. For example, a Korean reader will likely view the data region first and the title last.

When the findings were examined in detail, a bimodal selection pattern was identified as regards clicking the Y-scale prior to the Legend or the Legend prior to the Y-scale. Since almost none of the line graphs in the textbooks investigated included a Legend, this was eliminated as a component. Therefore, when analyzing the results without the Legend, one typical viewing order was identified: Data region, Y-scale, X-scale, Y-label, X-label, and Title. Also worth noting was that most subjects clicked the Y-scale prior to the X-scale and the Y-label prior to the X-label. That is, either the scale components or the label components, indicating that the subjects unconsciously viewed the dependent variables prior to the accompanying independent variables.

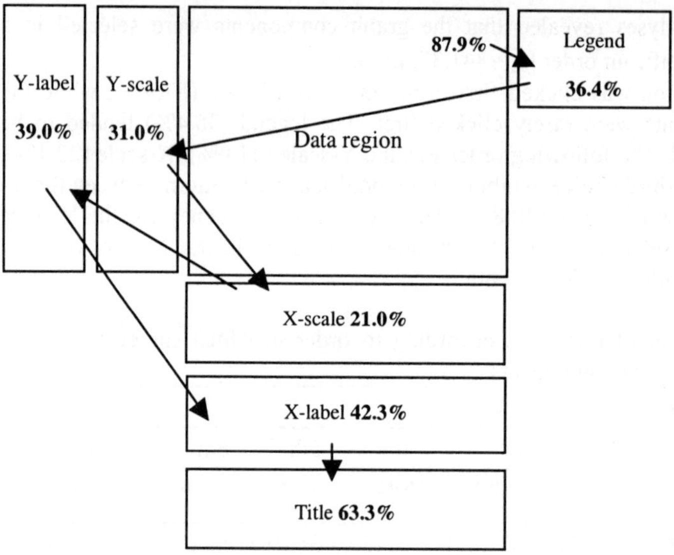

Figure 6. Typical viewing order

3.3 *Time Spent Viewing Each Component*

The difference in time spent studying each of the seven components (ie, Title, Data region, X-scale, Y-scale, X-label, Y-label, and Legend) was determined by means of an analysis of variance. Table 6 shows the results of the ANOVA.

Table 7. ANOVA results for time spent viewing each component

	SS	df	MS	F
Between groups	24927.5	6	4154.583	19.841
Within groups	570180.1	2723	209.394	
Total	595107.6	2729		

p< .05

Table 6 reveals statistically significant differences in the length of time subjects spent reading each line graph component. A series of post hoc Bonferroni analyses were also conducted to identify the exact source of these differences. Table 7 provides a summary of the Bonferroni analyses results. The values in each column (i - j) represent the values (ie, j) in the first column subtracted from those in the base row (ie, i). An asterisk indicates a significant difference and the unit of Table 7 is milliseconds.

Table 8. Results of Bonferroni analyses (i - j) (unit : ms)

j \ i	Title	Y-label	X-label	Y-scale	X-scale	Data region
Y-label	4726*	-				
X-label	-1202	-5928*	-			
Y-scale	5788*	1062	6990*	-		
X-scale	5194*	468	6396*	-594	-	
Data region	7613*	2887	8815*	1825	2419	-
Legend	1742	2985	2943	-4046*	-3453*	-5871*

* $p < .05$,

The time spent viewing the title of the line graphs was significantly longer than that for the other components, except for the X-label and Legend. The subjects viewed the X-label much longer than the Y-label. Similarly, the X-scale was viewed much longer than the Y-scale. In addition, a statistically significant difference was found between the time spent viewing the Legend and the time spent viewing the X-scale, Y-scale, and Data region. Therefore, the results showed that the subjects spent significantly more time viewing the Legend, Title, and X-label than the other components.

4. Conclusions and Implications

Line graphs play an important role in all fields of education, as they display the relationship between two continuous variables. Recently, researchers such as Fisher (1991, 1992) [25] [26], Pinker (1983, 1990) [22] [28], and Winn (1994) [30] realized the need for models to explain the student cognitive processes for line graphs. However, relatively little empirical research has been conducted on student cognitive processes for line graphs to support such models. Accordingly, the goal of the current study was to determine whether or not there was a significant difference in the viewing order of each component and the time spent viewing each one. Then, if there was any difference, what were the implications.

The tests were conducted after analyzing various kinds of high school science-related textbooks, Jones' revised program, and the subjects' TOGS scores.

As regards the viewing order for each component in a line graph, most students were separated into two groups: one group clicked the scale components prior to the Legend, while the other group clicked the Legend prior to the scale components. When the analysis results were considered without the Legend (as in most line graphs in Korean textbooks), this represented one route for Korean students to read line graphs (Figure 6).

In addition, most of the subjects clicked the Title last. This is certainly an area worthy of further study, as the Title functions as an excellent clue to understanding the contents of a line graph and can decrease the time spent reading all the other components. Therefore, where did the tendency come from? There are at least two possible explanations. One is that the subjects were influenced by the Korean custom of reading from top to bottom and left to right. In the current study, the Titles were located at the bottom of the line graphs, as in Korean textbooks, whereas in Jones' study the Titles were located at the top. The other possible explanation is that, since the Titles were so short, they could not express the content of the line graphs. The Titles of line graphs in Korean textbooks usually only take up one line and do not include any details, whereas graphs in scientific journal articles provide many resources in the captions that can assist readers in directing their attention and interpretive efforts.

As regards the time spent reading each component in a line graph, the ANOVA results revealed statistically significant differences. ($F=19.841$, $p< .05$) Most time was spent reading the X-label, Title, and Legend, in that order. As stated in the above paragraph, since the Title was usually clicked last, the time spent viewing the Title component may have actually been the time interval between the moment the graph was entirely exposed and the moment the subject clicked to answer the question (Jones et al., 1999) [27]. As such, if the Legend is eliminated, most time was clearly spent viewing the X-label. Therefore, the analysis results of the viewing order and time spent viewing each component revealed that Korean students have a tendency to ignore the Title component when reading a line graph and focus on the X-label for a long time.

Since line graphs are an efficient tool for making sense of information and used as an integral part of communication information in many contexts, it is important that they become cognitively available to students in many fields. Accordingly, the current study provided various insights into the cognitive strategies used by students to read line graphs based on investigating the viewing order and time spent viewing each component in a line graph.

References

1. C. A. Berg and D. G. Phillips, Investigation of the Relationship between Logical Thinking Structures and the Ability to Construct and Interpret Line Graphs, *Journal of Research in Science Teaching*, vol. 31, pp. 323-344, 1994.
2. C. A. Berg and P. Smith, Assessing Students' Abilities to Construct and Interpret Line Graphs: Disparities between Multiple-Choice and Free-Response Instruments, *Science Education*, vol. 78, pp. 527-554, 1994.
3. D. F. Jackson, B. J. Edwards, and C. F. Berger, Teaching the Design and Interpretation of Graphs Through Computer-Aided Graphical Data Analysis, *Journal of Research in Science Teaching*, vol. 30, pp. 483-501, 1993.
4. D. L. McKenzie and M. J. Padilla, The Construction and Validation of the Test of Graphing in Science (TOGS), *Journal of Research in Science Teaching*, vol. 23, pp. 571-579, 1986.
5. G. L. Lohse, A Cognitive Model for Understanding Graphical Perception, *Human-Computer Interaction*, vol. 8, pp. 353-388, 1993.

6. G. Leinhardt, O. Zaslavsky, and M. K. Stein, Functions, Graphs, and Graphing: Tasks, Learning, and Teaching, *Review of Educational Research*, vol. 60, pp. 1-64, 1990.
7. H. M. Brasell, Graphs, Graphing, and Graphers, *What Research Says to the Science Teacher - The Process of Knowing*, chap. 5, 1990.
8. H. M. Brasell, The Effect of Real-Time Laboratory Graphing on Learning Graphic Representations of Distance and Velocity, *Journal of Research in Science Teaching*, vol. 24, pp. 385-395, 1987.
9. H. M. Brasell and M. B. Rowe, Graphing Skill among High School Physics Students, *School Science and Mathematics*, vol. 93, pp. 63-70, 1993.
10. J. R. Mokros and R. F. Tinker, The Impact of Microcomputer-Based Labs on Children's Ability to Interpret Graphs, *Journal of Research in Science Teaching*, vol. 24. pp. 369-383, 1987.
11. L. C. McDermott, M. L. Rosenquist, and E. H. van Zee, Student Difficulties in Connecting Graphs and Physics: Examples from Kinematics, *American Journal of Physics*, vol. 55, pp. 503-513, 1987
12. M. C. Linn, J. W. Layman, and R. Nachmias, Cognitive Consequences of Microcomputer-Based Laboratories: Graphing Skills Development, *Contemporary Educational Psychology*, vol. 12, pp. 244-253, 1987.
13. M. J. Wavering, B. Perry, L. J. Kelsey, and D. Birdd, Performance of Students in Grades Six, Nine, and Twelve on Five Logical, Spatial and Formal Tasks, *Journal of Research in Science Teaching*, vol. 3, pp. 321-333, 1986.
14. M. J. Wavering, Logical Reasoning Necessary to Make Line Graphs, *Journal of Research in Science Teaching*, vol. 26, pp. 373-379, 1989.
15. M. L. Rosenquist and L. C. McDermott, A Conceptual Approach to Teaching Kinematics, *American Journal of Physics*, vol. 55, pp. 407-415, 1987.
16. R. E. Reynolds and D. R. Baker, The Utility of Graphical Representations in Text: Some Theoretical and Empirical Issues, *Journal of Research in Science Teaching*, vol. 24, pp. 161-173, 1987.
17. R. J. Beichner, The Effect of Simultaneous Motion Presentation and Graph Generation in a Kinematics Lab, *Journal of Research in Science Teaching*, vol. 27, pp. 803-815, 1990.
18. W-M. Roth, G. M. Bowen, and M. K. McGinn, Differences in Graph-Related Practices between High School Biology Textbooks and Scientific Ecology Journals, *Journal of Research in Science Teaching*, vol. 36, pp. 977-1019, 1999.
19. W-M. Roth, M. K. McGinn, Graphing: Cognitive Ability or Practice?, *Science Education*, vol. 81, pp. 91-106, 1997.
20. J. Bertin, *Semiology of Graphics – Diagrams Networks Maps*, W. J. Berg (trans.), Chap. 1, part I, The University of Wisconsin Press, Wisconsin, 1983.
21. C-S. Moon, A Perceptual and Cognitive Process Model and Students' Error Patterns in the Interpretation and Comprehension of Line Graphs in Force and

Motion, Ph.D. thesis, Graduate School of Korea National University of Education, 1999.

22. S. Pinker, Pattern Perception and the Comprehension of Graphs, National Institute of Education Report. pp. 1-46, 1983.
23. D. D. Adams and J. W. Shrum, The Effects of Microcomputer-Based Laboratory Exercises on the Acquisition of Line Graph Construction and Interpretation Skills by High School Biology Students, Paper presented at the Annual Meeting of the National Association for Research in Science Teaching, Lake of the Ozarks, 1988.
24. D. L. McKenzie and M. J. Padilla, Effects of Laboratory Activities and Written Simulations on the Acquisition of Graphing Skills by Eighth Grade Students, Paper presented at the Annual Meeting of the National Association for Research in Science Teaching, New Orleans, 1984.
25. M. A. Fisher and C. C. McKnight, Features of Knowledge Representations for Data Presented Graphically in Populations with Diverse Mathematical Experience, Paper presented at the Annual Meeting of the American Educational Research Association, Chicago, 1991.
26. M. A. Fisher, Categorization, or Schema Selection in Graph Comprehension, Paper presented at the Annual Meeting of the American Educational Research Association, San Francisco, 1992.
27. R. W. Jones, J. W. Warner, and S. G. Fankhauser, Investigating Student Understanding of Graphs: A Successful Methodology and Results of a Study, Paper presented at the Conference of the National Association for Research in Science Teaching, Boston, 1999.
28. S. Pinker, A Theory of Graph Comprehension, in R. Freedle (Ed.), *Artificial Intelligence and the Future of Testing*, Chap. 4, Hillsdale, NJ: Lawrence Erlbaum Associates, Inc., 1990.
29. U. Maichle, Cognitive Processes in Understanding Line Graphs, in W. Schnotz and R. W. Kulhavy (Eds.), *Comprehension of Graphics*, Chap.13, North-Holland, Amsterdam, 1994.
30. W. Winn, Contributions of Perceptual and Cognitive Processes to the Comprehension of Graphics, in W. Schnotz and R. W. Kulhavy (Eds.), *Comprehension of Graphics*, Chap.1, North-Holland, Amsterdam, 1994.

STUDY OF STUDENTS' CONCEPTIONS OF PREDICTABILITY IN PHYSICS- EXPERIMENTAL PROGRAM FOR TEACHING BASIC NATURE OF CHAOS

INOUE TOKUYA & SHIMOJO TAKASHI

Dept. of Science Education, Faculty of Education, Tokyo Gakugei University, Koganei, Tokyo, Japan.

In a deterministic system, solving equations of motion with certain initial conditions gives a unique trajectory according to causality. In a periodic system, like a harmonic oscillator, the trajectory does not depend so much on the initial conditions, as a small change in the initial conditions only leads to a small change in the trajectory. However, in a chaotic system, a small change in the initial conditions significantly modifies the time evolution thereafter. The current paper reveals that few physics-major undergraduate students understand the importance of the initial conditions when solving equations of motion. In particular, no one could perceive the existence of a chaotic motion. Accordingly, an experimental program is proposed that demonstrates the basic nature of chaos using a double pendulum, a well-known chaotic oscillator. Observation of the motion intuitively teaches that its sensitive dependence on the initial conditions makes it difficult to predict a chaotic time evolution over a long period of time.

1. Introduction

Today, the term "chaos" is no longer a special technical word only understood among some physicists and mathematicians, but rather a common word used in everyday life. Chaos theory has also been applied to the field of social science, for example, economics. However, there are relatively few opportunities to study of the nature of chaos in the normal undergraduate physics curriculum.

Since the term "chaos" has become so familiar, it would seem common sense to learn its basic nature. In particular, it is very important to understand its sensitive dependence on initial conditions when reconsidering many physical theories described using deterministic equations, where the prediction of subsequent phenomena is based on solving equations of motion with certain initial conditions. However, in a chaotic system, it is impossible to predict future events because even a small change in the initial conditions, far beyond the resolving power of human eyes and measuring devices, can lead to a significant change in the subsequent phenomena.

Accordingly, the current study developed an educational program to demonstrate the basic nature of chaos using a double pendulum, a well-known chaotic system model, and conducted a trial lecture with junior, senior, and graduate students majoring in physics. First, the lecture highlights that a sensitive dependence on initial conditions allows the motion of the pendulum to be predicted. Then, the students are asked to consider the meaning of predicting phenomena in a deterministic system. Finally, the students are shown the importance of the initial conditions and the limitation of prediction through physical theories.

2. Prior Survey

Prior to the trial lecture, 58 sophomore students majoring in physics at the School of Science and Technology, Meiji University were surveyed on their level of understanding of the relationship between a small change in the initial conditions and the trajectory of motion in order to determine the current status of their cognition.

In the survey, the students were first shown three graphs of falling motion, a harmonic oscillator, and exponential motion. Each graph showed the relationship between the displacement and time. The students were then asked to draw the changed curves on each graph sheet respectively after a small change was made in the initial conditions.

2.1 *Problems*

Falling Motion

Fig. 1 shows the height of an object according to the time when the object falls to the ground after it has been thrown up vertically with an initial velocity of v_0 from a height of h at time $t = 0$.

Problem 1: The height from which the object is thrown, is shifted a little lower: $h - \Delta h$. Draw the motion on the original curve in Fig. 1.

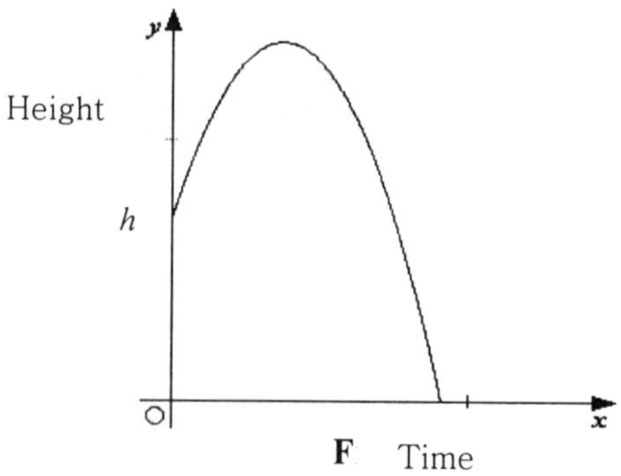

F

Harmonic Oscillator

Figs. 2 and 3 show the displacement of a weight according to time, where the weight is attached to the end of a vertically hanging spring extended by x_0 from its natural length at time $t = 0$.

Problem 2: The position of the weight is slightly lowered when the spring is initially released. The extension of the spring from its natural length becomes: $x_0 + \Delta x$. Draw the subsequent displacement of the weight on the original curve in Fig. 2.

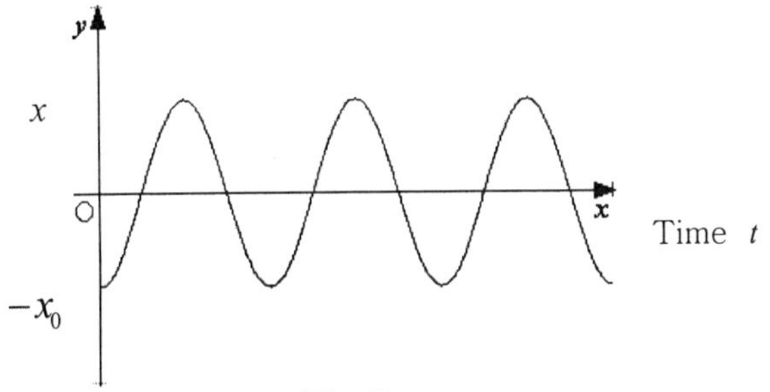

Fig. 2

Problem 3: The initial extension of the spring from its natural length reverts to x_0. An initial downward vertical velocity of v_0 is then added. Draw the subsequent displacement of the weight on the original curve in Fig. 3.

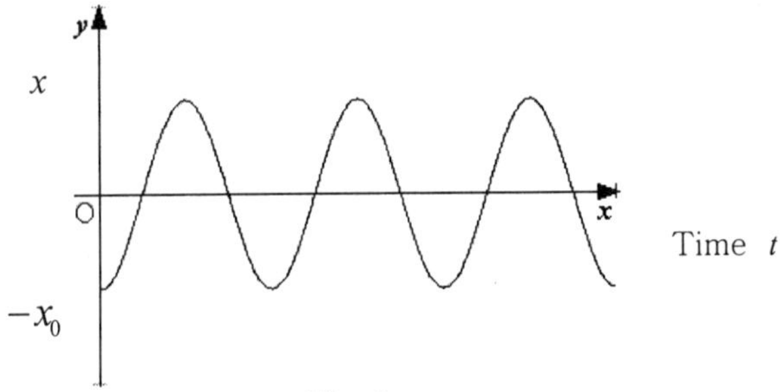

Fig. 3

Exponential Motion

Fig. 4 shows the displacement of a body. The displacement x changes with the passing of time t according to the equation: $x = x_0 e^{\lambda t}$, provided $\lambda > 0$.

Problem 4: The position of the body is slightly changed at time $t = 0$: $x_0 - \Delta x$. Draw the subsequent displacement on the original curve in Fig. 4.

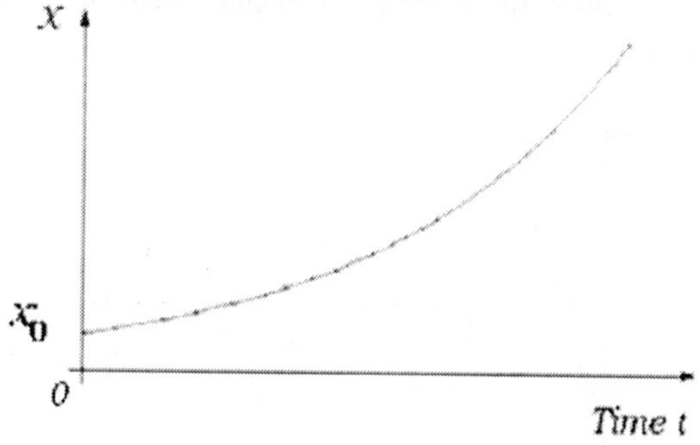

Fig. 4

2.2 *Results of Prior Survey*

The percentage of correct answers to Problems 1 and 2 was very high at 95% and 90%, respectively. However, the percentage of correct a nswers to problems 3 and 4 was lower at 40% and 17%, respectively.

Table 1 Percentages of Correct and Incorrect Answers (N=58)

	Correct Answers	Incorrect Answers
Problem 1	95% (55)	5% (3)
Problem 2	90% (52)	10% (6)
Problem 3	40% (23)	60% (35)
Problem 4	17% (10)	83% (48)

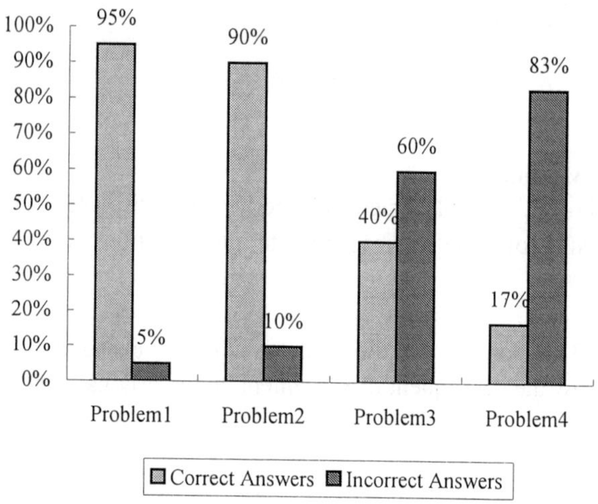

Fig. 5 Percentages of Correct and Incorrect Answers (N=58)

2.3 *Conclusion and Discussion of Prior Survey*

The reason for the high percentage of correct answers for problems 1 and 2 was most likely that the students had already studied falling motion and harmonic oscillation with a spring at high school, plus this type of motion is frequently seen in everyday life. Also, the change in the initial conditions was relatively simple so the resulting motion could be easily imagined.

Conversely, in problem 3, the change in the initial conditions was slightly more complicated, therefore, the resulting motion was harder to imagine even though this area was also already studied at high school. This was then reflected in the lower percentage of correct answers.

Meanwhile, for problem 4, it was difficult to imagine the trajectory change according to the change in the initial conditions, because this type of motion is unfamiliar in everyday life. Most of the incorrect answers (81%) merely moved the original curve downwards in parallel. This result implies that many students had difficulty imagining or theoretically conjecturing a motion where a small difference in the initial conditions expands with the passing of time, such as chaotic motions.

3. Trial Lecture

3.1 *Overview of Trial Lecture*

A double pendulum is a well-known model of a chaotic system. The major chaotic nature of a double pendulum is described below based on a comparison with the nature of a single pendulum. A trial lecture introducing the basic nature of chaos using a double pendulum was given to 16 junior, senior, and graduate students majoring in physics at the School of Science and Technology, Meiji University. All the students were already familiar with basic physics and mathematics, such as analytical mechanics and a Fourier transform. In the lecture, various graphs were shown on the basic nature of chaos mentioned and demonstrations conducted using a single and double pendulum made of aluminum.

3.2 *Analysis Methods*

The videotaped motion of the pendulum was captured by personal computer and analyzed with *Video Point* Ver. 2.0 (Motion Analysis Software: Lenox Softworks Inc., U.S. Imported by Shimadzu Rika Co., Japan.). At the same time, the motion was also analyzed through computer simulation using *Mathematica* Ver. 3.0 (Wolfram Research Inc., U.S.).

3.3 *Chaos and Random*

As regards disordered conditions, random is the same as chaos. However, in physics, random and chaos are considered as quite distinct. Chaos is the condition of a disordered system that can be described by a deterministic equation under a lower degree of freedom. Whereas, random is the condition of a system that can not be described by a deterministic equation, plus the values of expectation and their distribution can only be predicted statistically because the degree of freedom of the motions is too large.

3.4 *Equations of Motion*

First, the coordinate system of a double pendulum is defined as shown in Fig. 6. Let the mass of the first pendulum be m_1, its moment of inertia I_1, the distance between the original point and the center of gravity of the first pendulum d_1, the mass of the second pendulum m_2, its moment of inertia I_2, and the distance between the connection point of the two pendulums and the center of gravity of the second pendulum d_2.

The coordinates of the center of gravity for each pendulum are given as follows:

First Pendulum;
$$x_1 = d_1 \sin\theta_1$$
$$y_1 = d_1 \cos\theta_1$$

Second Pendulum;
$$x_2 = d_1 \sin\theta_1 + d_2 \sin\theta_2$$
$$y_2 = -d_1 \cos\theta_1 - d_2 \cos\theta_2$$

Provided the gravitational acceleration is given as g, the kinetic energy T_1 and potential energy U_1 of the first pendulum can be given as follows:

$$T_1 = \frac{1}{2} m_1 \left\{ \left(\frac{dx_1}{dt} \right)^2 + \left(\frac{dy_1}{dt} \right)^2 \right\} = \frac{1}{2} I_1 \dot{\theta}_1^{\,2}$$

$$U_1 = m_1 g y_1 = -m_1 g d_1 \cos\theta_1$$

Also the kinetic energy T_2 and potential energy U_2 of the second pendulum can be given as follows:

$$T_1 = \frac{1}{2} m_1 \left\{ \left(\frac{dx_2}{dt} \right)^2 + \left(\frac{dy_2}{dt} \right)^2 \right\} + \frac{1}{2} I_2 \dot{\theta}_2^{\,2}$$
$$= \frac{1}{2} m_2 \left\{ \left(d_1\dot{\theta}_1 \cos\dot{\theta}_1 + d_2\dot{\theta}_2 \cos\theta_2 \right)^2 + \left(d_1\dot{\theta}_1 \sin\dot{\theta}_1 + d_2\dot{\theta}_2 \sin\theta_2 \right)^2 \right\} + \frac{1}{2} I_2 \dot{\theta}_2^{\,2}$$

$$U_1 = m_2 g y_2 = -m_2 g d_1 \cos\theta_1 - m_2 g d_2 \cos\theta_2$$

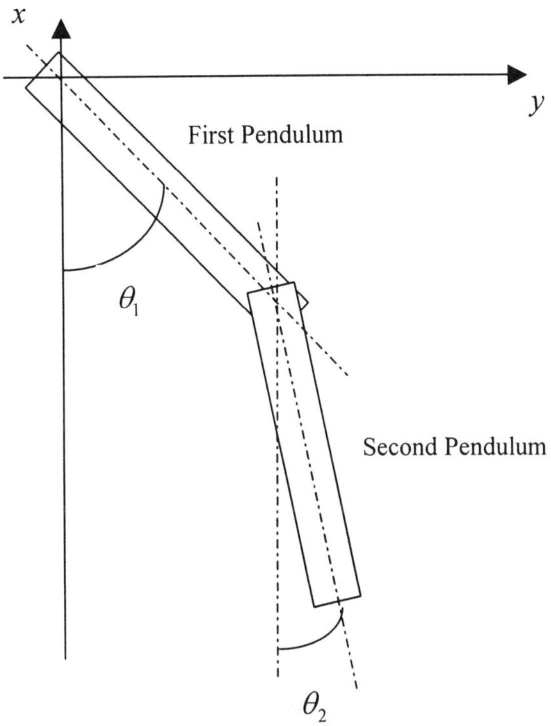

Fig. 6 Coordinates of Double Pendulum

Thus, the Lagrangian of the whole system can be given as follows:

$$L = T - U = (T_1 + T_2) - (U_1 + U_2)$$

$$= \frac{1}{2}\left(I_1 + m_2 d_1^2\right)\dot{\theta}_1^2 + \frac{1}{2}\left(I_2 + m_2 d_2^2\right)\dot{\theta}_2^2 + \left\{ m_2 d_1 d_2 \cos(\theta_1 - \theta_2) \right\}\dot{\theta}_1\dot{\theta}_2$$

$$+ (m_1 + m_2) g d_1 \cos\theta_1 + m_2 g d_2 \cos\theta_2$$

Then, the Lagrangian equations of motion for each pendulum can be given as follows:

First Pendulum:

$$\frac{d}{dt}\left(\frac{\partial L}{\partial \dot{\theta}_1}\right) - \frac{\partial L}{\partial \theta_1} = \left(I_1 + m_2 d_1^2\right)\ddot{\theta}_1 + \left\{ m_2 d_1 d_2 \cos(\theta_1 - \theta_2) \right\}\ddot{\theta}_2 +$$

$$\left\{ m_2 d_1 d_2 \sin(\theta_1 - \theta_2) \right\}\dot{\theta}_2 + (m_1 + m_2) g d_1 \sin\theta_1$$

$$= 0$$

Second Pendulum:

$$\frac{d}{dt}\left(\frac{\partial L}{\partial \dot{\theta}_2}\right)-\frac{\partial L}{\partial \theta_2}=\left(I_2+m_2 d_2^{\,2}\right)\ddot{\theta}_2+\left\{m_2 d_1 d_2 \cos(\theta_1-\theta_2)\right\}\ddot{\theta}_1+\left\{m_2 d_1 d_2 \sin(\theta_1-\theta_2)\right\}\dot{\theta}_2+m_2 g d_2 \sin\theta_2=0$$

As such, the time evolution of the angle and other variables were obtained by solving these equations through computer simulation.

3.5 *Orbit of Tip of Double Pendulum*

Since it is impossible to reproduce the movement of the double pendulum on paper, Fig. 7 indicates how the tip of the second pendulum moved during the first 5 seconds, which was obtained by computer simulation. Evidently, the movement is extremely complicated.

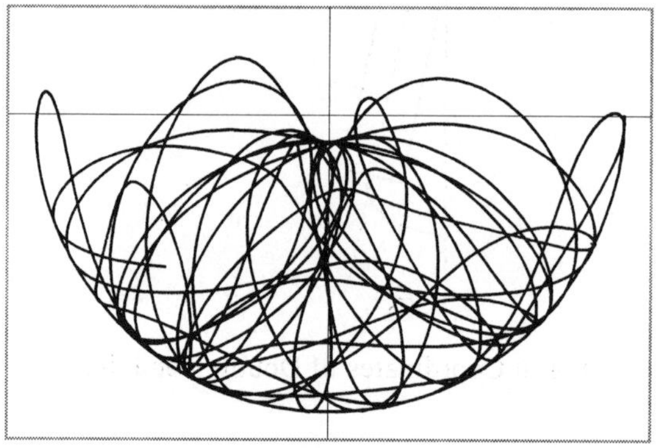

Fig. 7 Orbit of Tip of Double Pendulum

3.6 *Time Evolution*

When observing physical phenomena, generally there is some form of time evolution. However, in a chaotic system, a time evolution graph indicates a complicated curve without any apparent periodicity. As such, evaluating the time evolution is an effective method for judging whether a phenomenon is chaotic or not. Yet, there are two points that require attention. (1) Human beings evaluate phenomena through human eyes, thereby inevitably resulting in finite errors. (2) If a phenomenon occurs over a long time period and the observation lasts a shorter time period, it is impossible to find any periodicity from the time evolution. Nonetheless, evaluating the time evolution is often used to judge chaos, as it reveals the complexity and is easy to understand by intuition.

Fig. 8 shows the time evolution of the angle of the first pendulum, obtained by computer simulation. Fig. 9 shows the time evolution of the angle of the single pendulum, also obtained by computer simulation. The curve in Fig. 8 is more complicated than that in Fig. 9, which is characteristic of a chaotic motion.

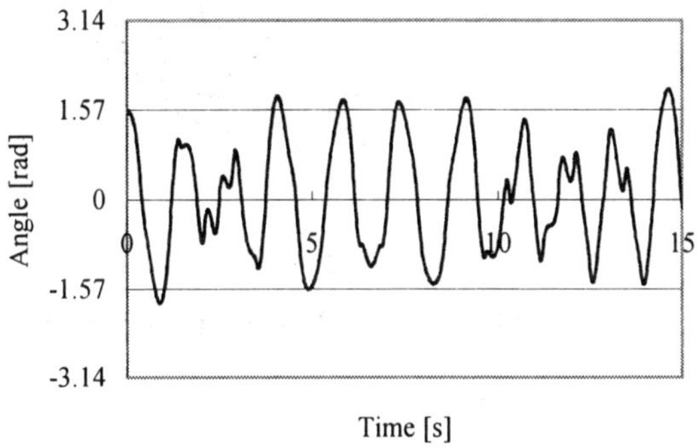

Fig. 8 Time Evolution of First Pendulum

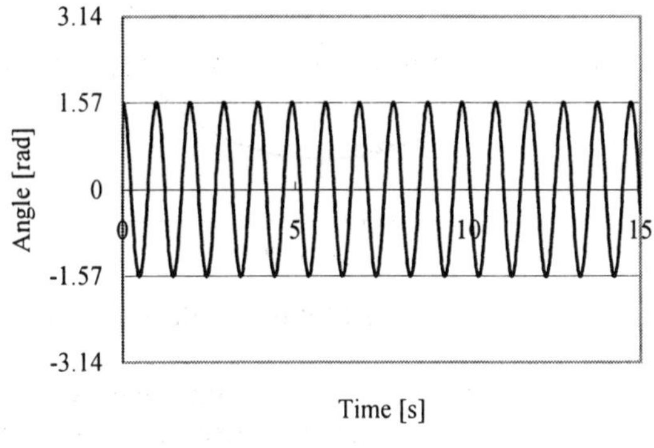

Fig. 9 Time Evolution of Single Pendulum

3.7 *Phase Space*

The phase space characterizes the motion in a dynamic system, while a point in the phase space indicates the status of the motion in the dynamic system at that moment. In a chaotic system, the phase space indicates a complicated curve, as with the time evolution.

Fig. 10 shows the phase space plot of the first pendulum, obtained by computer simulations. Fig. 11 shows the phase space plot of the single pendulum, also obtained by computer simulation. Similar to the time evolution, the nature of the chaotic motion shown by Fig. 11 is a simple closed curve, while the curve in Fig. 10 is extremely complicated.

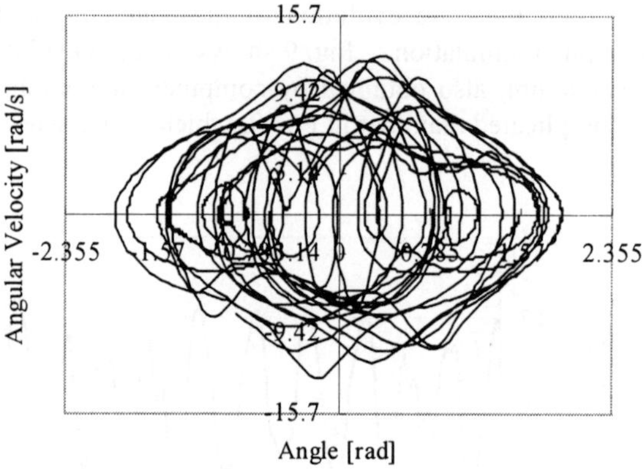

Fig. 10 Phase Space Plot of First Pendulum

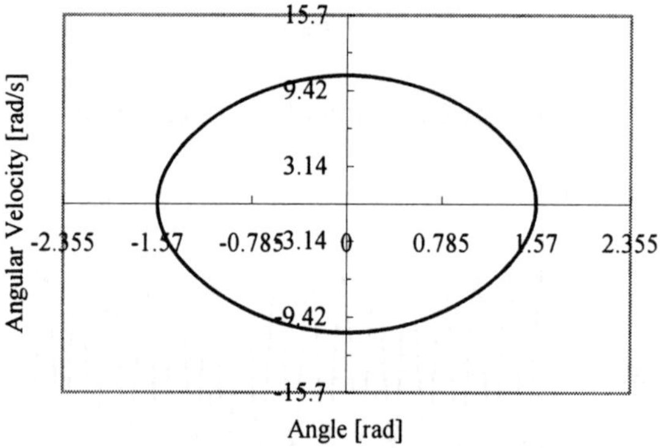

Fig. 11 Phase Space Plot of Single Pendulum

3.8 *Fourier Analysis*

No visible periodicity was identified in the time evolution of the first pendulum shown in Fig. 8. In a chaotic system, the time evolution does not reveal any periodicity. Therefore, evaluating the spectrum of the time evolution using an F.F.T. is an effective way of determining whether a system is chaotic or not.

Fig. 12 shows the analysis results of the time evolution of the first pendulum (Fig. 8) based on an F.F.T, while Fig. 13 also shows the analysis results of the time evolution of the single pendulum (Fig. 9) based on an F.F.T. Fig. 13 (single pendulum) reveals the existence of periodicity as sharp peaks can be seen at certain frequencies, whereas no periodicity is shown in Fig. 12 at any frequency because the curve is broad.

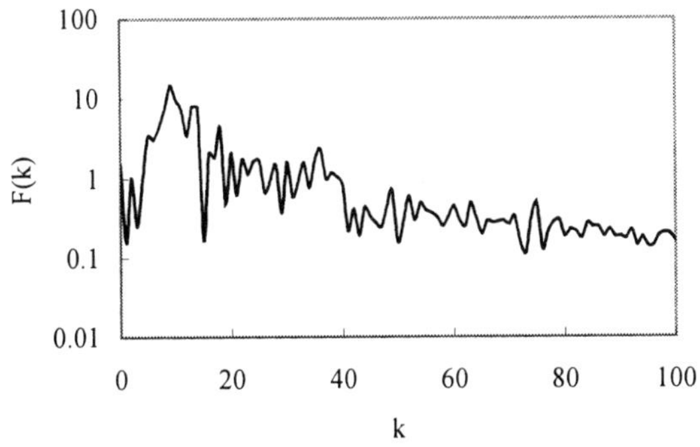

Fig. 12 F.F.T. of First Pendulum

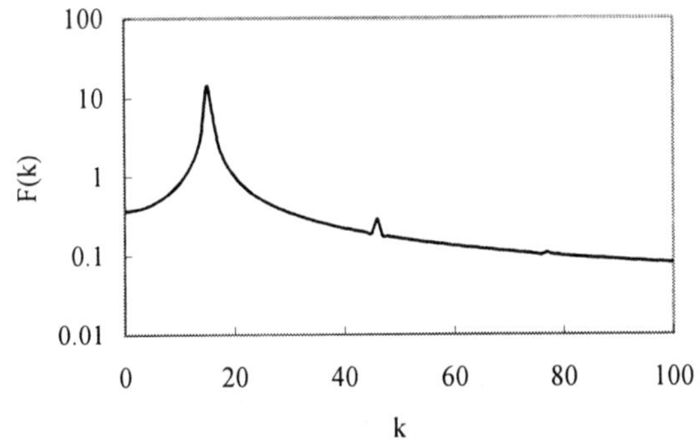

Fig. 13 F.F.T. of Single Pendulum

3.9 *Lyapunov Exponent*

The Lyapunov exponent is a quantity that indicates the sensitive dependence of a motion on the initial conditions and reveals the degree of separation over time of two different trajectories that started at the same time from two very close points. Based on the assumption that the difference between the two trajectories, $\theta(t)$, $\theta'(t)$, which are slightly separated at time $t=0$, satisfies $|\theta(t)-\theta'(t)|=e^{\lambda t}$, under the condition $\lambda>0$, the difference between the two trajectories will expand exponentially over time. However, under the condition $\lambda<0$, the difference comes close to 0 over time and the two trajectories converge to the same value. This λ is called the Lyapunov exponent, which acts as an index to show the sensitivity of the result to a change in the initial conditions. When the value of the Lyapunove exponent is positive, the system is usually determined as chaotic. If $\theta(t)$ is a discrete function,

$\theta(t_n)$, the Laypunov exponent λ is defined as follows.

$$\lambda = \lim_{n\to\infty}\frac{1}{n}\sum_{i=0}^{n-1}\log\left|\frac{d\theta(t_i)}{dt_i}\right|$$

Fig. 14 is the Lyapunov exponent for the angle of the first pendulum, obtained by computer simulation[1]. The figure reveals a tendency of convergence and the convergent value is positive, thereby indicating a high sensitive dependence on the initial conditions in the time evolution for the first pendulum.

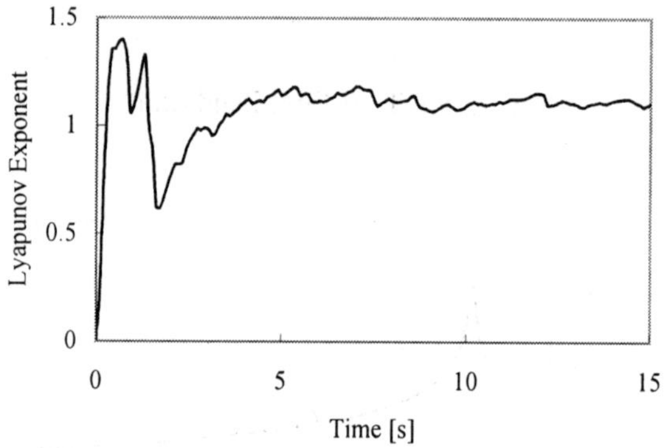

Fig. 14 Lyapunov Exponent of First Pendulum

3.10 *Unpredictability*

Fig. 15 shows the absolute value of the difference in the time evolution of the first pendulum when the initial angle of the second pendulum is slightly changed by $10^{-10}[rad]$. It is clear that the difference began to increase drastically around the time $t = 7[s]$.

The value "$10^{-10}[rad]$" is in an area that is far beyond the ability of measurement by devices and human eyes. The fact that such a small difference in the initial conditions results in such a large difference is evidence of the sensitive dependence on the initial conditions of the double pendulum, which was considered based on the Lyapunov exponent. It also means that it is virtually impossible to predict subsequent phenomena, even though the system is deterministic, due to the sensitive dependence on the initial conditions.

[1]Here, we need to pay attention that Lyapunov exponent λ is not a function of time. The Lyapunov exponent λ is a constant defined as a limited value under $n \to \infty$, which means $t \to \infty$.

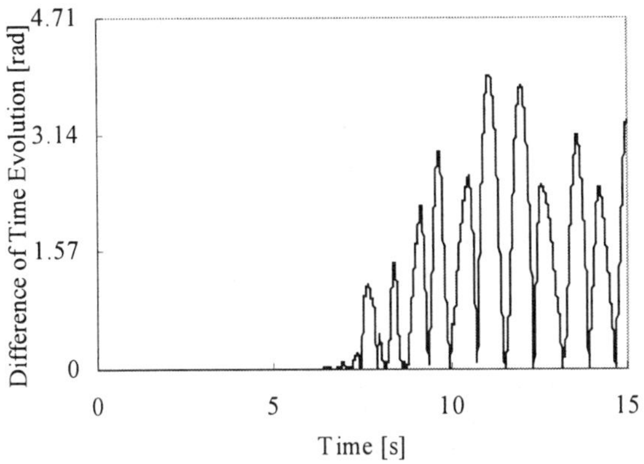

Fig. 15 Difference in Time Evolution with Simulation

In advance of the trial lecture, the following experiment was conducted. The motion of the pendulum was analyzed twice by measuring the initial angle using an eye measurement. Fig. 16 shows the absolute values of the difference in the time evolution of the first pendulum, where a significant difference is evident from the beginning.

During the trial lecture, in addition to showing Fig. 15 and Fig. 16, the students were allowed to observe the motion of two sets of double pendulums where the initial angles were set as the same based on eye measurements.

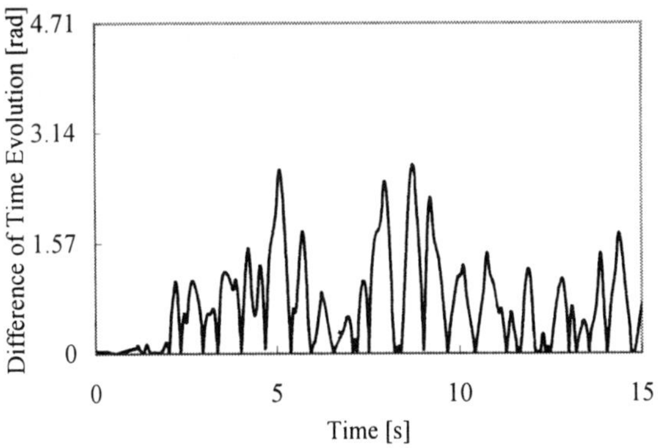

Fig. 16 Difference in Time Evolution with Measurements

In a chaotic system, it is important to consider that a theoretical prediction will not agree with experimental results no matter how precise the measuring device is. However, this does not mean that the theory is imperfect. Rather, in a chaotic system, it is virtually impossible to predict subsequent phenomena because of the sensitive dependence on the initial conditions.

3.11 ***Student Impressions***

After the trial lecture, the students were asked to complete a questionnaire related to the lecture. The following are some sample answers.

Student A: The lecture was easy to understand because it was not just an explanation of equations.

Student B: The lecture was very easy for me to understand because I could see the actual motion of pendulums. In the comparison between a single and double pendulum, we could see the difference clearly. It was a very interesting lecture.

Student C: It was a good lecture because we could see that the motion of the double pendulum was quite different from the periodic motion of the single pendulum. I could deepen my understanding more than usual lectures because I could see the actual motion of the pendulum.

Fig.17 Trial Lecture (1)

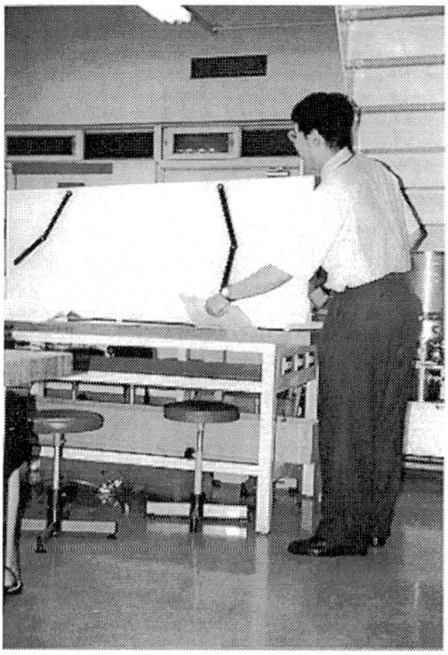

Fig.17 Trial Lecture (2)

3.12 *Student Reactions*

After the trial lecture, the students were asked to fill out a questionnaire: "Please write your opinion about the meaning of prediction in physics." 13 of the 16 students answered that they recognized the importance of the initial conditions in predicting phenomena and the limits of prediction through physical theories. The following are some typical responses. In particular, Student 3 mentioned a very important point on the significance of prediction based on calculations.

Student 1: Although equations can identify a motion, even a small factor can drastically change the results. In this sense, I think that predicting may be meaningless in chaos, yet it is important to learn what happens theoretically. A desire to understand is essential to studying physics. In physics, it is possible to produce predicted results by conducting experiments, however, this is impossible in economics and politics, thus prediction is necessary.

Student 2: A small change in the values, less than the resolving power, seriously affected the results. In other words, even if we simulate using a computer or advanced and precise techniques, we can't simulate real phenomena accurately. Thus, in physics, prediction is only an ideal theory. But this doesn't mean that prediction is meaningless, because not all things in the real world are as chaotic as a double pendulum.

Student 3: When there is a large difference between the results of prediction based on a calculation and those of an actual experiment, how can I identify which is right?

Fig.18 Trial Lecture (3)

4. Conclusion

According to the results of the prior survey, it is clear that even undergraduate students majoring in physics have difficulty in indicating the change of trajectory with a small change in the initial conditions for unimaginable motions. Moreover, it is obvious that very few students can imagine or theoretically conjecture the existence of motion where a small difference in the initial conditions expands drastically over time, such as chaotic motions.

The questionnaire survey after the trial lecture revealed that actual observation of the motion of a pendulum helped students to clearly understand the basic nature of chaos.

The questionnaire survey also implied that learning about the basic nature of chaos helped students recognize the importance of the initial conditions in predicting subsequent phenomena and the limitations of prediction based on physical theories.

Acknowledgement

The authors would like to express their appreciation to Mr. Maki Tchikawa, Dept. of Physics, School of Science and Technology, Meiji University, for his kind help and suggestions related to this research.

The authors would also like to express their appreciation to the students of the Dept. of Physics, School of Science and Technology, Meiji University for their kind corporation.

Finally, the authors would like to thank Ms. Yasuko Sugiura for translating this paper into English from Japanese.

References

1. Tokuya Inoue, A Chaotic Nature of Double Pendulum, Bachelor's Thesis Meiji Univ., 1998.
2. Mitsuo Suzuki, Kenji Masuda, Chaotic Motion of a Double Pendulum, *Journal of the Physics Education Society of Japan*, Vol.48, No.1, pp.1-5, 2000.
3. T. Shinbrot, C. Grebogi, J. Wisdom and J.A. Yorke, Chaos in a Double Pendulum, *American Journal of Physics*, Vol.60, No.6, pp.491-499, 1992.
4. R. B. Levien, S. M. Tan, Double pendulum: An experiment in chaos, *American Journal of Physics*, Vol.61, No.11, pp.1038-1044, 1993.
5. R. L. Kautz, Chaos in a computer-animated pendulum, *American Journal of Physics*, Vol.61, No.5 pp.407-415, 1993.
6. Takashi Shimojo, Introduction to Chaos Dynamics Third Edition, Kindai-Kagakusya, 1995.

RELATIONSHIP BETWEEN LEARNERS' EPISTEMOLOGICAL BELIEFS, LEARNING BEHAVIOR AND CONCEPTIONS OF MECHANICAL ENERGY

SEONGSOOK MOON, HANYONG CHOI, SANGOH SEO AND JAESOOL KWON

Dept. of Science Education, Korea National University of Education, Chungbuk, 363-791, Korea

Although it is difficult to understand the role that students' knowledge about the nature of science plays in their daily learning of science in school, the current study focuses on the important interplay between students' scientific epistemological beliefs, their learning behavior, and conceptions of mechanical energy. To explore this topic, the epistemological beliefs of six sophomores on the nature of physics knowledge in textbooks and their beliefs on the nature of the larger body of scientific knowledge were investigated through interviews. All participants were majoring in physics education and all thought that the larger body of scientific knowledge could be changed. Three categories were then introduced that classified how the students' understanding of the nature of scientific knowledge was linked with their learning behavior: 'receiving physics knowledge based on authority', 'receiving physics knowledge based on extrinsic motivation', and 'authentic understanding'. These categories were concerned with the acquisition of personal conceptions of mechanical energy. In particular, those students who pursued a superficial consistency and held a belief concerned with a definition of energy experienced a cognitive conflict with the conceptions of negative mechanical energy and a standard position. They received physics knowledge from textbooks based on authority.

1. Introduction

The development of adequate student conceptions on the nature of science has been a perennial objective of science instruction, regardless of the currently advocated pedagogical or curricular emphases. It has also been an area of prolific research characterized by several parallel, yet distinct, lines of investigation [1].

In the beginning of the twentieth century, the drastic changes in the ideas of modern physics started to undermine the belief that science could make valid 'absolutist ' claims. Philosophical, psychological, and logical arguments led to the proposal that ' knowledge' is not discovered, but rather a human construction and subjective. Thus, constructivism replaced the 'absolutism' of the empiricist and rationalist tradition [2]. Constructivists assert that learners build meaning around the objects and events with which they come into immediate and direct contact [3].

Philosophical views on science then influence science education [2]. As such, there is a view of science that emphasizes the tentative and revisionary nature of science education, and the interaction of social and cognitive processes in its development [4]. Dutschl et al. asserted 'that changes in scientific knowledge provide a better model for the development of a curriculum that seeks to bring about conceptual change in learners' [5].

A few studies concerning students' knowledge about the nature of science illuminate possible relationships between classroom practice and the development of knowledge. Edmonson [6] illustrated the dynamic relationship between students' epistemological commitment and their orientation to learning science. Students identified as logical positivists tended to be rote learners oriented towards grades. Hammer [7] described two ways in which students can perceive the role of conceptual knowledge in introductory physics. Students' understanding of what science is about, how it is presented, and their expectations as to what goes on in a science course, can play a powerful role [8].

Hogan [9] introduced two categories on the nature of science : distal knowledge and proximal knowledge. She suggested that these two kinds of structures can guide future research on students' knowledge about the nature of science as a mediator, not just an outcome, of their science experiences in school.

In Korea, most previous studies on knowledge about the nature of science have been related to students' and teachers' conceptions of the nature of science [10, 11, 12, 13, 14]. Korean high school students do not appear to adequately understand the nature of science. Although they understand the idea of empirical testing, they do not understand that scientific knowledge can also be the product of human creativity [12]. Park [15] showed that Korean secondary science teachers' views of science and learning were closely related to each other., yet elementary pre-service teachers were found to have a constructivist view of learning as distinct from their concept of the nature of science [16].

It follows from what has been said that many researchers are concerned about the role that student's knowledge about the nature of science plays in their daily learning of science in school.

Accordingly, the purpose of the current study was to qualitatively explore the following three topics :

First, how do learners' epistemological beliefs about the nature of scientific knowledge influence their learning behavior?

Second, what are their conceptions of mechanical energy?

Third, how do their conceptions of mechanical energy change ?

2. Methods

2.1 *Participants*

To explore these topics, a sample group of undergraduates majoring in science education and taking an introductory physics course was selected. The university was situated on the outskirts of a Korean city. 73 students responded to a questionnaire on philosophical views on science and attitudes towards physics learning in Nov. 2000. The questionnaire on attitudes towards physics(Chronbach's alpha = 0.9765) was a modified version Lee's questionnaire on attitudes towards science [17]. Whereas, the questionnaire on philosophical views on science was developed by J. Stewart and translated into Korean by Ban et al. [14] These questionnaires were chosen as it was considered that students' beliefs may influence their middle or high school learning.

About 4 months later, in Mar. 2001, six sophomores majoring in physics education were selected as the final participants. The reason for their selection was their rapport with one of the researchers based on a physics problem solving class. As such, it was considered easier to study them using a qualitative method. Plus, their distribution as regards their philosophical views on science was not very different from that of the originally sampled students

A paper-and–pencil questionnaire was constructed to probe the relationship between the students' beliefs and their conceptions of mechanical energy. The construction of the questionnaire was guided by physics education specialists. The participants responded to the questionnaire on basic conceptions of mechanical energy. Table 1 gives details of the results.

Table 1. Participants' philosophical views on science and attitudes towards physics

Student	Gender	Philosophical views on science	Attitudes towards physics	
			Learning	Experiment
K	Male	Eclectic	Like	Like
J	Female	Eclectic	Neutral	Dislike
D	Male	Empirical	Really dislike	Really dislike
B	Female	Rational	Like	Like
H	Male	Constructive	Neutral	Like
Y	Female	Eclectic	Really like	Really like

2.2 *Data collection and analysis*

The current study was interpretive and used multiple data sources to achieve a triangulation of data. The participants were interviewed 5 to 8 times over four months. Within this period, the participants learned about Newtonian mechanics. In-depth and half-structured interviews were conducted along with seven class observations. The interviews were all audio-recorded for transcription and analysis.

3. Results and Discussion

Whereas Hogan [9] introduced the two terms distal knowledge and proximal knowledge, the current study used the larger body of scientific knowledge and physics knowledge in textbooks. The larger body of scientific knowledge means the students' conceptions of comprehensive and general scientific knowledge. All participants considered that the larger body of scientific knowledge could be changed (Table 2). Three categories were introduced that classified how students' understanding of the nature of science was linked with their learning behavior.

Table 2. Confirmation of scientific knowledge and physics knowledge in textbooks and using metacognitive strategy

Student	Confirmation		Using metacognitive strategy in class
	Scientific knowledge	Physics knowledge in textbooks	
K	Can be changed over time	Confirmatory	A few
J	Can be changed over time	Can be changed over time	A few
D	Can be changed over time	Can be changed over time	Never
B	Can be changed over time	Can be changed over time	A few
H	Can be changed over time	Can be changed over time	Using naturally
Y	Can always be changed	Correct until now /Can be changed now	Using naturally

3.1 *Receiving Physics Knowledge Based on Authority*

'Receiving physics knowledge based on authority' means students accept the physics knowledge currently approved by many scientists.

Three participants (K, J, & D) attempted to understand the contents of lecture, yet their understanding was just linking pieces of physics knowledge approved by scientists. The participants considered physics knowledge to be objective and approved. Although they did not try to memorize physics knowledge, they first accepted the lecture contents, then considered it. Plus, they pursued a superficial consistency in learning:

> Interviewer: J, what is your role in a physics class?
> J (student): My role in physics class?
> Interviewer: Yes.
> J (student): I just intake objective knowledge.
> Interviewer: ..., what does it mean that you intake objective knowledge?

J (student); When other friends learn something, they can link it to daily life. But in my case, I cannot link, knowledge is just knowledge, knowledge comes into my head, just like a formula. That is how it is, it is just for the test.

....

Interviewer: .... then, do you have any thoughts concerning the knowledge you should intake?
D (student): I think after I intake
Interviewer: Ah, after you intake knowledge, you think about it, right?
D (student): Yes.
Interviewer: Why? Why do you do that?
D (student): No idea.

The participants (K, J, D) who received physics knowledge based on authority only had a superficial understanding of the mechanical energy concepts.

In short, since they only understood the definition of energy superficially, they thought that mechanical energy should not be negative. They thought that energy is the ability to perform work, thus the value of energy should always be positive:

J (student): Ah, I read it before and generalization means that it is always valid. Isn't that right?
Interviewer: No, it is important to have your own idea.
J (student): It seems that the standard position should be at one place.

......

J (student): Therefore, the value of potential energy at one place is multiplied with different standard positions. Therefore, it was wrong to make the mechanical energy's value absolute.

......

Therefore, it is contextually valid that the standard position must always be at a one place.
Interviewer: Ah, it is contextually valid.

...

Interviewer: Jinsu's solution, not Yeojin's?
J(student): I think that is wrong. I think the negative energy is wrong. Although the measurement was wrong, the standard position was wrong too.

Table 3 shows the conceptual change process for basic mechanical energy experienced by participant J. J held an alternative concept about energy. The reason for this alternative concept was the definition that energy is the ability to do work. J was explained the mean of $F(x)=-dU(x)/dx$ to illustrate the relationship between gravitational force and potential energy on the ground. After the explanation J commented that her previous concept of energy was superficial and agreed with the idea of negative mechanical energy.

Table 3. Process of conceptual change experienced by participant J

Items	In questionnaire	Interview about questionnaire	After taking mechanics class	Object under the ground	Explaining F=-dU/dx: gravitational force
Negative potential energy	Agree	Refuse	Refuse	Refuse	Agree
Negative mechanical energy	Agree	Refuse	Refuse	Refuse	Agree
Standard position	-	Just on the ground	Just on the ground	Under the object	Convenient position

At first participant (D) rejected the idea of negative potential energy and negative mechanical energy. However, after he read a description of the potential energy and mechanical energy of a ball under the ground, he experienced a cognitive conflict about negative mechanical energy. The reason for this was that h is understanding of the definition of energy was only superficial. In short, he was confused.

In the case of negative potential energy, he explained the relative velocity and displacement by saying that the measured value varied when the observation position was changed. Thus, he agreed with the idea of negative potential energy. A month later participant (D) came to think that negative energy could work based on his mechanics class.

In particular, participant K thought that scientific concepts should not be produced according to convenience. Therefore, he could not understand that it is convenient to locate a standard position at infinity:

Interviewer: What do you think about describing scientific concepts based on convenience?

...

K(student): Authors think that it is too difficult for students to understand.... Therefore they write like that.

Interviewer: What about scientists? Scientists?

K (student): No.

Interviewer: Um, should it be explained with verification or universal validity among scientists?

K (student): Yes.

Afterwards, when K was interviewed again, he said that he attained physics knowledge by explaining natural phenomena based on convenience. As such, he came to understand about convenience as a result of his dialogue with the interviewer. Therefore, he came to understand authentically, not mathematically, that it is convenient to locate a standard position at infinity.

3.2 *Receiving Physics Knowledge Based on Extrinsic Motivation.*

'Receiving physics knowledge for the extrinsic motivation' means that the student admits approved physics knowledge for the extrinsic motivation. Although she tried to understand the contents of the lectures, her main motivation was to just discern the important contents of the lecture so that she could become a teacher who could impart knowledge to her students. As such, she considered it a waste of time to have her own ideas that might differ from the textbook contents:

> Interviewer: Um, and then in your case, when you study, do you ever think of adding your own ideas?
>
> ....
>
> B (student): Um, I just accept. I don't think like that.
>
> ....
>
> Interviewer : Why are you serious about learning?
>
> B (student) : ho ho ho, I want to be a teacher, so I need to learn and then teach what I learned. This is my goal, and I only think about my goal.
>
> Interviewer: Oh, what a goal!
>
> B (student): I intake scientific knowledge and bring my thoughts into line with my future life.
>
> Interviewer: You said that you did not have any counter arguments.
>
> B(student): Yes.

Participant (B) who received physics knowledge based on extrinsic motivation initially refused the concept of a negative value for potential energy and mechanical energy. Yet B's conceptions of energy value were changed after the interview about the questionnaire on mechanical energy, and she thought about the relativity of displacement. Yet 20 days later, B became confused and thought that the definition of energy was the ability to do work (Table 4):

> B(student): However......if Jinsu is above the ground and the ball is under Jinsu's position and then goes down, the potential energy's value is negative.
>
> Interviewer: Do you agree with the negative value of the potential energy?
>
> B (student): Because the position is a negative distance, it can have a negative value.

I: Ah!

....

(20 days later)

B (student): I think that energy's value is always positive.

Interviewer: Why do you think it should have a positive value?

B (student): I don't know.

......

B (student): (Energy is) the ability to do work, but energy's value is minus...

I: Um, you don't agree with the negative value?

B: No.

Table 4. Process of conceptual change experienced by participant B

Items	On the Question-naire	Interview about question-naire	20 days later	After learning in mechanics class	Explaining F=-dU/dx: gravitational force
Negative potential energy	Rejected	Rejected Agreed	→Confused	Confused	Agreed
Negative mechanical energy	Rejected	Rejected Agreed	→Confused	Confused	Agreed
Standard position	-	Wherever	-	Wherever	Wherever

The definition that energy is the ability to do work created a cognitive conflict for the participant.(Table 4). When she was told the meaning of negative potential energy in the mechanics class, she also became confused. Later, when B interacted with one of the interviewers, it was explained that the formula F=-dU(x)/dx could be linked with the gravitational force and potential energy. Then she adjusted her mechanics notes and agreed with the concept of negative mechanical energy.

Therefore, the definition of energy made the participant think about the value of energy unscientifically. However, appreciation of the formula F=-dU(x)/dx resolved the conflict about the value of energy.

3.3 *Authentic Understanding*

'Authentic understanding' means that some participants naturally use a metacognitive strategy. This understanding is related to their approval of the diversity of individual ideas. Plus, since they believe that scientific knowledge is the product of *multiple* ideas, this makes them use a metacognitive strategy. One participant (H) thought that both the larger body of scientific knowledge and physics knowledge could be changed. He thought that scientific knowledge is what helps humans explain natural phenomena and is not absolute truth. In addition, he thought that the reason for being is personal.

> Interviewer: The property of the paradigm that science is not absolute truth is linked to your learning method. Does this affect your attitude in class?
> H(student): Last year when I learned about electric and magnetic fields, I thought that my own idea was better than the professor's. By the way, later on I realized that the professor's method was better.
> Interviewer: What does 'better' mean?
> H(student): What?
> Interviewer: What does 'better' mean?
> H(student): The more I learned, the more my idea was overturned.
> ....
> Interviewer: ...Um, I see, do you have any opinion?
> H: any opinion? I don't make my own theory when learning. But others follow the ideas of the professor. You said that I had a very strong personality. I think that if I think like others, there is no reason for being. ....

Another participant (Y) thought that the larger body of scientific knowledge could be always changed. However, she said that physics knowledge in textbooks was merely confirmation. In addition, she also said that she would not be surprised if physics knowledge changed:

> Interviewer: Yes, it can be changed and it is not absolute. When you attend a physics lecture, does this thought influence your learning behavior?
> Y(student): Of course. When I am faced with certain circumstances, if there is no flexibility, I become stressed. There are always good alternatives.
> .....
> Interviewer: The larger body of scientific knowledge can always be changed, but?
> Y(student): The knowledge that the professor instructs is approved by many scientists.
> Interviewer: Is the knowledge in textbooks also approved?
> Y(student): All the knowledge I learn is approved
> Interviewer: Any doubt about it? Please express your opinion.
> Y(student): hoho, explanation

Interviewer: Don't you suspect it?
Y(student): I have met strange things, but
...
Interviewer: What does 'strange things' mean?
Y(student): It seemed that there was something wrong.
Interviewer: And so, are there any strange things in textbooks?
Y(student): It seems that there are a few errors.
Interviewer: (Contents) In textbooks?
Y(student): I think so
......
Interviewer: By the way, because you think that human thought is unlimited, ...you may have different emotions or learning behavior. What do you think about that? Do you think you learn in a normal way?
Y(student): Um, I can think that why isn't this method possible? When I confirm something the professor has explained, I think about the following. Isn't this method possible, or a different way?
Interviewer: Ah, you apply your own idea
Y(student): I just think....When there is any error, I draw a line, and adjust it.

These examples from interviews with two participants indicate that the following views made the participants more active in class and naturally use a metacognitive strategy in their physics class (Table 2): human thoughts are very diverse and science is the product of human ideas, plus the physics knowledge in textbooks is not absolute.

The two participants (H, Y) did not agree to the definition of energy superficially and literally. They thought that negative mechanical energy was the ability to do work and the important aspect was the transformation of energy.

4. Conclusions and Implications

When the participants learned physics knowledge in class, they were basically passive due the fact that the physics knowledge in the textbook was still approved by many scientists. When they had different ideas to the professor, the mostly gave up on their own thoughts. They had all experienced that the scientific knowledge in textbooks is always right in daily life since they went to elementary school. Therefore, since the scientific knowledge in textbooks was never doubted, textbook knowledge was found to have the strongest power. The professor also had a strong power. Furthermore, the participants felt they needed a lot of science knowledge. When they learned physics knowledge in class, they just received the knowledge based on the authority of the textbook or professor. In short, the belief that physics knowledge was objectively approved made the students passive in class. In addition, when the participants learned physics knowledge in class, extrinsic motivation also made them passive.

Although the authority of physics knowledge in textbooks and knowledge of professors are both powerful, there was another factor that made the participants active in learning. This factor was when the students recognized that scientific knowledge is the product of diverse and infinite human thought. Students who understood this fact were very active. And it was natural for them to use a metacognitive strategy. Stenhouse [18] asserted that the notion of "understanding a concept" viz. that it equated with "knowing the rules for the use of a word", could be seen as relativistic rather than absolute, elastic, and open-ended. As such, it would appear that his assertion considers scientific ideas to be diverse. In fact, participant (H) amended a 'period' in a physics textbook to understand the context.

In conclusion, first, it is important for students to understand about the flexibility and diversity of scientific knowledge.

Second, the definition of energy made the students hold alternative concepts about energy. Those students who understood physics knowledge based on authority maintained their alternative concepts and rejected the concept of a negative value for energy. Of course, it is not easy to develop a student's epistemological belief on the nature of physics knowledge in a textbook as well as their epistemological belief on science knowledge.

However, when they learned about the concept of mechanical energy it was necessary to make them recognize the following:

The value of potential energy and mechanical energy at certain positions is meaningless. These values are measured by convenience and are necessary to emphasize the difference in the potential energy between two positions. As such, it was important for the students to understand the transformation relationship of energy.

The thought that some scientific concepts can be created not only logically but also 'conveniently' to explain natural phenomena was also found to influence the student's acquisition of energy conceptions

Third, the appreciation of a formula was able to resolve the students' conflict. Although all the participants thought that many of the formulas in physics could be linked to natural phenomenon, some participants were poor at creating such links. As such, it is extremely important to reflect on the meaning of formulae.

References

1. N. G. Ledermann, Students' and Teachers' Conceptions of the Nature of Science: A Review of the Research, *Journal of Research in Science Teaching,* vol.29, pp. 331-359, 1992.
2. J. Nussbaum, Classroom Conceptual Change: Philosophical Perspectives, *The History & Philosophy of Science Teaching,* Florida State Univ.: Florida, pp.278-291, 1989.

3. J. H. Wandersee, J. J. Mintzes and J. D. Novak, Research on Alternative Conceptions in Science, In D. L. Gabel (Ed.), *Handbook of Research on Science Teaching and Learning*, chap.5, pp.177-210,1994.
4. P.Thagard, Mind, society and the Growth of Knowledge, *Philosophy of Science*. vol 61, 629-645, 1994.
5. R. A. Duchl, R. J. Hamilton and R. E. Grandy, Psychology and Epistemology: Match or Mismatch When Applied to Science Education? In R. A. Duschl & R.J. Hamilton (Eds), *Philosophy of science, cognitive psychology, and educational theory and practice,* pp.19-47, Albany, NY:State University of New York Press, 1992.
6. K. M. Edmodson, The Interplay of Scientific Epistemological Views, Learning Strategies, and Attitudes of College Students, *Journal of Research in Science Teaching,* vol.30, pp. 547-559, 1993.
7. D. Hammer, Students' beliefs about conceptual knowledge in introductory physics, *International Journal of Science Education,* vol. 16, pp. 385-403, 1994.
8. E. F. Redish, J. M. Saul, and R. N. Steinberg, Student Expectations in Introductory Physics, *American Journal of Physics,* vol 66, pp.212-224, 1998.
9. K. Hogan, Exploring a Process View of Students' Knowledge about the Nature of Science, *Science Education,* vol.24, pp.51-70, 2000.
10. J. W. Song and S.K. Kwon, The Changes of Postgraduate Students' Conceptions Towards the Nature of Science through the Course of Philosophy of Science, *Journal of Korean Association for Research in Science Education,* vol.12, pp. 1-9, 1992.
11. S. C. Lim, Elementary Teachers' Conceptions of the Nature of Science, master's thesis, Graduate School of Korea National University of Education, 1994.
12. S. M. Seo, The Analysis of High School Students' understanding of the nature of Science, master's thesis, Graduate School of Seoul National University , 1994.
13. W. J. Kim, Students' Conceptions about the Nature of Scientific Knowledge, Master's thesis, Graduate School of Korea National University of Education, 1996.
14. E. K. Ban, S. K. Lee, U. H. Kim and H. J. Park, A Study of High School Students' Philosophical Views on Science, Students' Conceptions about the Nature of Scientific Knowledge, *Journal of Korean Association for Research in Science Education,* vol.20, pp. 881-100, 2000.
15. Y. B. Park, Secondary Science Teachers' Views on Science and Learning, *Journal of Korean Association for Research in Science Education,* vol.20, pp. 244-249, 2000.
16. S. K. Kwon and S. J. Park, Elementary Preservice Teachers' Conceptions about and its Changes in the Nature of Science and Constructivist' View of Learning, *Journal of Korean Association for Research in Science Education,* vol.15, pp. 104-115, 1995.

17. K. H. Lee, Construction of a Structural Equation Model on Attitudes Science using LISREL, Ph. D. Dissertation, Graduate School of Korea National University of Education, 1996 .
18. Stenhouse, Conceptual Change in Science Education: Paradigms and Language-Games", *Science Education*, vol.70, pp.413-425, 1986.

DEVELOPMENT OF COURSE ON PRACTICE OF COGNITIVE CONFLCIT STRATEGY FOR PHYSICS TEACHERS*

HYUKJOON CHOI, YEOUNSOO KIM, YOUNGJICK LEE & JAESOOL KWON

Department of Physics Education, Korea National University of Education, Chungbuk, Korea

The purpose of the current study was to develop a course on the practice of the cognitive conflict strategy for physics teachers. As such, a graduate-level course was designed and twenty-one physics teachers were enrolled. During the course, each teacher developed two cognitive conflict materials for demonstration along with teaching plans to apply the materials. To improve the course, the participants were given a questionnaire and interviewed to discern their opinions about the cognitive conflict strategy. Data was collected based on the questionnaires, interviews, and videotapes. Most physics teachers confirmed that the cognitive conflict strategy is important and has a good effect on changing students' conceptions when learning physics. However, difficulties were experienced when developing cognitive conflict materials and applying then to a classroom context. Accordingly, the current article suggests certain implications for the science pedagogical curriculum including the development and application of cognitive conflict materials.

1. Introduction

Over the last 20 years, researchers in science education have been interested in understanding students' conceptions of various scientific areas along with the processes of conceptual changes. Initially, science educators investigated the patterns and characteristics of students' alternative conceptions and found that students use their everyday experiences and intuition to develop their conceptions about phenomena. Plus, it was revealed that students' preconceptions are extremely resistant to instructional change, with some students persisting in their misconceptions even after an extensive amount of instruction [1, 2, 3, 4].

Since then, the major concern of science educators has been how to change students' misconceptions of scientific concepts. As such, various theoretical models and instructional situations have been developed to facilitate conceptual change. In particular, many researchers have emphasized that students need to be confronted with an incongruous event where they cannot explain certain phenomena using their own preexisting conceptions [5, 6, 7, 8, 9].

---

* This work was supported by the Brain Korea 21 in 2001.

In previous research on cognitive conflict situations, there have been many studies on methods of representing cognitive conflict materials, the content of cognitive conflict materials, and the degree of difficulty [10, 11, 12]. Based on these research results, the characteristics of students' responses to cognitive conflict situations have been studied [13]. Lee [14] and Druyan [15] identified differences in conceptual change according to the method of representing cognitive conflict materials and their content. Various researchers also found that students confronted with cognitive conflict situations exhibited various patterns and characteristics in their responses; ignoring, rejecting, excluding, abeyance, reinterpreting, peripheral change, theory change [16, 17, 18, 19, 20, 21].

Furthermore, several studies have focused on instructional strategies to facilitate conceptual change in science teaching [22, 23, 24]. However, an excellent instructional strategy is only a vacant theory if it cannot be applied in a classroom context. Accordingly, it is important to develop practical instructional strategies along with instructional materials related to the learner's characteristics.

2. Foundational Study

As a foundational study for the current study, Kim et al. [25] investigated science teachers' understanding of the cognitive conflict learning strategy and the practice of learning strategies in science classes. A total of 40 secondary school science teachers majoring in science education at graduate school participated in the study.

Even though all the science teachers had already attended graduate-level lectures on 'science pedagogy' and 'science learning and instruction: theory in practice', 30% of them stated that they did not understand the cognitive conflict learning strategy well.

In response to two questions on the necessity of cognitive conflict for conceptual changes in students and the efficiency of the cognitive conflict strategy in facilitating conceptual changes in students, more than 70% answered positively. Yet only 45% answered that they had made an effort to apply the cognitive conflict strategy in their instruction. The reasons given for not attempting to apply the cognitive conflict strategy were : a poor understanding of the cognitive conflict strategy, too few teaching hours, and a lack of teaching/learning materials, especially ideas related to anomalous data.

The solutions suggested for such difficulties were the development of materials arousing cognitive conflict, special teacher training (51%), and the establishment of a system for sharing developed materials(21%).

Therefore, the conclusions of the foundational study highlighted the need for the development of a curriculum for colleges or teacher training facilities focused on the practice and application of a concrete conceptual change learning strategy.

3. Development of Course

The need for a course on the practice and application of a concrete conceptual change learning strategy was established by the foundational study. Accordingly, the current study developed, conducted and monitored the outcome of such a course.

Twenty-one secondary school physics teachers majoring in science education at graduate school enrolled on the developed course. The course was composed of 48 classes over 12 days, i.e. 4 classes a day. For subsequent evaluation of the course, all the classes were videotaped with the participants' permission. The course consisted of 4 steps with 12 classes over 3 days.

1st step:

After a brief introduction to the course, the participants were asked to answer a questionnaire on their use of the cognitive conflict learning strategy in class. Several lectures were then given on the necessity of anomalous data arousing cognitive conflict and various materials introduced to demonstrate actual phenomena.

2nd step:

The participants were divided into groups and asked to consider different materials for development, plus various demonstration resource books were introduced to help in this selection. Based on class discussions, the materials selected by each group were estimated, modified, and substantialized. Each group explained the scientific concept related to their selected material and the process involved in making it. The rest of the class then asked questions, especially related to the feasibility of making it.

3rd step:

The materials were made and demonstrated in class. One or two days were allowed to make the materials outside class.

4th step:

Based on the developed materials, the participants created teaching plans and explained them. Finally, the participants were asked to complete a questionnaire reflecting on their experience in the class.

4. Results of Course

1st step:

The teachers participating on the course initially exhibited a lower understanding of the cognitive conflict strategy than the respondents in the foundational study. 11% of the participants responded that they understood it well, whereas 41% of the respondents in the foundational study answered that they understood the strategy well or very well as they had already taken graduate-level courses on 'science pedagogy' and 'science learning and instruction: theory in practice'. The lectures and demonstrations of previously developed teaching materials were found to help the participants understand the cognitive conflict strategy and develop their own cognitive conflict materials and teaching plans. It was also emphasized that students' alternative conceptions needed to be substantially revealed in the teaching plans.

2nd step:

The participants were divided into three groups; mechanics, electrodynamics, and light and wave, then asked to consider more than two materials for developing. The materials were limited to materials for demonstrating actual phenomena. Resource books, for example, 'Hands-on Physics Activities' and ‘Physics Demonstration Resource Book' [26, 27], were supplied to help in developing the teaching materials. The materials considered were presented to the class and modified. On the whole, the materials presented did not create a cognitive conflict, but rather introduced the concept and helped in understanding it. As such, the authors proposed that effective materials should create a cognitive conflict and present the students with a contradictory demonstration. The feasibility of making the materials was also discussed.

3rd step:

The participants selected one of the materials discussed in step 2, then created and demonstrated it. While creating the teaching materials, various unpredicted problems were encountered. Thereafter, the participants shared their experiences when making the materials.

4th step:

Each participant prepared a teaching plans for two materials, then conducted a lesson applying the cognitive conflict strategy using one of the materials created to demonstrate an actual phenomenon. It was found that the materials developed did not necessarily facilitate the students' cognitive conflicts if the students' alternative conceptions were not reflected in the teaching plan. Furthermore, the conflict was not resolved unless a concrete method of resolving the conflict was presented.

5. Analysis of Questionnaire

Questionnaires were conducted in the first and last class. Over 80% of the respondents answered positively or neutrally to the questions on the necessity of cognitive conflict for changing students' conceptions and the effectiveness of the teaching plans for first and last classes (Figs. 1 & 2). In the first class, 11% of the respondents answered positively to the question on understanding the cognitive conflict strategy, while 48% answered negatively, whereas 58%

answered positively and 11% answered negatively in the last class (Fig 3). Yet, it was still disappointing that only 58% answered positively are finishing the course. One possible reason was that none of the participants in the course had taken any previous classes on science education or had any learning experience on the cognitive conflict strategy. Therefore, to make the course more effective for teachers who have not previously studied about the cognitive conflict strategy, more theoretical lectures focusing on the cognitive conflict strategy need to be included in the first step.

In the first class, none of the respondents answered positively to the question on previous efforts to apply the cognitive conflict strategy to instruction, while 42% answered negatively., whereas, in the last class, no one answered negatively to applying the cognitive conflict strategy in their future teaching and 79% answered positively (Fig 4).

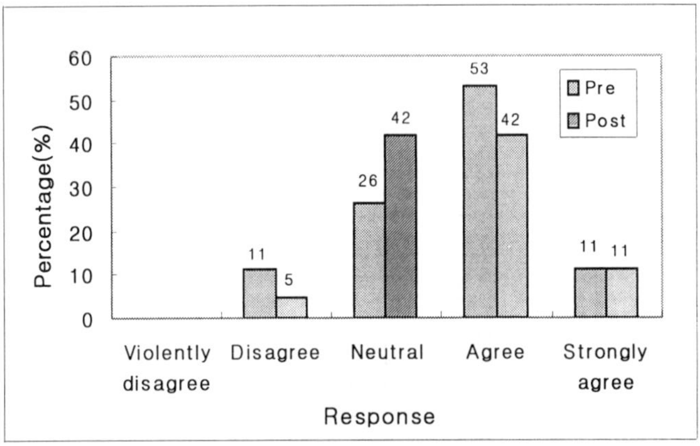

Figure 1. Necessity of cognitive conflict for changing students' conceptions

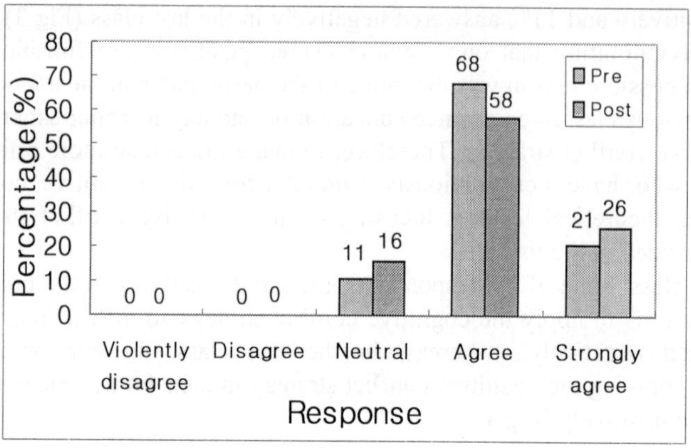

Figure 2. Efficiency of cognitive conflict strategy in changing students' conceptions

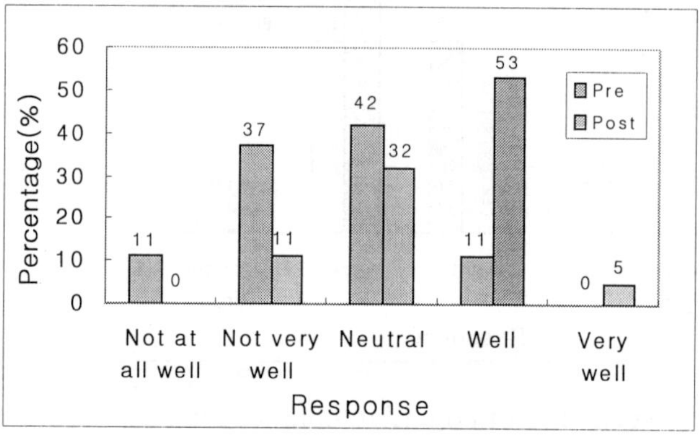

Figure 3. Understanding of cognitive conflict

As regards anomalous data, in the first class, only 5% of the participants answered they had a lot, while 68% answered they had very little, whereas 21% answered positively and 26% answered negatively in last class (Fig 5). Although the positive responses increased and negative responses decreased after the course, there were still more negative responses than positive ones. It is believed that this will change if more anomalous that is developed.

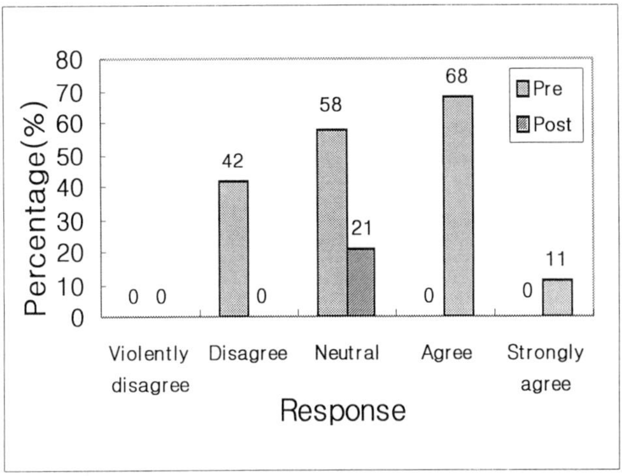

Figure 4. Previous effort to apply cognitive conflict strategy in teaching

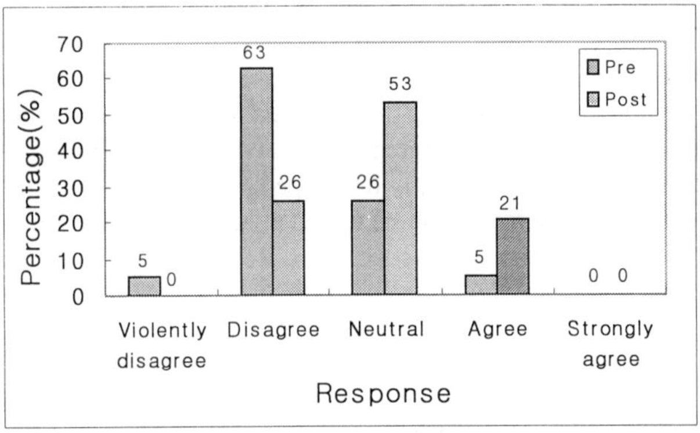

Figure 5. Ideas on anomalous data

6. Conclusions and Implications

The number of materials developed for demonstrating actual phenomena were 51, including the materials presented by the current authors: 28 on mechanics including thermodynamics, 13 on electrodynamics, and 10 on light and waves. The responses of the participants as regards the merits of the course are presented in Fig 6.

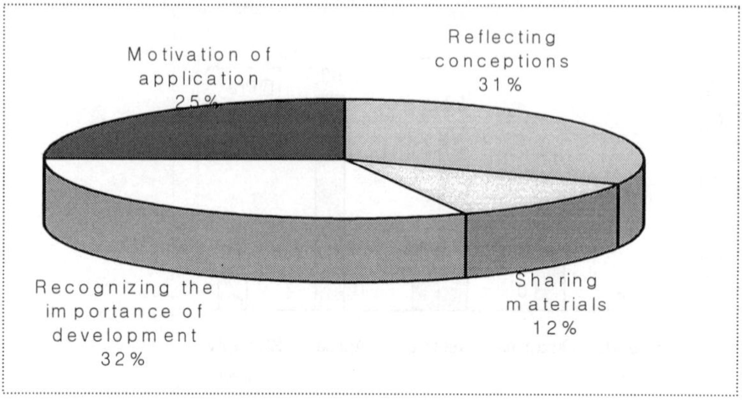

Figure 6. Merits of lectures

Through the course, the participants were able to improve their understanding of the cognitive conflict strategy and clarify many of the concepts they had already taught in school. Plus, the determination and self-confidence to apply the cognitive conflict strategy were strengthened, while sharing the materials developed to demonstrate actual phenomena. A significant difference was found between the theory and practice of developing materials for demonstrating actual phenomena, yet the value of such materials for use in school was underscored.

To improve the course, additional theoretical lectures on the cognitive conflict strategy are needed in the first step for teachers who have not taken previous courses on the cognitive conflict strategy.

Although the proposed course was originally developed as a 3-week graduate-level course, it is anticipated that it could also be included in various educational programs for science teachers.

References

1. Y. Lee and J. Kwon, The index of the stability of misconceptions, *Journal of the Korean Association for Research in Science Education*, vol. 13(3), pp. 310-316, 1992.
2. M. Hashweh, Toward an explanation of conceptual change, *European Journal of Science Education*, vol. 8, pp. 229-249, 1986.
3. G. Posner, K. Strike, P. Hewson, and W. Gertzog, Accommodation of a scientific conception: Toward a theory of conceptual change, *Science Education*, vol. 66, pp.221-227, 1982.
4. P. Hewson, A conceptual approach to learning science, *European Journal of Science Education*, vol. 3, pp. 383-396, 1981.

5. J. Kwon, A comparative discussion on the instructional procedure and strategies on learning scientific concepts, *Journal of the Korean Association for Research in Science Education*, vol. 12(2), pp. 19-29, 1992.
6. J. Park, Theoretical model about cognitive conflict, *Journal of Science Education Chonnam National University*, vol. 16, pp. 17-35, 1992.
7. J. Kwon, The necessity of cognitive conflict strategy in science teaching, A paper presented at *the International Conference on Science Education*: Globalization of Science Education, May 26-30, Seoul, Korea, 1997.
8. C. Chinn and W. Brewer, The role of anomalous data in knowledge acquisition: A theoretical framework and implications for science instruction, *Review of Educational Research*, vol. 63, pp. 1-49, 1993.
9. K. Strike and G. Posner, Conceptual change view of learning and understanding, in L.H.T. West and A.L. Pines (eds.), *Cognitive structure and conceptual change*, pp. 211-230, Academic Press, 1985.
10. J. Kim, Y. Lee and J. Kwon, The analyses of the change process of students' physics conceptions by the types of conflict situations, *Journal of the Korean Association for Research in Science Education*, vol. 20(1), pp. 77-87, 2000.
11. L. Lee, Y. Lee and J. Kwon, The effects of conflict situation types on inducing students' cognitive conflicts in Newton's laws, *Journal of the Korean Association for Research in Science Education*, vol. 18(4), pp. 473-483, 1998.
12. B. Kim and J. Kwon, The influence of the types of scientific concepts and the patterns of cognitive conflict on the change of students' conceptions, *Journal of the Korean Association for Research in Science Education*, vol. 15(4), pp. 472-486, 1995.
13. G. Lee and J. Kwon, Students' responses confronted with discrepant situation patterns about inertia concept, *Journal of the Korean Association for Research in Science Education*, vol. 19(4), pp. 516-527, 1999.
16. Y. Kim and J. Kwon, The patterns of students' conceptual changes on force by age, *Journal of the Korean Association for Research in Science Education*, vol. 20(2), pp. 221-233, 2000.
17. J. Park and M. Park, The analysis of students' response about demonstrations for cognitive conflict on the force and motion, *Journal of the Korean Association for Research in Science Education*, vol. 17(2), pp. 149-162, 1997.
18. D. Shepardson, The role of anomalous data in restructuring fourth graders' frameworks for understanding electric circuits, *International Journal of Science Education*, 21, 77-94, 1999.
19. P. Tao and R. Gunstone, The process of conceptual change in force and motion during computer-supported physics instruction, *Journal of Research in Science Teaching*, vol. 36, pp. 859-882, 1999.
20. C. Chinn and W. Brewer, An empirical test of a taxonomy of responses to anomalous data in science, *Journal of Research in Science Teaching*, vol. 35, pp. 623-654, 1998.

21. C. Chan, J. Burtis, and C. Bereiter, Knowledge building as a mediator of conflict in conceptual change, *Cognition and Instruction*, vol. 15, pp. 1-40, 1997.
22. H. Kim, Theoretical analyses and classroom applications of science teaching models, Ph. D. thesis, Korea National University of Education, 1995.
23. M. Niaz, Cognitive conflict as a teaching strategy in solving chemistry problems: A dialectic-constructivist perspective, *Journal of Research in Science Teaching*, vol. 32, pp. 959-970, 1995.
24. A. Dreyfus, E. Jungwirth and R.Eliovitch, Applying the "cognitive conflict" strategy for conceptual change: Some implications, difficulties, and problems, *Science education*, vol. 74, pp. 555-569, 1990.
25. Y. Kim, S. Seo, G. Lee, H. park and J. Kwon, A survey on the practice of cognitive conflict learning strategy on secondary science education, *Journal of the Korean Association for Research in Science Education*, vol. 21(2), pp. 400-410, 2001.
26. J. Cunningham and N. Herr, *Hands-on physics activities with real-life applications*: easy-to-use labs and demonstrations for grades 8-12, West Hyack, NY, 1994.
27. Korean Gifted Society, *Physics Demonstration Resource Book*, 1

STUDENTS' ATTRIBUTIONS AND CHARACTERISTICS OF COGNITIVE CONFLICT RESOLUTION IN PHYSICS LEARNING*

YEOUNSOO KIM & JAESOOL KWON

Department of Physics Education, Korea National University of Education, Chungbuk, Korea

The purpose of the current study was to investigate the relationship between students' attributions and cognitive conflict, and identify the characteristics of cognitive conflict resolution in physics learning. Twenty-nine college students attending a basic general physics course took an attribution test and conceptual pre-test related to action-reaction concepts. As a result, twenty students with alternative conceptions were selected. Next, these students were confronted with a discrepant demonstration that contradicted their existing conceptions, then given a cognitive conflict level test (CCLT), post-test, and delayed post-test to determine their conceptual change. Those students who experienced high levels of cognitive conflict were also interviewed to find out what kinds of attribution and motivational belief affected their resolution of the conflict and successful/unsuccessful understanding of the concepts. When confronted with a discrepant event, the students who attributed successful outcomes to "effort" seemed to experience higher levels of cognitive conflict, compared to those students who attributed successful outcomes to "task difficulty". The different attribution profiles exhibited for resolving conflict and successful/unsuccessful understanding of physics concepts were found to have a strong influence on the resolution of cognitive conflict and process of conceptual change. In addition, the profiles also appeared to be related to metacognitive and volitional strategies. High levels of cognitive conflict did not always produce conceptual change. For the conflicts to lead to change, students needed to have a perspective on effort that implied the use of a self-regulated learning strategy. Accordingly, the current findings highlighted the importance of a self-regulated cognitive conflict strategy (SRCCS) or management strategy of cognitive conflict (MSCC), indicating the need for a model of a self-regulated cognitive conflict strategy for a classroom context

1. Introduction

Student's alternative conceptions have been a dominant area of research in science education for more than two decades. Many researchers of constructivism in science education argue that cognitive conflict is a very important aspect of conceptual change in learning [1-8], as such, cognitive conflict strategy-guided research has influenced instructional practices in science learning for many years. Nonetheless, many studies have also indicated that counter-evidence alone does not necessarily change students' personal conceptions in many teaching and learning approaches that emphasize cognitive conflict [9-12]. Moreover, research has shown that some students maintain their concepts or subsequently revert back to their alternative concepts away from scientific concepts even after experiencing high levels of cognitive conflict [13].

While previous studies have mainly focused on the resistance of students' scientific conceptions to change even when students experience high levels of cognitive conflict, few studies have investigated why and how this happens. Consequently, science education research in the area of cognitive conflict has focused on the supposed underlying epistemological aspects (logical structure, rational process etc), and paid minimal attention to the affective aspects or motivational constructs in students' science learning [1, 14-18]. The current study is interested in the way student's motivational beliefs about themselves as learners and the roles of individuals in a classroom learning community facilitate or hinder conceptual change [17]. The cognitive model of motivation based on cognitive psychology emphasizes that internal events, such as cognitive conflict, causal attribution, expectations about success (or failure), and memory of another person's success, have a mutual effect on students' motivation to learn [19,20]. In particular, the cognitive conflict and causal attribution that affect future behavior may play an important role in student conceptual change. However, little research has attempted to study the relationship between student's attributions and cognitive conflict in conceptual change.

Accordingly, the purpose of current study is to investigate the relationship between students' general attributions and the level of cognitive conflict, and identify what kinds of attribution affect the resolution of cognitive conflict and successful/unsuccessful understanding of basic physics concepts. Based on an analysis of the results, the current study suggests the necessity of a model of self-regulated cognitive conflict strategy in the classroom context.

2. Cognitive Conflict and Attribution Theory: Cognitive Approaches to Motivation

Some researchers have suggested that a student's motivation to learn affects the level of cognitive conflict and conceptual change in physics learning [1, 22]. Motivation is usually defined as an internal state that arouses, directs, and maintains behavior. Motivation is a vast and complicated subject encompassing many theories. Historically, there are four kinds of approaches to motivation; the behavioral view, humanistic view, cognitive view, and sociocultural view [23,24]. Among these, cognitive theories of motivation have mainly been developed as a reaction to the behavioral view. One of the central assumptions in cognitive approaches is that people do not respond to external events or physical conditions, such as hunger, but rather to their subjective interpretations of these events. As such, cognitive theories emphasize intrinsic motivation. People are seen as active and curious, searching for information to solve personally relevant problems. Bernard Weiner's attribution theory is a good example. This cognitive explanation of motivation begins with the assumption that we all ask "Why?" in our attempts to understand our successes and failures. Students may ask themselves, "Why did I understand this physics concept so well?" "What's wrong with my problem solving?" Consequently, students can attribute their successes and failures to ability, effort, mood, knowledge, luck, help, interest, clarity of instructions, the interference of others, unfair policies, and so on. Attribution theories of motivation describe how the individual's explanations, justifications, and excuses influence motivation.

Bernard Weiner is one of the main educational psychologists responsible for relating attribution theory to school learning. According to Weiner, most of the attributed causes for success or failure can be characterized in terms of three dimensions [25], for example, luck is external (locus), unstable (stability), and uncontrollable (responsibility). Weiner suggests that these three dimensions have important implications for motivation. The internal/external locus seems to be closely related to a feeling of self-esteem, while the stability dimension is closely related to expectations about the future.

If people feel that they are not in control of their own lives, their self-esteem is likely to be diminished. Likewise, students' perceptions of their attributions for resolving cognitive conflict and acquiring a successful/ unsuccessful understanding of physics concepts can affect the process of conceptual changes in learning physics. The cognitive model of motivation based on cognitive psychology emphasizes that internal mediating events, such as cognitive conflict, causal attributions for what led to success or failure in reaching goals of interest in the past, expectations about success or failure with respect to reaching future goals, and memories of what others did before reaching a goal, play a mutually important role in students' motivation to learn [19,21].

3. Research Methods

The current study started with quantitative research. After investigating the first question, i.e. the relationship between students' attributions and the characteristics of cognitive conflict, an interpretative research design was used based on principles of naturalistic inquiry [25] to identify what kinds of attribution affected the resolution of cognitive conflict and successful/ unsuccessful understanding of physics concepts.

3.1 *Participants and Procedure*

The participants were 29 college students who were attending a general physics course. As shown in table 1, the students took an attribution test [27] and scientific concept test at the end of the semester. All students took a test to examine their conceptions of action-reaction in mechanics and 20 students who held alternative views of action-reaction were selected. Next, these 20 students were confronted with a discrepant demonstration ("Can a fan move a cart?", as shown in table 1) that contradicted their existing conceptions about action and reaction, then given a cognitive conflict level test (CCLT) [28], post-test, and delayed post-test to determine their conceptual change.

Table 1. Procedure used in current study

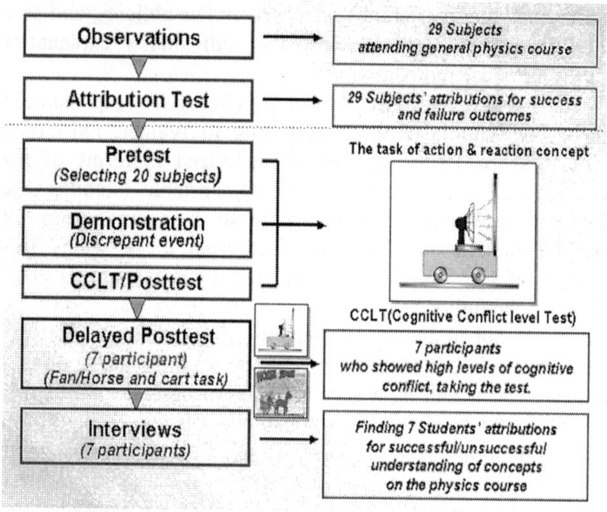

The post-test was administered after the CCLT, and the delayed post-test was administered one month later to 7 students who experienced high levels of cognitive conflict. As a case study, the 7 students were interviewed to find out what kinds of attribution affected their resolution of cognitive conflict and successful/ unsuccessful understanding of physics concepts.

4. Data Collection and Analysis

4.1 *Attribution Test and Cognitive Conflict Level Test (CCLT)*

The attribution test was based on the Rotter's criterion [29] to determine the internal and external locus of control. The test consisted of 40 items that questioned the attributions for success and failure, such as ability, effort, task difficulty, and luck.

To determine the level of the students' cognitive conflict, the cognitive conflict level test (CCLT) developed by Lee and his colleagues [28] was conducted. This instrument included four components of cognitive conflict: recognition of contradiction, interest, anxiety, and cognitive reappraisal. The higher the total test score, the greater the level of cognitive conflict. A correlation analysis and ANOVA were used to investigate the relationship between the attributions and the cognitive conflict.

4.2 *Pre-test, Post-test and Delayed Post-test*

The pre-test was designed to assess the students' existing alternative conceptions about action and reaction (Newton's 3rd law). The test consisted of predict - explain tasks using an electric fan and cart related to action and reaction. The post-test was a CCLT and consisted of a predict-observe-explain (POE) [30] task using the fan and cart. The POE task was designed to provide cognitive conflict and facilitate conceptual change.

The delayed post-test was designed to assess the students' conceptual change and administered one month later to 7 students who experienced high levels of cognitive conflict. The delayed post-test consisted of the fan task and a horse and cart task and was conducted based on an interview about instance (IAI) [30]. When a student correctly explained both instances using scientific conceptions, they were regarded as having changed their scientific conception from their original alternative conception.

4.3 *Participant Observations, Interviews, Document Analyses*

Data was collected from several sources for triangulation during the semester [31]. First, participant observations were conducted throughout the general physics course to capture the students' learning style and learning volition, etc. Second, documents, such as notes, journals, and exams, written by the students as they took the course were collected. Third, semi-structured and in-depth interviews were conducted to find out what kinds of attributions affected the students' resolution of cognitive conflict and successful/ unsuccessful understanding of physics concepts, plus their perspectives of the significance of attributions in the process of conceptual change. The students were asked to answer questions and explain the reason for their answers. Triangulation on the data collection and analyzing was implemented [31].

5. Results

5.1 *Relationship Between Cognitive Conflict and Students' Attributions for Successful Outcomes*

The first purpose of the current study concerned the relationship between students' general attributions and cognitive conflict in an anomalous situation. The inclination to attribute success to effort ($r= .74$, $p < .01$) and ability ($r= .59$, $p < .05$) was found to be correlated with high levels of cognitive conflict.

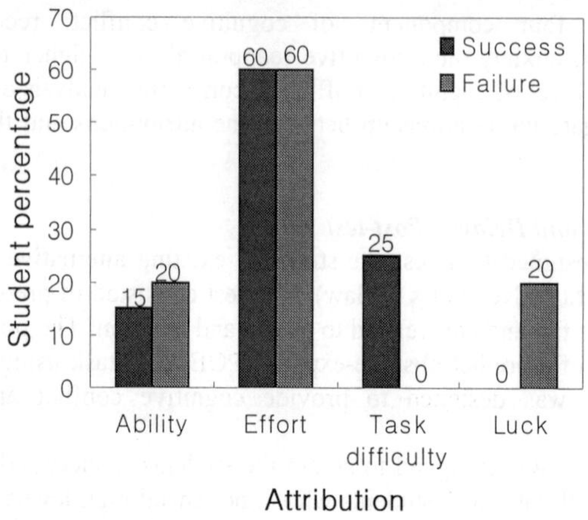

Figure 1. Distribution of students' attributions for success and failure

Figure 2. Level of cognitive conflict relative to students' attributions for successful outcome

Plus, the inclination to attribute success to effort (r= .46, $p < .05$) was also found to be correlated with conceptual change in the physics course.

As shown in Figure 1, the distribution of the students' attributions for success was as follows: 60% of the students believed that they succeed based on their effort; 15% based on their ability; and 25 % based on the easiness of the task. As shown in Figure 1, many students attributed successful/unsuccessful outcomes to 'effort'. There is evidence of cultural differences in beliefs about the causes of achievement outcomes. According to some research, Asians attribute outcome more to effort and less to ability than Americans [33]. This emphasis is consistent with the traditional Asian philosophy that assumes malleability in humans and stresses the importance of striving for improvement. Culturally different beliefs about the causes of achievement outcomes suggest that these perceptions are to some degree socialized. Thus parents and teachers can influence students' perceptions of the cause of achievement outcomes, and affect students' perspectives of 'effort' to resolve cognitive conflict when confronted with discrepant events.

As shown in Figure 2, the result of an ANOVA analysis on the level of cognitive conflict among the three groups with different attributions to success showed significant differences ($F=6.89$, $p< .01$). In particular, the result indicated that the group who attributed successful outcomes to "effort" experienced higher levels of cognitive conflict than the group that emphasized "task difficulty". Thus, it was considered important to find out how those who experienced higher levels of cognitive conflict changed/maintained their alternative conceptions about action and reaction after they were confronted with a discrepant demonstration that contradicted their existing conceptions.

5.2 *Characteristics of Students' Attributions for Cognitive Conflict Resolution and Conceptual Change in Pre-test and Delayed Post-test*

The second purpose of the current study was concerned with the kinds of student attribution for cognitive conflict resolution and successful/unsuccessful understanding of physics concepts, and how these attributions affect the process of conceptual change.

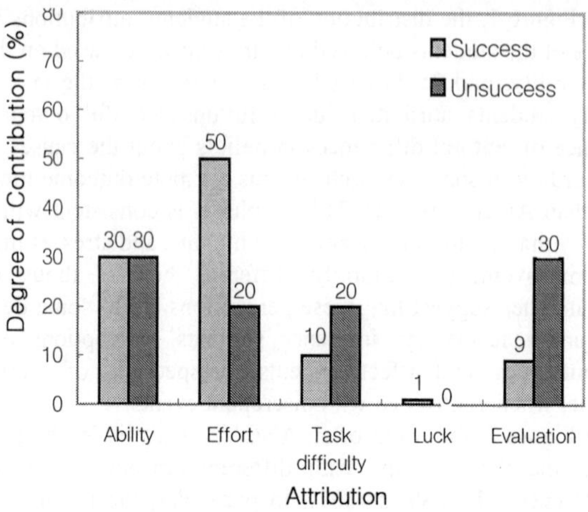

Figure 3. Banna's attributions

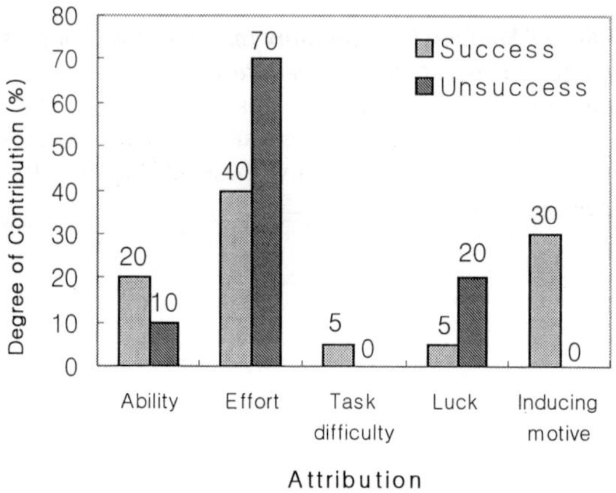

Figure 4. Choo's attributions

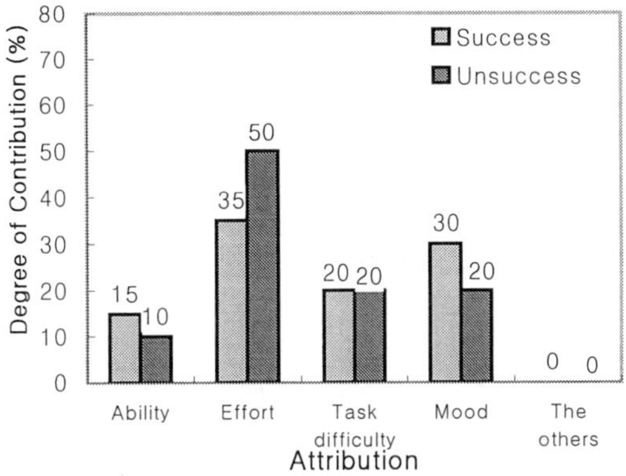

Figure 5. Jana's attributions

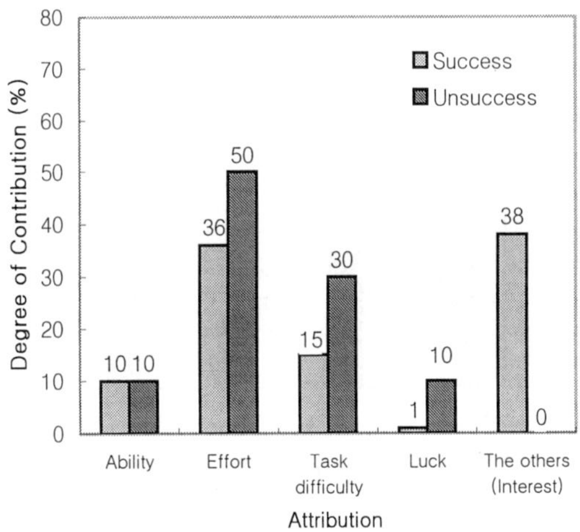

Figure 6. Leena's attributions

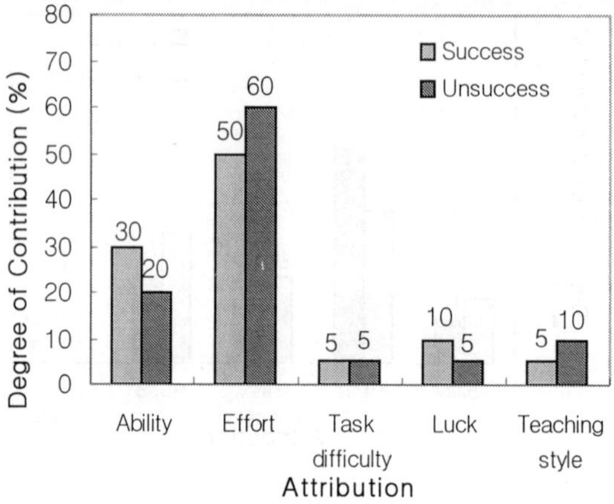

Figure 7. Nam's attributions

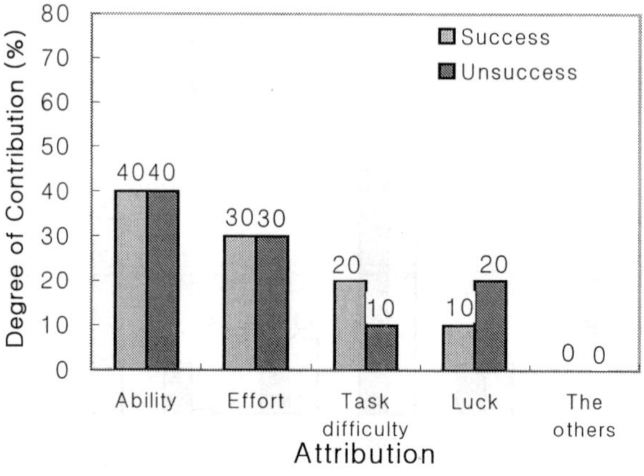

Figure 8. Sami's attributions

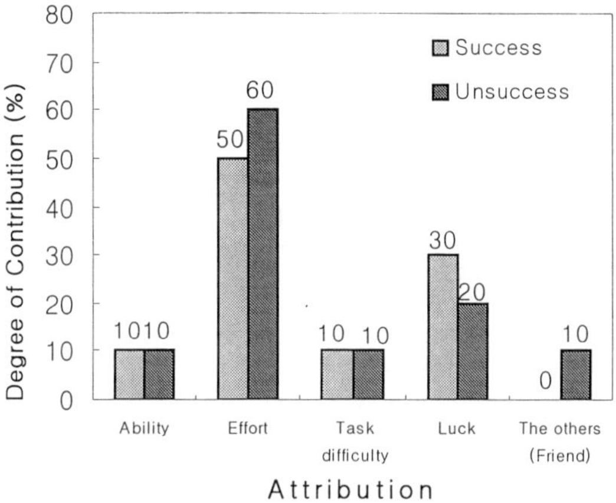

Figure 9. Young's attributions

As a case study, 7 students who showed high levels of cognitive conflict were selected and interviewed to find out what kinds of attribution influenced their resolution of cognitive conflict and successful/unsuccessful conceptual understanding of physics concepts.

As shown in Figure 8, Sami attributed a successful /unsuccessful understanding of physics concepts and resolution of conceptual conflict to 'ability'. Except for Sami, the other six students argued that "effort" played a very important role in the successful resolution of cognitive conflict and conceptual change in learning physics. However, as a result of informal interviews with the students, it was found that each student had a different perspective on the effort required to resolve conceptual conflict and understand physics concepts. For example, as shown in Table 1, Sami's idea of making an effort to successfully understand physics concepts and resolve conceptual conflict in physics learning was "doing many exercises before the test".

In contrast, Bana's perspective of "effort" was "studying in preparation for each class", "rewriting and summarizing, then trying to understand it", "collecting notes related to each class and using them to understand physics concepts deeply independent of preparing for an exam", and "using a personal learning strategy: reading and understanding the text first; closing the text and reflecting on what I understand, rewriting the content to be learned on a sheet of paper, and comparing this with the text".

Table 2. Results of interviews, students' perspectives on "effort" for resolving conceptual conflict and successfully understanding physics concepts, and conceptual change in pre-test and delayed post-test.

Student*	Student's perspectives on effort	Posttest	Delayed posttest
Bana (Fig. 3)	•Doing study for the each day's lesson. •Rewriting and summarizing first, and then trying to understand it. •Collecting my portfolios related to the day's lesson each day and using them to understand scientific concepts deeply independent of preparing for an exam. •Using my own learning strategy: Reading and understanding the text first, closing the text and reflecting what I understand; writing the content to be learned on a sheet of paper and comparing them with the text.	Δ	O
Choo (Fig. 4.)	•Before the day of the test, taking a look at the text and solving the problems. •(When I can't solve the problem) Always checking a correct answer with the solution.	◑	⊠
Jana (Fig. 5)	•Concentrating my attention to what teacher says in instruction, but I am not accepting all what teacher explained, rather I reflect what it means and apply it to the real life. •Often thinking deeply on the concepts learned.	Δ	O
Leena (Fig. 6.)	•Studying the concepts very hard and trying to understand them thoroughly. •Often reflecting what they mean.	Δ	O
Nam (Fig. 7)	•Before the test, studying what teacher explained in the instruction. •Memorizing only the proofs and formulas to be taught. •Solving only the problems that teacher addressed on. •Listening carefully to what teacher explained	◑	⊠
Sami (Fig. 8)	•Only doing many exercises before the test.	◑	⊠
Young (Fig. 9)	•Investing more times than another students in understanding the concepts and making up my mind to study sincerely.	◑	⊠

Posttest

◑ : Changing the alternative conceptions, but not explaining the cause of the result of the demonstration.

Δ : Not changing the alternative conceptions, but accepting the result of demonstration.

Delayed Posttest

⊠ : Having the alternative conceptions of action & reaction.

O : Having the scientific conceptions of action & reaction.

***These students showed high levels of cognitive conflict, when confronted with a discrepant event (demonstration)**

Effort is at the heart of self-regulated learning [34]. One goal of physics teaching is to free students from the need for a teacher so that students can continue to learn independently about natural science and the environment throughout their lives. The ability to be a self-regulated learner requires academic learning skills and the will to learn [35,36]. Knowledge, motivation, and volition (self-discipline) influence skill and will. In particular, knowledge means metacognitive learning strategies. Volition, an old-fashioned word for will power, means knowing how to protect yourself from distractions and cope with such feelings as anxiety, drowsiness, or laziness [37,38]. Where motivation represents commitment, volition represents follow-through. As shown in Table 2, similar to Bana, Jana, and Leena, some students' perspectives on effort implied applying metacognitive, motivational, and volitional strategies to resolve cognitive conflict and understand physics concept.

Banas' perspectives on effort and his conceptual change

In particular, Bana's perspective on effort included three factors of self-regulated learning. He said that the effort required to resolve conceptual conflict involved trying to understand scientific concepts deeply independent of preparing for an exam, rewriting, and summarizing first, then trying to understand it and collecting his notes related to each class. These perspectives exhibited an intrinsic motivation to seek out and conquer challenges as a personal interest and exercise his capabilities [39], as such he was a good learner and motivated to learn physics.

As to the factor of metacognitive strategies, Bana thought that when/after being confronted with discrepant events, the effort required to understand physics concepts and resolve conceptual conflict was to apply his personal learning strategy; first, reading and understanding the concept in the text, reflecting on what he understood, writing the content to be learned on a sheet of paper, and comparing this with the text. What was most important was that he learned these strategies from his math teacher when he was in high school. These learning tactics were related to metacognitive strategies [40,41]. Plus, as shown by his statements, metacognitive strategies that support self-regulated learning can be taught by a teacher. Some researchers recommend embedding instruction on metacognitive skills into regular instruction, like physics, rather than as a separate curriculum [42].

As regards the factor of volition, Bana contended that he could make an effort any time if he wanted to work. He recognized 'effort' as a dimension of stability, regardless of the surrounding environment. As such, he exhibited a strong will power. According to action-control theory [43], volition plays a mediating role between the intention to learn and the use of learning strategies [37,43].

In the post-test, although Bana had alternative conceptions, he accepted the result of the demonstration. However, in the delayed posttest, Bana demonstrated a change in his alternative conceptions to scientific conceptions through self-regulated learning. As shown in Table 2, Jana and Leena also showed similar perspectives on effort and exhibited a comparable process of conceptual change.

Choo's perspectives on effort and her conceptual change

As a result of the interviews, Choo, Nam, Sami, and Young were found to have similar perspectives on the effort needed to resolve conceptual conflict and successfully understand physics concepts. As shown in Table 2, Choo contended that making an effort is taking a look at the text and solving the problems the day before a test, and always checking the correct answer when she could not solve the problem. Similar to Choo, Nam said that effort meant studying what the teacher explained in class before a test, memorizing the proofs and formulas, and only solving the problems the teacher covered. Young and Sami also focused on a quantitative effort. In the post-test, those students who only focused on a quantitative effort seemed to accept the scientific conceptions, however, one month later in the delayed post-test, they reverted to their alternative scientific conceptions, as shown in Table 2.

Most educational psychologists portray students as being motivated to learn, willing to put in effort, and capable of monitoring their own effort. Unfortunately, as seen in the results of the interviews, this is not realistic. Although effort is clearly at the heart of self-regulated learning, researchers do not agree on its conceptualization. According to Boekaerts [33], Alexander made a distinction between mindful and mindless effort [44], while Weinert et al. distinguished between qualitative and quantitative effort. The former type of effort refers to the quality of the strategy used [45], i.e. the extent to which effort is expended to process the material extensively (deep-level processing) and in a context-sensitive way. Meanwhile, the latter refers to time allocation (e.g., sitting in front of one's books, or cramming before an exam in a surface-level processing style). Mindful and qualitative effort is associated with using metacognitive and volitional strategies. In the delayed post-test, as with Bana, Jana, and Leena, those who expended a high qualitative effort achieved a scientific understanding, however, as with Choo, Nam, Sami, and Young, those who expended a more quantitative effort reverted to alternative conceptions.

Therefore, in physics instruction, to facilitate a student to become a self-regulated learner and able to manage cognitive conflict, physics teachers must be aware of student's perspectives on the effort needed to resolve conceptual conflict and understand physics concepts.

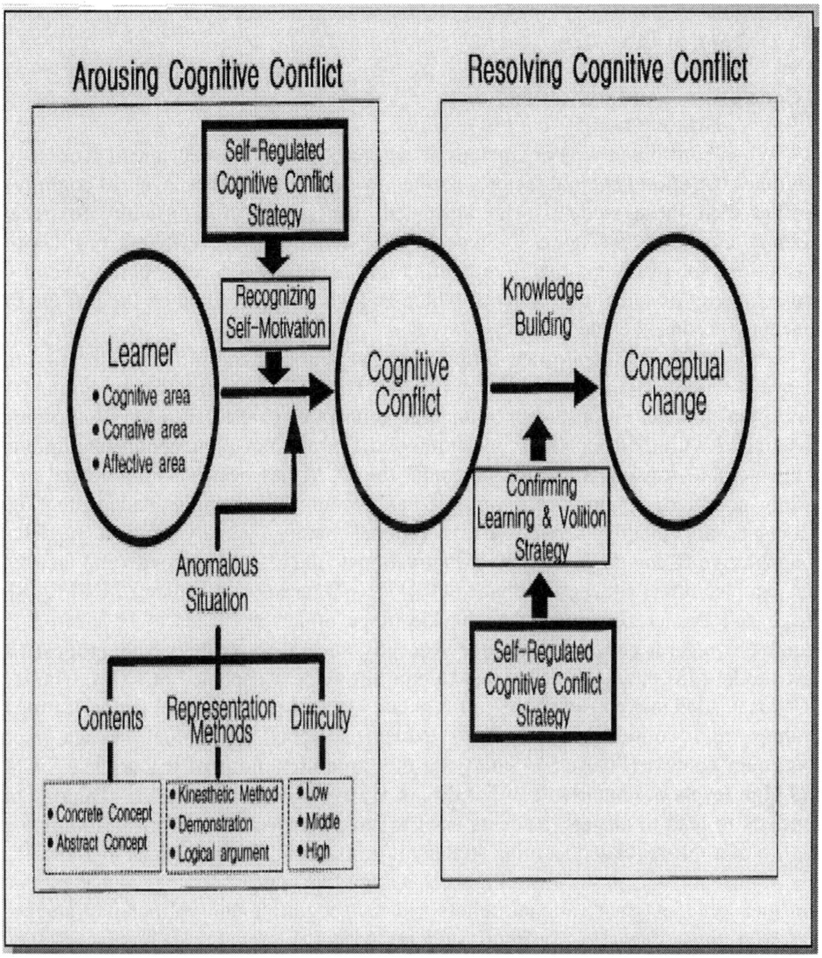

Figure 10. Need to consider self-regulated factor in conceptual change

In summary, even though the students experienced high levels of cognitive conflict, not all of them changed their alternative conceptions into scientific conceptions. However, only those students who had a perspective on effort that implied the use of a self-regulated learning strategy exhibited a conceptual change in the delayed post-test, as shown in Table 2.

6. Conclusions and Implications

When confronted with anomalous situations, students who attributed successful outcomes to "effort (internal factor)" seemed to experience higher levels of cognitive conflict than those students who attributed success to "task difficulty (external factor)", as shown in Figure 2. Students were found to have different profiles of attributions for resolving cognitive conflict and acquiring a successful/unsuccessful understanding of scientific concepts, which had a strong influence on the process of conceptual change (Table 2).

In the early 1980s, Posner, Strike, Hewson, and Gertzog [46] were the first to suggest a model of conceptual change (CCM). The CCM specifies the conditions for conceptual change: For a student to undergo conceptual change, they have to become dissatisfied with their existing conception and find the new conception intelligible, plausible, and fruitful. Since its inception, the CCM has been very influential and widely accepted, yet in recent years it has become increasingly inadequate. The criticisms are mainly leveled at its rational nature; namely, that it neglects noncognitive factors (e.g., co-native, motivational, and classroom contextual factors) that can also affect conceptual change [17,47]. In a further explication of CCM [18], Strike and Posner contended that a wide range of factors needs to be taken into account as regards conceptual change. Recently, several researchers have proposed a new model [48], however, it has yet to gain the wide acceptance of CCM. Thus, a CCM is needed that considers the self-regulated factors (motivational factor, learning strategy, and volitional strategy) to explain students' conceptual change in a classroom context (Figure 9). Satisfying the three conditions of the original CCM and high levels of cognitive conflict do not always produce conceptual change. For conflicts to lead to change, students need to have perspectives on "effort" implying the use of a self-regulated learning strategy (For example, Bana and Jana in Table 2).

Consequently, as shown in Figure 9, a cognitive conflict strategy is needed that considers students' motivational beliefs and metacognitive and volitional strategies. Before a physics teacher confronts students with a discrepant demonstration, they should present students with ways to increase their intrinsic motivation for learning physics. "One can lead a horse to water, but you can't make it drink." What is important is increasing the likelihood of drinking, for example, by feeding the horse salt before offering it water. Similarly, before students are confronted with a discrepant event, it is important that they understand why they are studying physics concepts. Based on this understanding, students can then actively attempt to resolve cognitive conflict. After confronting students with a demonstration, the new scientific concept needs to be taught along with metacognitive strategies to resolve the conflict, thereafter, volitional strategies [49] are required to develop and retain the scientific conception. This model of strategies, including motivational, metacognitive, and volitional strategies, is called a self-regulated cognitive conflict strategy (SRCCS) or management strategy of cognitive conflict (MSCC). However, further studies on SRCCS are needed in a classroom context.

This strategy seems to facilitate students' conceptual change in physics learning and help them gain perspectives on "effort" implying the use of a self-regulated learning strategy to become self-regulated learners of physics.

References

1. G. H. Lee, The effects of cognitive conflict, learning motivation and learning strategy on high school students' conceptual changes in physics, Ph.D. thesis, Korea National University of Education, 2000.
2. Y. J. Lee, The effect of cognitive conflict on students' conceptual change in physics. Ph.D. thesis, Korea National University of Education, 1998.
3. S. Druyan, Effect of the Kinesthetic Conflict on Promoting Scientific Reasoning, *Journal of Research in Science Teaching*, vol. 34 (10), pp. 1083-1099. 1997.
4. J. S. Kwon, The necessity of cognitive conflict strategy in science teaching, Paper presented at *the International conference on Science Education: Globalization of Science Education*, May 26-30, 1997, Seoul, Korea, 1997.
5. J. S. Kwon, A cognitive model of conceptual change in science learning, *Physics Teaching* (written in Korean) vol. 7 (1), pp. 1-9. Korean Physics Society, 1989.
6. M. Z. Hashweh, Toward an Explanation of Conceptual Change, *European Journal of Science Education*, vol. 8 (3), pp. 229-249, 1986.
7. P. W. Hewson and M. G. A., Hewson, The role of conceptual conflict in conceptual change and the design of science instruction, Instructional Science, vol.13, pp. 1-13, 1984.
8. C. Chan, J. Burtis, and C. Bereiter, Knowledge building as a mediator of conflict in conceptual change, *Cognition and Instruction*, vol. 15, pp. 1-40, 1997.
9. P. K. Tao & R. F. Gunstone, The process of conceptual change in force and motion during computer-supported physics instruction, *Journal of Research in Science Teaching*, vol. 36, pp. 859-882, 1999.
10. Chinn, and W. F. Brewer, An empirical test of a taxonomy of responses to anomalous data in science, *Journal of Research in Science Teaching,* vol. 35, pp. 623-635, 1998.
11. Chinn, and W. F. Brewer, The role of anomalous data in knowledge acquisition: A theoretical framework and implications for science instruction, *Review of Educational Resecrch,* vol. 63, pp.1-49, 1993.
12. M. Niaz, Cognitive Conflict as a Teaching Strategy in Solving Chemistry Problems: A Dialectic-Constructivist Perspective, *Journal of Research in*

Science Teaching, vol. 32 (9), pp. 959-970, 1995.
13. J. S. Kwon, and Y. J. Lee, The effect of cognitive conflict on students' conceptual change in physics, Paper presented *at the Annual Meeting of the National Association for Research in Science Teaching,* Boston, Massachusetts, March 31, 1999.
14. N. J. Kwon, The influences of learners' characteristics on scientific conceptual changes by cognitive conflict in Korean middle schools. Ph.D. thesis, Korea National University of Education, 2000.
15. S. Vosniadou, and C. Ioannides, From conceptual development to science education: a psychological point of view, *International Journal of Science Education*, vol. 20, pp. 1213-1230, 1998.
16. S. Demastes, R. Good, and P. Peebles, Patterns of conceptual change in evolution, *Journal of Research in Science Teaching,* vol. 33, pp. 407-432, 1996.
17. P. R. Pintrich, R. W. Mark, and R. A. Boyle, Beyond cold conceptual change: The role of motivational beliefs and classroom contextual factors in the process of conceptual change, *Review of Educational Research*, vol.63, pp. 167-199, 1993.
18. K. A. Strike, and G. J. Posner, A revisionist theory of conceptual change, in R. Duschl and R. Hamilton (eds.), *Philosophy of science, cognitive psychology, and educational theory and practice* (pp. 147-176). Albany, NY: SUNY Press, 1992.
19. D. Stipek, *Motivation to learn: from theory to practice* (3rd ed), Allyn & Bacon, 1998.
20. Weiner, Integrating social and personal theories of achievement striving, *Review of Educational Research*, vol. 64, 557-573, 1994.
21. D. Gagné, C. W. Yekovich, and F. R. Yekovich, *The cognitive psychology of school learning,* Addison-Wesley, 1993.
22. L. Barlia, & M. E. Beeth, High school students' motivation to engage in conceptual change learning in science, Paper presented at *the annual meeting of the National Accociation for the Research in Science Teaching*, Boston, Massachusetts, March 31, 1999.
23. S. Graham and B. Weiner, Theories and principles of motivation. In D. Berliner & R. Calfee (Eds.), *Handbook of educational psychology* (pp. 63-84), New York: Macmillan, 1996.
24. Woolfork, *Educational Psychology* , 8th ed, Allyn and Bacon, 2001.
25. Weiner, The role of affect in rational (attributional) approaches to human motivation, *Educational Researcher,* vol. 9, pp. 4-11, 1980.
26. R. R. Sherman and R. B. Webb, *Qualitative research in education: focus and methods*, The Falmer Press, 1988.
27. Y. J. Lee, The study of the attributional style to success and failure, *Educational Research*, vol. 9, Kongju University, 1993.
28. G. H. Lee, J. S. Kwon, S. S. Park, J. W. Kim, H. G. Kwon, and H. K. Park, The

Development of an Instrument for the Measuring of Students' Cognitive Conflict Levels, Paper presented *at the Annual Meeting of the National Association for Research in Science Teaching,* Boston, Massachusetts, March 31, 1999.
29. J. Rotter, Generalized expectancies for internal versus external control of reinforcement, *Psychological Monographs*, 1 (Whole 609), 1966.
30. R. Osborne and P. Freyberg, *Learning in science: The implications of children's science,* Auckland: Heinemann. 1985.
31. S. Mathison. Why triangulate? *Educational Researcher,* vol, 17(2), pp. 13-17, 1988.
32. Chen and H. Stevenson, Motivation and mathematics achievement: A comparative study of Asian-American, Caucasian-American, and East Asian high school students, *Child development*, vol. 66, pp. 1215-1234, 1989.
33. M. Boekaerts, Self-regulated learning: a new concept embraced by researchers, policy makers, educators, teachers, and students, *Learning and Instruction*, vol. 7(2), pp. 161-186, 1997.
34. L. McCombs, B. L. and R. J. Marzano, Putting the self in self-regulated learning: The self as agent in integrating skill and will, *Educational Psychologist*, vol. 25, pp. 51-70, 1990.
35. P. K. Murphy and P. A. Alexander, A motivated exploration of motivation terminology, *Contemporary Educational Psychology*, vol. 25, pp. 3-53, 2000.
36. L. Corno, The best-laid plans: Modern conceptions of volition and educational research, *Educational researcher*, vol. 22(2), pp. 14-22, 1993.
37. R. E. Snow, L. Corno and D. Jackson, Individual differences in affective and cognitive function, in D. Berliner and R. Calfee (Eds.), *Handbook of educational psychology* (pp. 243-310), New York: Macmillan, 1996.
38. J. P. Raffini, *150 ways to increase intrinsic motivation in the classroom*, Allyn and Bacon, 1996.
39. Schunk and B. Zimmerman, *Self-regulated learning: From teaching to self-reflective practice*, Guilford, 1998.
40. P. R. Pintrich, and De Groot, Motivational and self-regulated learning components of classroom academic performance, *Journal of Educational Psychology*, vol. 82, pp. 33-40, 1990.
41. R. Brown and M. Pressley, Self-regulated reading and getting meaning form text: The transactional strategies instruction model and its ongoing validation, in D. Schunk and B. Zimmerman (Eds.), *Self-regulation of learning and performance* (pp. 101-124). Hillsdale, NJ: Lawrence Erlbaum Associates, 1994.
42. J. Kuhl, Volitional mediators of cognition-behavior consistency: Self-regulator processes and action versus state orientation, in J. Kuhl and J. Beckman (Eds.), *Action control: From cognition to behavior* (pp. 101-128), New York: Springer-Verlag, 1985.
43. T. Garcia, E. J. McCann, J. E. Turner, L. Roska, Modeling the mediating role of

volition in the learning process, *Contemporary Educational Psychology*, vol. 23, pp. 392-418, 1998.

44. P. A. Alexander, Superimposing a situation-specific and domain-specific perspective on an account of self-regulated learning, *Educational Psychologist*, vol. 30(4), pp. 189-193. 1995
45. E. Weinert, F. W. Schrader, and A. Helmke , Quality of instruction and achievement outcomes, *International Journal of Educational Psychology*, vol. 13(8), pp. 895-912, 1989.
46. J. Posner, K. A. Strike, P. W. Hewson, and W. A. Gertzog, Accommodation of a scientific conception: Toward a theory of conceptual change, *Science Education,* vol. 66, pp. 211-227, 1982.
47. Lee and C. W. Anderson, Task engagement and conceptual change in middle school science classrooms, *American Educational Research Journal,* vol. 30, pp. 585-610. 1993.
48. S. Vosniadou, Capturing and modeling the process of conceptual change, *Learning and Instruction*, vol. 4, pp. 45-69, 1994.
49. E. J. McCann and T. Garcia, Maintaining motivation and regulating emotion: Measuring individual differences in academic volitional strategies, *Learning and Individual Differences*, vol. 11, pp. 259-279, 1999.

CONCEPTUAL CHANGE ACCORDING TO DEGREE OF CONFLICT BETWEEN PRECONCEPTION AND EXPERIMENTAL DATA

I.KIM, I.YANG , D.YANG AND S.JEONG

School of Science Education, College of Education, Chungbuk National University, Cheongju, Korea

The current study investigates the rate of conceptual change according to the level of conflict between students' preconceptions and given experimental data about a simple electric circuit. As such, a pre-test was given to determine students' preconceptions as regards current changes in an electric bulb when the applied voltage is increased. Those students who answered that the current would increase linearly as the applied voltage increased due to Ohm's law were selected and divided into six groups. Each group was then given different virtual experimental data with a specific mean value of percent error, thereby creating six conflict levels. The students were asked to plot the current as a function of the voltage using the given data, then answer questions about changes in their preconception after interpreting the graph. In general, the rate of conceptual change gradually increased as the conflict level increased, plus a threshold value for the conflict level in relation to the conceptual change rate was also observed.

1. Introduction

One of the most important themes in recent research regarding conceptual change is cognitive conflict. Cognitive conflict occurs before conceptual change [1], [4]-[7] and many studies have reported that cognitive conflict-based strategies are more effective than traditional instruction for conceptual change. [6],[11],[13] However, it is questionable whether cognitive conflict is a necessary and sufficient condition for conceptual change, although the importance of cognitive conflict is still undeniable. [2],[4],[5],[9] The reasons why cognitive conflict by itself cannot lead to conceptual change have already been shown in previous research as students can resolve cognitive conflict using various psychological strategies instead of changing their conceptions. [7] Examples include confirmation and biased-assimilation. There are also some cases where unusual data is treated as a special case or exceptional phenomenon rather than evidence of an alternative theory. [7],[8] Another reason why conceptual change rarely occurs is that students interpret natural phenomena directly and intuitively rather than using a theoretical base. [3]

It is important for science teaching conceptions to recognize the fact that conceptual conflict does not necessarily lead to conceptual change. As such, a detailed investigation is necessary of students' responses on given conflict situations. Kim and Park [10] reported that some middle school students modified experimental data when faced with a conflict in the experimental process. They also reported that the frequency of modification increased as the process continued. Another study reported different types of student response to observing conflict. Whenever students changed their preconceptions, they almost always introduced a new idea to explain the contradictory observations. As such, the process of conceptual change is seemingly closely related to a protective belt of auxiliary conceptions, initial conditions, or related assumptions. [14] This means that when an alternative concept is unavailable, students resolve a conflict by modifying their scientific knowledge related to the concept rather than changing their concept. Therefore, these results indicate that cognitive conflict is not a necessary and sufficient condition for conceptual change. Thus, studies on how students respond to cognitive conflict are important for understanding conceptual change.

Cognitive conflict needs to be studied from two perspectives, qualitative and quantitative. Previous research has only focused on the qualitative aspects of student responses to conflict situations. Accordingly, the current study is interested in the quantitative aspect of cognitive conflict. Qualitative conflicts are concerned with the types, presenters, or contexts of conflict, whereas quantitative conflicts are concern with the strength of a conflict. There is also the question of whether there is a threshold value of cognitive conflict for student conceptual change. Consequently, the appropriate way of providing conflict will help students shift from their own concepts to scientific ones.

Definition of terms

In the context of this study the terms below are used with a limited meaning as follows:

Cognitive rate: Some students realized the fact that there was a difference between their thoughts and the experimental data even though the others did not. Cognitive rate means the percentage of students who recognized the conflict.

Discrepancy rate: The level of conflict the students experienced was expressed as a ratio, based on the assumption that the student's thought (preconception) was the true value and the experimental values had a relative error. As such, if the discrepancy rate is 10%, the curve fitting with the relation of the voltage and current as linear gives a 10% discrepancy. This is also related to the threshold conflict defined below.

Threshold conflict: Students recognized the fact that there was a difference between their thoughts and the given information, including the experimental data, observation, and explanation. However, not all students changed their concepts as a result of the conflict. Sometimes they disregarded the conflict and instead introduced a variety of strategies or other knowledge. Whether or not there was any conceptual change depended on the strength of the cognitive conflict. Thus, the threshold conflict means the threshold value of conflict that was strong enough for the students to make a conceptual change, i.e. when alternative strategies were no longer introduced to persist with their original concepts.

2. Research Goals and Methods

2.1 *Research Goals*

The current study aimed to investigate how students respond to experimental results inconsistent with their predictions, especially in relation to voltage and current, plus, how the rate of conceptual change depends on the conflict level of the difference between the original concept and the experimental results.

2.2 *Test for Selecting Subjects*

Subjects were needed with the same strong yet physically unsound ideas about electricity. Therefore, a paper and pencil test was developed to select subjects. The task was measuring the resistance of an electric bulb, which is a common problem used in Korean textbooks and thus familiar to students (Figure 1). The students were asked how the current in a bulb would change when the applied voltage was increased by changing the variable resistance. In general, students thought that the current would be linearly proportional to the applied voltage based on their memorization of Ohm's law. When the resistance of a bulb remains constant, the voltage and current are proportional to each other. Yet, when the voltage applied to a bulb increases, the bulb's temperature also increases, thereby causing the resistance of the bulb to increase. Therefore, as the voltage increases, the rate of the current increase is reduced.

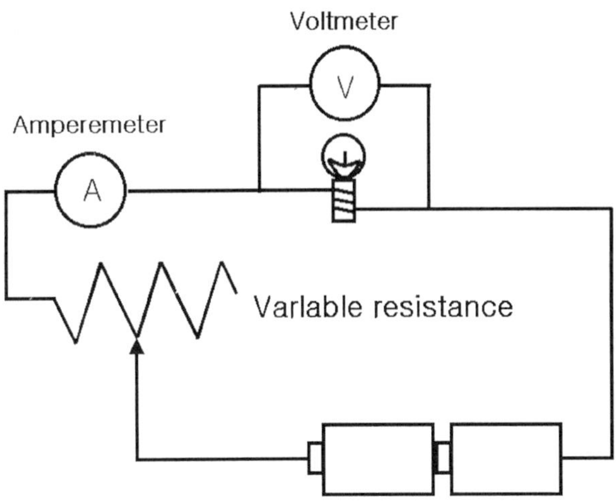

Figure 1. A series circuit

As a result, the voltage and current in the bulb are not linearly proportional but parabolic. Accordingly, a test was devised to investigate the student's thoughts about this topic. The students were asked to answer the following questions: First, what is the relationship between the current and the applied voltage? Second, draw a graph of the relation between the voltage and the current. Third, write an explanation of your idea.

2.3 *Selection of Subjects Using Initial Test and Creation of Conflict Levels*

Selection of subjects: A paper and pencil test was given to 300(6 classes) 9th grade students in Cheong-ju city. The test asked about the relation between the current and the voltage in relation to the abovementioned experiment. Those students who answered that the relation was linear and produced a consistent graph and explanation were selected .

Creating conflict: The conflict was created using a set of virtual experimental data about the same topic. Each set consisted of 10 measured current and voltage values. When the students plotted the data, the relationship between the applied voltage and the current was not linearly proportional yet parabolic. As such, the result was not consistent with the students' preconceptions, thereby creating a cognitive conflict. Although the students did not measure the results themselves, they had already performed the experiment as part of regular class. The students were asked to plot and draw a graph of the voltage vs. the current based on the given data, then asked two questions. (1) What is the relationship between the current and the applied voltage? (2) After analyzing the experimental results, have you changed your ideas in any way?

Levels of conflict: To differentiate the levels of conflict, six sets of data were prepared and each set had a different mean value of the percent error. In the current study, the error means the difference between the students' preconception and the given data. When the given data (a curved line) and students' preconception (a linear line) were plotted as two lines in the same graph, the difference between two points on the two lines was the error, assuming that the points on the linear line (preconception) were theoretical values and the points on the curved line (results of experiment) were measured values. Also, the error was calculated as a percentage. There were percent errors and the mean value of the percent errors for each set was different according to the curvature. As such, the conflict levels were determined by the mean value of the percent errors for each set of data. The lowest mean value of the percent error was 8%, followed by 15%, 20%, 25%, 30%, and 35%. Some of these are shown in the graph in fig. 2.

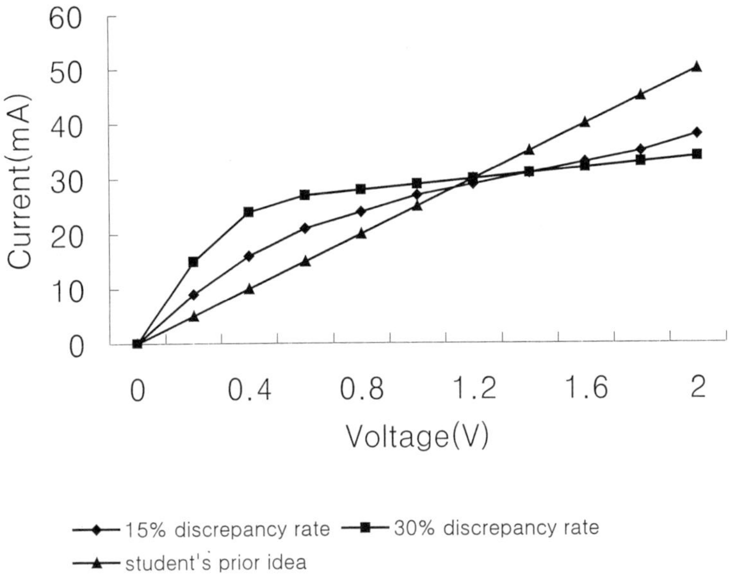

Figure 2. Students' Preconception

Identifying students' responses about cognitive conflict: The subjects selected from the initial test were divided into six groups, A to F. Each group received a different level of conflict data, then plotted the given data into a graph and compared the results with their preconception after curve fitting. The students' responses from each group with a different level of conflict were analyzed based on two research questions: First, according to the conflict level, 'how many students recognized the conflict?' This was related to the cognitive rate. Second, according to whether or not the conflict was recognized, 'how many students changed their preconception?' This was related to determining the threshold conflict level.

3. Analysis of Results

3.1 *Student's Preconceptions.*

From the initial test, 8 types of student preconception were identified, as shown in Fig. 3. The most frequent was ①, 75%, 226 students, the second was ②, 12%, 37students, followed by ③, 7, ④, 7, ⑤, 6, ⑥, 7, ⑦, 6, and ⑧, 3. These results were similar to a previous study on college students [12] where type ① was the most frequent (52%) followed by type ② with a frequency of 18%.

When the students were asked to write a reason for their answer, most students answered "because of Ohm's law " or " the voltage and current are proportional to each other due to Ohm's law". Some responded that the current and resistance were inversely proportional.

Ohm's law is an important topic in Korean textbooks. However, students seemed to perceive Ohm's law as a simple formula of V=I*R. Many test questions prepared in schools also ask for numerical calculations using Ohm's law. The type ① students were chosen as the subjects for investigating the effect of the degree of conflict

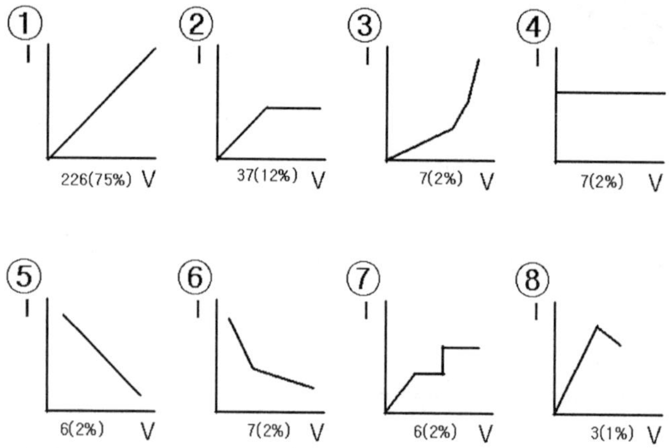

Figure 3. Students' Responses in Pretest

3.2 *Responses to Different Levels of Cognitive Conflict*

The following are the results according to the different levels of cognitive conflict, i.e. discrepancy rate.

(1) 8% discrepancy rate

A set of data with an 8 % discrepancy rate was given to group A. The responses of these 31 students after conducting the experimental task were as follows.

Question 1: When the voltage increases, how does the current change?

22 students, 71%, answered 'increases linearly' and 8 students, 29%, answered 'increases yet not linearly'. This means that 29% of the students recognized the cognitive conflict

Question 2: After analyzing the experimental results, have your thoughts changed in any way?

26 students, 84%, answered 'the same as before the experiment'. Whereas, 5 students, 16%, answered 'since my idea differed from the experiment results, I changed my thought.' Therefore, among the 8 students that recognized the fact that

the experimental results differed from their previous conception and answered 'increases, yet not linearly', only 5 students actually changed their concepts. As such, 3 students, 37.5%, did not change their ideas even after recognizing the discrepancy between the experimental data and their concepts.

(2) 15% discrepancy rate

Data with a 15 % discrepancy rate was given to group B. The responses of these 29 students after conducting the experimental task are as follows.

Question 1: When the voltage increases, how does the current change?

13 students, 45%, answered 'increases linearly', 15 students, 52%, answered 'increases yet not linearly', and 1 student, 3 %, answered 'increases and becomes even'. This means that 55% of the students recognized the cognitive conflict.

Question 2: After analyzing the experimental results, have your thoughts changed in any way?

21 students, 72%, answered 'the same as before the experiment'. Whereas, 8 students, 28%, answered that 'since my idea differed from the experiment results, I changed my thought.' Among the 16 students that recognized that the experimental results differed from their previous conception and answered 'increases, yet not linearly', only 8 students actually changed their ideas. As such, 8 students, 50%, did not change their ideas even after recognizing the discrepancy between the experimental data and their thoughts.

(3) 20% discrepancy rate

Data with a 20 % discrepancy rate was given to group C. The responses of these 31 students after conducting the experimental task were as follows.

Question 1: When the voltage increases, how does the current change?

8 students, 26%, answered 'increases linearly', 21 students, 68%, answered 'increases, yet not linearly', and 2 students, 6 %, answered 'increases and becomes even'. This means that 74% of the students realized the cognitive conflict.

Question 2: After analyzing the experimental results, have your thoughts changed in any way?

20 students, 65%, answered 'the same as before the experiment'. Whereas, 11 students, 35%, answered that 'since my idea differed from the experiment results, I changed my thought'. Among the 23 students that recognized that the experimental results differed from their previous conception and answered 'increases, yet not linearly', only 11 students actually changes their concepts. As such, 12 students, 52%, did not change their ideas even after recognizing the discrepancy between the experimental data and their thoughts.

(4) 25% discrepancy rate

Data with a 25 % discrepancy rate was given to group D. The responses of these 32 students after conducting the experimental task were as follows.

Question 1: When the voltage increases, how does the current change?

5 students, 16%, answered 'increases linearly', 21 students, 66%, answered 'increases, yet not linearly', and 6 students, 19 %, answered 'increases and becomes even'. This means that 85% of the students recognized the cognitive conflict.

Question 2: After analyzing the experimental results, have your thoughts changed in any way?

19 students, 59%, answered 'the same as before the experiment'. Whereas, 13 students, 41%, answered that 'since my idea differed from the experiment results, I changed my thought'. Among the 27 students that recognized that the experimental results differed from their previous conception and answered 'increases yet not regularly', only 13 students actually changes their concepts. As such, 14 students, 52%, did not change their ideas even after recognizing the discrepancy between the experimental data and their thoughts.

(5) 30% discrepancy rate

Data with a 30 % discrepancy rate was given to group E. The responses of these 33 students after conducting the experimental task were as follows.

Question 1; When the voltage increases, how does the current change?

4 students, 12%, answered 'increases linearly', 15 students, 45%, answered 'increases, yet not linearly', and 14 students, 42 %, answered 'increases and becomes even'. This means that 87% of the students realized the cognitive conflict.

Question 2: After analyzing the experimental results, have your thoughts changed in any way?

18 students, 55%, answered 'the same as before the experiment'. Whereas, 15 students, 45%, answered that 'since my idea differed from the experiment results, I changed my thought'. Among the 29 students that recognized that the experimental results differed from their previous conception and answered 'increases, yet not linearly', only 15 students actually changed their concepts. As such, 14 students, 48%, did not change their ideas even after recognizing the discrepancy between the experimental data and their thoughts.

(6) 35% discrepancy rate

Data with a 35 % discrepancy rate was given to group F. The responses of these 35 students after conducting the experimental task were as follows.

Question 1: When the voltage increases, how does the current change?

2 students, 6%, answered 'increases linearly', 6 students, 17%, answered 'increases, yet not linearly', and 27 students, 77%, answered ' increases and becomes even'. This means that 94% of the students recognized the cognitive conflict.

Question 2: After analyzing the experimental results, have your thoughts changed in any way?

8 students, 23%, answered 'the same as before the experiment'. Whereas, 27 students, 77%, answered that 'since my idea differed from the experiment results, I

changed my thought.' Among the 33 students that recognized that the experimental results differed from their previous conception and answered 'increases, yet not linearly', 27 students changed their concepts. As such, only 6 students, 18%, did not change their ideas even after recognizing the discrepancy between the experimental data and their thoughts.

(7) Summary of Results

Several characteristics according to the students' thoughts and discrepancy rate can be compared as follows:

Cognitive rate: the relation between the discrepancy rate and the student cognitive rate is shown in Figure 4. In general, the student cognitive rate increased as the discrepancy rate increased. At a low discrepancy rate, 8% to 20%, the cognitive rate increased linearly. However, after the discrepancy rate reached 25%, the cognitive rate became gradually saturated. Most students recognized the conflict at a 30% discrepancy rate. Yet, a relatively high number of students, 29%, also recognized the cognitive conflict at an 8% discrepancy rate.

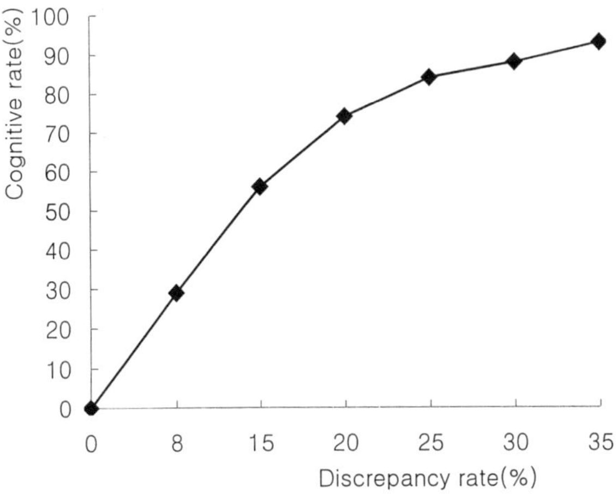

Figure 4. Discrepancy rate vs. student cognitive rate

Conceptual changes according to discrepancy rate: The number of students in each group who changed their concepts is shown in Table 1. The results are also shown in Fig. 5. In general, the rate of conceptual change increased with the discrepancy rate, yet not linearly. The number of students who changed their concepts increased with an increase in the discrepancy rate from 8% to 25%. Thereafter, there was a sudden increase when the discrepancy rate was 35%, indicating that the threshold value for the discrepancy rate in relation to the students' conceptual changes was between 30% and 35%.

Table 1. Students' conceptual changes (%) according to discrepancy rate.

Discrepancy rate(class) / Student responses	8% (A)	15% (B)	20% (C)	25% (D)	30% (E)	35% (F)
I did not change my concept (%)	26/31 (84)	21/29 (72)	20/31 (65)	19/32 (59)	18/33 (55)	8/35 (23)
I changed my concept (%)	5/31 (16)	8/29 (28)	11/31 (35)	13/32 (41)	15/33 (45)	27/35 (77)

Conceptual changes in students who recognized conflict: A critical discrepancy rate was found in relation to conceptual change. The responses of the students who recognized the cognitive conflict were analyzed to confirm the threshold value for conceptual change. Among the students who recognized the conflict, some changed their concepts, while others did not. Although the number of subjects was insufficient to generalize this result, it is still important to understand why some students changed their ideas, while others did not. The ratios of the non-changing rate of the students who recognized a conflict are shown in Table 2 and Figure 6. Among the students that recognized the cognitive conflict, the number of students who did not change their idea remained comparatively constant around 50% with discrepancy rates from 8% to 30%, see Figure 6.

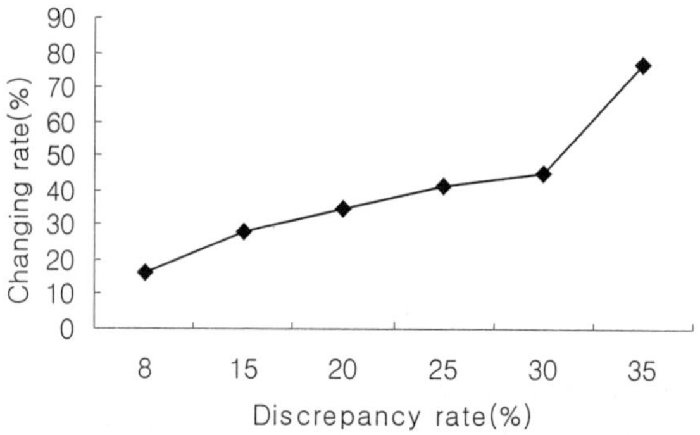

Figure 5. Student conceptual change rate (%) according to discrepancy rate

Thereafter, there was a drop to 18% when the discrepancy rate became 35%. In other words, there would seem to be a conflict threshold as regards yielding a conceptual change.

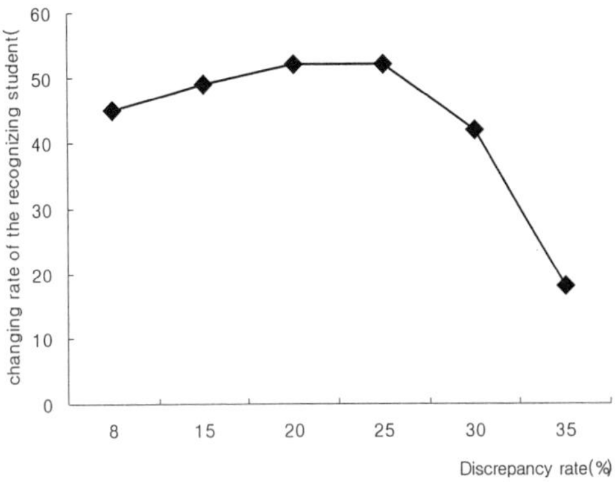

Figure 6. Conceptual change rate (%) for students who recognized conflict

Table 2. Non-changing rate of students who recognized cognitive conflict

Discrepancy rate (%)	8	15	20	25	30	35
Non-changing rate of students who recognized	45	49	52	52	42	18

4. Summary and Discussion

The current study was aimed at investigating student responses to experimental results inconsistent with their preconceptions. The process of student conceptual change can be a very complex one. However, the present investigation focused on a simple process of conceptual change where all the subjects had the same simple preconception, then specially designed inconsistencies between the students' preconception and experimental results were introduced using the same patterns yet different strengths.

Based on the results of a paper-pencil test, students were selected that had the same preconception as regards the relation between a current and an applied voltage. The paper-pencil test was related to a simple electric circuit, i.e. a bulb connected to a battery, which was chosen as most students are familiar with the task and were expected to simply apply Ohm's law, as previously memorized. However, since the actual results exhibit a non-linear relation, this creates a cognitive conflict. The selected students anticipated that the relation between the two variables (current and applied voltage) would be linear, i.e. the current through the bulb would increase linearly with an increase in the applied voltage. The students were given experimental results that showed a non-linear (parabolic) relation between the two variables. The sets of data provided were designed with different discrepancy rates from the students' preconception based on adjusting the mean value of the percent error. The students were then asked to draw a I-V graph using the given data and compare the results with their preconception.

The student responses were investigated and two main issues were highlighted. The first was related to the students' recognition of cognitive conflict and the second was related to the rate of conceptual change. Cognitive conflict is one of the most important factors in conceptual change and often used by science teachers.

From the current results, it was found the rate of recognition of cognitive conflict increased with the discrepancy rate, with a threshold value between 30% and 35%. This means the higher the discrepancy rate, the more students that recognized the cognitive conflict, yet there was a limit. Some students still did not recognize the conflict even with a high discrepancy rate, whereas other students were able to recognize the conflict at a low discrepancy rate. Further research on 'why some students do not recognize conflict with a strong discrepancy and some students recognize conflict with a weak discrepancy?' is needed to understand this aspect of conceptual change.

In general, it was found that the student conceptual change rate increased with the discrepancy rate. Plus, a critical (threshold) value for the discrepancy rate was identified based on a sudden increase in the conceptual change rate. Attention was focused on the responses of the students who recognized the cognitive conflict to clarify the existence of a threshold value. The conceptual change rate for those students who recognized the cognitive conflict was about 50% and remained stable with a discrepancy rate from 8% to 25%. However, the rate suddenly increased to 92% when the discrepancy rate reached 35%, implying a threshold value for cognitive conflict as regards student conceptual change.

Accordingly, cognitive conflict was found to be an important condition, yet not absolutely necessary or sufficient for conceptual change. Some of students did not change their ideas even though they recognized the conflict. A threshold value of discrepancy was also identified, which has implications for both recognizing cognitive conflict and conceptual change. Plus, the strength of cognitive conflict was highlighted as an important factor in conceptual change.

References

1. Stay & Berkowitz, Cognitive conflict as a basis for teaching quantitative aspect of the concept of temperature. *Science Education*, .64(5), .679-692., 1980
2. Champagne, A. B., Klopfer, L. E. and Anderson, J. H, Factors influencing the learning of classical mechanics. *American Journal of physics*, 48(12), 1074-1079,1980.
3. McCloskey, M, Intuitive physics. *Scientific American*, .248, 122-238,1983.
4. Posner, G. J., Strike, K. A., Hewson, P. W. and Gertzog,W. A, Accommodation of a scientific conception: Toward a theory of conceptual change. *Science Education*, .66(2), 211-227,1982.
5. Pines, A. E. and West, L. H. T, A framework for conceptual change with special reference to misconceptions. in Helm, H. and Novak, J. (eds.) Proceedings of the misconceptions is science and mathematics, Cornell University, Ithaca, pp. 47-52, 1983
6. Hewson , P. W., & Hewson, M. G. The role of conceptual conflict in conceptual change and the design of science instruction. *Instructional Science*, 13, 1-13,1984.
7. Hashweh, M. Z, Toward an Explanation of Conceptual change. *European Journal of Science Education*, .8(3), 229-249, 1986.
8. Kuhn, D., Amsel, E. and O'Loughlin, M, The development of scientific thinking Skills. Academic Press, Inc.,1988.
9. Kim, I., College students' conceptual change about force and acceleration through critical discussion of the rival concepts based on evidences and reflective thinking, Ph.D. thesis, Seoul National University, Seoul, Korea, 1991.
10. Kim, I. and Park, J. The student' responses on the conflict observation, data and result in electricity experiment. Paper presented at the annual meeting of the Australasian Science Education Research Association(ASERA), Bendingo, Australia,1995
11. Niaz, M., Cognitive conflict as a teaching strategy in solving chemistry problems : A dialectic-constructivist perspective. *Journal of Research in Science Teaching*, 32(9), 959-970,1995
12. Kim, I., College students' responses to conflicting results from electricity experiments, Mulli Kyoyuk, *The Korean Physical Society*, 15(2), 88-99,1997
13. Druyan, S., Effect of the kinesthetic conflict on promoting scientific reasoning. *Journal of Research in Science Teaching*, 34(10), 1083-1099, 1997
14. Park, J., Kim, I., Kim, M., Lee, M., Analysis of the students' process of confirmation and falsification. *Journal of the Korean Association for Research in Science Education,* 18(3), 283-296,1998.

ANALYZING COGNITIVE OR NON-COGNITIVE FACTORS INVOLVED IN PHYSICS PROBLEM SOLVING IN EVERYDAY CONTEXT

JONGWON PARK & LIMOOK LEE
Department of Physics Education, Chonnam National University, Kwangju, KOREA, 500-757

All physics educators emphasize the importance of context, especially an everyday context, in physics learning, teaching, and problem solving. However, do students really want to learn physics in an everyday context? Do students think physics problems in an everyday context are better than de-contextualized problems, or prefer the former type of problem to the latter in a physics test? What do physics teachers or researchers think about the above questions? Accordingly, the current study was initiated from these questions. Students were asked to solve two types of physics problem, i.e. problems in an everyday context and de-contextualized problems, then an analysis was made of the additional cognitive or non-cognitive factors involved in the process of solving the two types of problem. Finally, it was ascertained whether physics teachers considered the additional factors identified in the above analysis to be important in physics learning or not. The main results and implications for physics teaching and learning in an everyday context are discussed in this paper.

1. Introduction

Nowadays, context as well as physics content and inquiry processes are being emphasized in physics teaching, learning, and problem solving (Keeves & Aikenhead, 1995). For instance, the APU(Assessment Performance Unit) science assessment framework consists of three dimensions: concepts, scientific processes, and contexts. Plus, in the 7th Korean National Curriculum, the importance of context in learning physics is also described: "Physics should be learned in everyday contexts that can arouse students' interest and curiosity, for instance, in the context of sports, transportation, amusement parks, music, household electric appliances, communication, medical treatment …."

When learning physics, students prefer specific contexts. Song and Choi (1994) examined the degree of students' preferences as regards different contexts in learning basic concepts of mechanics, and ranked the order of the preferred contexts as follows: sports > military weapons > everyday life > living things > natural phenomena > laboratory. In addition, Betty & Fred (1996) developed science materials with a technological approach, and observed that contextualized activities were highly appreciated by students.

Context can have an influence on recognizing the purpose of physics learning. Reif & Larkin (1991) found that many students had erroneous conceptions of scientific goals. For instance, the goals of students for everyday life are mainly to lead a good life, whereas their central goal of science is to achieve optimal explanation and predication based on parsimoniously structured theoretical knowledge. Plus researchers have pointed out that students' difficulties in learning science may be increased due to the importation of goals that are effective in everyday life yet inadequate in science.

It has also been found that students' performances with science problems differ according to the context. For example, Schofield et al. (1989) observed that the performance of APU was higher for everyday context problems than for scientific context ones. This means that the processes or strategies of the solving program differ according to the context of the problem. Saunders & Jesunathadas (1988) also observed that students did better in solving problems familiar to them rather than unfamiliar ones. Plus Song and Black (1992) noted an interactive effect between the concepts required to solve problems and the problem context; students exhibited a better performance with problems requiring no concept application in an everyday context, while they showed no difference between everyday and scientific contexts in problems requiring concept application.

Accordingly, based on the above research on the effect of context on physics teaching, learning, and problem solving, the current study focuses on the area of physics problem solving in relation to the following general questions: Do students really want to learn and solve physics problems in an everyday context? Do students think physics problems in an everyday context are better than de-contextualized ones, or prefer the former type of problem to the latter as a physics performance test in school? What do physics teachers or researchers think about the above questions?

Reif and Larkin (1991) argued that the inferences or ways of thinking in everyday life are different from those in science. That is, the main methods used by students to solve everyday life problems rely heavily on perceptual processes, pattern recognition, and qualitative reasoning, however, such methods alone are inadequate to solve physics problems. From this perspective, it can be inferred that the processes used to solve problems in an everyday context should be different from used to solve de-contextualized problems, that is, problems in a pure scientific context.

Therefore, the second focus of the current study is to compare the process of problem solving in an everyday context with that in a pure scientific context. To do this, students were asked to solve two types of physics problem, problems in an everyday context and de-contextualized problems, then a comparison was made of the additional cognitive or non-cognitive factors involved in solving the two types of problem. In particular, if a student failed to obtain the correct answer for an everyday context problem, it was determined whether or not the reason for the student's failure was related to the context.

Finally, an assessment was made of whether physics teachers considered the additional factors identified in the above analysis were important in physics teaching and learning or not. If the importance of these additional factors, for instance, reading skills or context-based judgement, is disregarded due to the fact that they are not included in traditional mainstream physics teaching and learning, then these factors will not be dealt with appropriately in formal school learning even though they may play an important role in solving everyday context problems. Therefore, the last research focus was to determine the implications for teaching, learning, and problem solving in an everyday context.

2. Procedures

The current study consisted of 3 steps: (1) investigating the general recognition of physics problems in an everyday context compared to de-contextualized problems, (2) exploring the characteristics of the processes of solving problems in an everyday context and identifying the cognitive/non-cognitive factors involved in the process of solving physics problems in an everyday context and de-contextualized physics problems, and (3) investigating whether physics teachers attach importance to the additional factors involved in the processes of solving problems in an everyday context.

As the first step, eight problems were developed consisting of 4 everyday context problems (designated as 'E' problems) and 4 de-contextualized problems (designated as 'D' problems). As such, there were 4 pairs of problems: E1-D1, E2-D2, E3-D3, and E4-D4. Each pair of problems incorporated the same concepts. The contexts of the 'E' problems were related to 'safety in everyday life', with two on 'auto safety' and the others on the 'safe use of home electricity'. One pair of problems (E1and D1) is described in Appendices 1and 2.

After reading and solving the problems, the subjects were asked questions to investigate their general recognition of the physics problems in an everyday context. The questions used in the first step of the study were designated as the 'A' questions and are described in Tables 1and 2.

Table 1. Number of Subjects in First Step

Subjects	Number
High School Students	20
High School Physics Teachers	10
University Physics Educators	9

Table 2. 'A' Questions used in First Step

Q. No.	'A' Questions
A-I	Can you easily distinguish the E problem from the D problem in this pair of problems?
A-II	Which type of problem do you think is better for physics problems?
A-III	Which type of problem would you prefer to solve as a physics test ?
A-IV	Which type of problem can you solve better ?
A-V	Which type of problem would you prefer to learn in a physics class ?

Subjects responded to each question (in Table 2) by marking a check on a five-point scale, comprised of –2, -1, 0, 1, and 2. For question A-I, the value of -2 corresponded to 'strongly no', while +2 corresponded to 'strongly yes'. For questions A-II, A-III, A-IV, and A-V, -2 corresponded to 'de-contextualized problem' and +2 corresponded to 'everyday context problem'. The 'A' questions were asked for each pair of problems.

As the second step, another 20 high school students were asked to solve the problems using the 'think aloud' method. Ten students solved 4 problems, that is, E1, E2, D3, and D4, while the other 10 students solved the remaining 4 problems, E3, E4, D1, and D2. It took each student about twenty minutes to solve 4 problems. If any student failed to obtain the correct answer, they were interviewed to explore the reasons for the failure. The major concerns of this second step interview are described in Table 3.

Table 3. Major Concerns in Second Step

No.	Concerns
I	Characteristics of processes used to solve E problems compared to D problems
II	Reasons for failure to obtain correct answers to E problems.
III	Additional cognitive/non-cognitive factors involved in processes of solving E problems compared to D problems

Based on observing how the students solved the given problems plus the ensuing interviews, a sequence of the problem solving processes was compiled, and the additional cognitive/non-cognitive factors involved in the processes of solving the problems in an everyday context identified and compared to those involved in solving the de-contextualized problems.

As the last step, another 16 physics teachers were asked to investigate what they thought of the additional factors identified in the second step. The questions used in this step were designated as the 'B' questions and are described in Table 4.

Table 4. 'B' Questions used in Third Step

No.	'B' Questions
B-I	Do you think students should have abilities related to these factors to solve physics problems ?
B-II	Do you think these factors should be taught in the physics classroom ?

From the results of these final questions, various educational considerations for teaching and learning physics are suggested.

3. Results

3.1 *General Recognition of Physics Problems in Everyday Context*

Table 5 summarizes the results of the 'A' questions used in the first step (see Table 2).

Table 5. General Recognition of Physics Problems

Q. No. (N=10)	Students			Teachers (N=10)	Educators (N=9)
	Male (N=10)	Female (N=20)	Sum		
A-I	0.85	1.30	1.08	1.25	0.94
A-II	0.68	0.35	0.51	1.1	1.34
A-III	0.95	-0.03	0.46	0.68	1.47
A-IV	-1.23	-1.08	-1.12	-1.33	-1.08
A-V	0.78	0.53	0.65	1.28	1.42

For question A-I, a positive value for the responses indicates 'yes', because the maximum value for the response is '2'. Therefore, the positive values for question A-I ranging from 0.85 to 1.25 mean that the given pair of problems could be appropriately discriminated.

For the other questions (A-II, A-III, A-IV, and A-V), a positive value for the responses indicates 'problems in an everyday context', while a negative value indicates 'de-contextualized problems'. Therefore, the responses for question A-II mean that the subjects generally thought that problems in an everyday context were better for physics problems than de-contextualized ones. The subjects also answered that they preferred to solve everyday context problems for physics tests in school (question A-III).

However, the general opinion was that de-contextualized problems were easier to solve than problems in an everyday context (question A-IV). The reasons given for this were that "too much time is needed to understand the situation in the E problems", "we are more familiar with D type problems", and "E type problems require us to apply physics concepts to a real situation, whereas D type problems are more simple and straightforward as they only require the use of the relevant formulas".

To compare their predicted and real performances with physics problems, 93 students, including the 20 students that participated in the first step, were randomly sampled and asked to solve the pairs of problems.
For each pair of problems, the students were asked to find the correct answers and describe how they came to those answers. If a student found the correct answers, yet could not clearly describe how they came to their conclusions, their responses were classified as 'undetermined' and excluded from the analysis.
Table 6 shows the results of the students' performance

Table 6. Percentage of Correct Answers

Everyday	Percentage	de-contextualized	Percentage	P
E1(N=47)	31.9	D1(N=46)	63.0	.002
E2(N=41)	87.8	D2(N=45)	88.9	.877
E3(N=40)	47.4	D3(N=47)	50.3	.251
E4(N=43)	72.1	D4(N=44)	50.0	.035

According to Table 6, for the first pair of problems E1-D1, more students gave a correct answer for D1 than E1, and this difference was statistically significant (p=.002). This result matches the prediction given for question A-IV in Table 5. However, there were no significant differences between the performances for E2 and D2, and E3 and D3. Moreover, for the last pair of problems, more students gave a correct answer for E4 than D4.

The result for the last pair is also consistent with other previous studies. Rennie & Parker (1996), who used two sets (real-life problems and abstract ones) of matched physics problems, observed that seven out of eight students performed better with the real-life problems compared to the abstract ones. Plus, as mentioned earlier, the APU performance was higher for the everyday context problems than the scientific context problems (Schofield, et al., 1989). However, not all research results show the same tendency that everyday context problems are easier to solve. For example, Song & Black (1992) observed that there was no performance difference between the use of an everyday context and scientific context in physics problems when concept application was required.

Therefore, it would seem that the effect of context on problem solving in physics is not consistent, i.e. context is not the only factor that determines student performance between everyday and scientific context problems.

From the responses to question A-V, it was found that many students, teachers, and educators wanted to teach and learn physics in an everyday context. The reasons given for this were : " because learning physics in an everyday context can closely connect physics to students' lives", "teaching in an everyday context can evoke students' interest in physics", and "we (students) can understand the basic concepts through D type problems, however, we should be able to apply these concepts to everyday life by ourselves".

3.2 *Processes of solving physics problems in everyday context*

Based on observing the procedures used by the 20 students to solve the given problems (two E problems and two D problems), the detailed processes involved in solving the problems were explored. Figure 1 shows a summary of the process used by a student to solve the E1 problem.

Student (male) reads the problem

→ underlines the key information (for instance, 19.4m/s, car B decreases speed, after 2 seconds)

→ rewrites the underlined information in the given figure

→ calculates the average velocity

→ obtains the correct result (that is, the speed of car B is above the speed limit)

→ however, selects answer '(3)', which is the wrong answer, that states that car A should have taken proper action because driver A could recognize that car B was driving above the speed limit.

Figure 1. Process of Solving E1 Problem

According to the above process, it could determined that the student was able to extract the important information from the problem situation, apply the correct mathematical process, and obtain the result that the speed of car B was above the speed limit. However, the student did not conclude that car B was wrong in the traffic accident due to a personal judgment that "car A should have taken appropriate action because driver A could recognize that the speed of car B was above the speed limit" As a result, the student chose the wrong answer that "car A was wrong".

From this analysis, an additional factor involved in the process of solving E type problems compared to D type problems was identified, i.e. "*some students try to import personal judgments which are not presented in the problems*". Personal judgments imported by problem solvers were also frequently observed in other cases. Figure 2 is another example of a process used to solve the E1 problem.

In this case (Fig. 2), several factors were identified as blocking the student from finding the correct answer. First, the student did not understand the physical concepts of average velocity and instantaneous velocity. However, even after becoming aware of the meaning of these concepts, there was still a problem understanding the situation or context of the problem. Therefore, only after listening to several explanations about the situation was the student able to solve the problem correctly.

When the researcher asked the student why this problem was so difficult to solve, the answer was, "the situation of the traffic accident was vague" and "it was difficult to understand the meaning of the sentence '*instantaneous velocity will exceed 20m/s*', because no exact value was indicated."

Student (female) reads the problem
→ underlines the numerical values
→ responds "I don't understand what the problem means"
→ (after being encouraged) asks for an explanation of the meaning of average velocity and instantaneous velocity
→ (after listening to a brief explanation)
→ responds again "I can't grasp the situation in which two cars collide with each other"
→ (after listening to an explanation about this situation)
→ calculates the two velocities and obtains the correct results
→ chooses the correct answer

Figure 2. Another Process for Solving E1 Problem

From this analysis of the process used to solve the problem, plus the interview with the problem solver, two additional factors involved in the process of solving an E type problem were identified; first, "*some students did not read the long sentences of the problem carefully, or did not understand the meaning of the situation or context of the problem*", and second, "*some students did not understand the information given, because the numerical values of the information included a range or were not in a simple form, such as integer*".

The first factor identified in Figure 2 was also easily observed in other cases, for example, some students responded that, "Long sentences and complex situations strain my nerves", or "Due to the long sentences, it was hard to make sense of the situation related to the problem."

As regards the second factor, it is generally difficult to determine an exact value for certain variables in real life situations. For example, the electric resistance of the human body naturally varies according to the body's condition, therefore, the value of the current flowing through the human body usually has a certain range rather than an exact value. However, when solving the E2 problem where the context is related to '*electric shock*', some students responded that, "Here (indicating the problem), I read that '*1-2mA of electric current flows through the human body when it is shocked by 220V of electricity for 0.1 seconds*', however, I could not decide exactly how much current flowed through the human body."

After exploring all the processes used by the students to solve the problems, six additional factors, including the 3 factors mentioned earlier, were identified when solving the everyday context problems, as indicated in Table 7.

Table 7. Cognitive/Non-cognitive Factors Involved in Problem Solving in Everyday Context

No.	Factor
I	Excluding personal judgments not presented in the problem.
II	Reading sentences carefully, and understanding the meaning of the situation or context of the problem.
III	Treating variable values effectively, for example, values with a range or not in a simple form, such as an integer.
IV	Extracting important descriptive information from long sentences.
V	Neglecting information irrelevant to solving the problem.
VI	Being able to deal with a new type of problem, which is not in the usual form used for a school performance test.

Explanations have already been given above for Factors I, II, and III (related to Figures 1 and 2). When students failed to solve the problems correctly, they were allowed to try again. This time, the students generally recognized the information missing from the earlier attempt, and obtained the correct answer. For instance, as an example related to factor IV, some students responded that, "(during their second attempt to solve the E1 problem) I missed '*that car B decreased its speed!*'", while other students said that, "I did not realize that the sentence ('--- *to use electricity safely*') was important when I tried to solve it the first time."

Problems in a scientific context are usually idealized, therefore, the conditions or assumptions are simple and explicit. However, in an everyday context, many inexplicit conditions are included. Moreover, irrelevant information to problem solving is also included. In fact, several students that participated in the current study asked about irrelevant aspects while solving the problems. For instance, when solving the E1 problem, some students asked "What happened to the two cars after they collided with each other?", and for the E3 problem, which is related to safe driving on an expressway veiled in mist, some students asked , "Why did they step on the brake hard?" Therefore, the 5th factor was identified from these responses.

Finally, some students commented that, "this problem was difficult for me, because the style of the problem was unfamiliar", or "I was afraid to solve the problem, because it was different from the usual problems in school tests." As such, the 6th factor was identified from these responses.

3.3 *Teacher Recognition of Additional Cognitive/Non-cognitive Factors*

Table 8 shows the responses to the 'B' questions, described in Table 4. As mentioned earlier, the percentage values range from –2 to +2, and the values +2, 0, and -2 represent 'strongly yes', 'neutral', and 'strongly no', respectively.

Table 8. Teacher Recognition (N=16) of Additional Cognitive/Non-cognitive Factors

Factor No.	Question No.	Percentage	p[1)]	p[2)]
I	B-I	1.13	.000	.252
	B-II	0.75	.003	
II	B-I	1.69	.000	.013
	B-II	0.75	.041	
III	B-I	1.50	.000	.167
	B-II	1.06	.000	
IV	B-I	1.75	.000	.061
	B-II	1.13	.002	
V	B-I	1.06	.002	.122
	B-II	0.37	.270	
VI	B-I	1.63	.000	.043
	B-II	1.13	.000	

1) whether the value of each percentage differed from 0.
2) whether the two values for the percentage of B-I and B-II differed from each other.

In Table 8, percentage of responses to question B-I related to Factor I was 1.13 and the first probability was 0.000. This means that the physics teachers generally thought that problem solvers should exclude personal judgements when solving problems in an everyday context.

For question B-II related to Factor I, the percentage value of 0.75 indicates that the teachers agreed that Factor I should be taught in the classroom. Even though it would appear that the values of 1.13 and 0.75 are different, they are not as the second probability is 0.252. Therefore, it was concluded that the physics teachers thought that students should have the ability described in Factor I, and also thought that Factor I should be taught in the classroom.

According to the percentages for all the B-I questions in Table 8, almost all the physics teachers thought that students should have the abilities related to the 6 additional factors involved in solving the everyday context problems. Plus, the order of importance of these factors was considered to be IV>II>VI>III>I>V. Similarly, for Factors III and IV, the physics teachers thought that students should be able to treat complex forms of variable values in an everyday context (Factor III) and extract important descriptive information from long sentences (Factor IV). However, not all teachers agreed that those factors should be taught in the physics classroom.

Whereas, for Factor V, although the physics teachers thought that students should be able to ignore irrelevant information when problem solving, they gave a neutral response when asked, "Do you think that this factor should be taught in the physics classroom?" (percentage=0.37 and p=0.270).

Finally, for Factors II, V, and VI, the physics teachers thought that students should have the abilities related to these factors, and also said that these factors should be taught in the physics classroom. However, the tendency between the two answers was significantly different, that is, the degree of response about student abilities related to these factor was stronger than the responses about teaching it.

4. Educational Implications and Conclusion

The current study found that the student subjects: (1) could discriminate everyday context problems from de-contextualized problems, (2) thought that everyday context problems were better problems, (3) preferred to solve everyday context problems as a school test, (4) wanted to learn physics in an everyday context, and (5) believed de-contextualized problems were easier to solve. However, according to the real performance results, no significant difference was noted related to the problem context. Second, based on observing the problem solving processes used by the students, six additional factors were identified as influencing the problem solving in an everyday context. These factors can be summarized as follows: even though the form of the problem in an everyday context may be unfamiliar to students, students need to read the problem carefully, understand the meaning of the situation in the problem, extract important information while neglecting irrelevant information, be able to treat complex variables that have ranges and are not in a simple form, such as an integer, and finally draw a conclusion without personal judgements.

As regards the above factors, most of the physics teachers questioned thought that these factors were important in problem solving in an everyday context, and that students should have abilities related to these factors. However, not all the teachers agreed that these factors should all be taught in the physics classroom. For instance, although the physics teachers agreed that students should be able to neglect information irrelevant to solving the problems, they did not feel that this ability should be taught in the physics classroom.

Accordingly, the current study was able to clarify the opinions of students, physics teachers, and physics educators at university on the use of everyday contexts in physics problem solving, while also identifying specific factors required to improve physics problem solving in everyday contexts. Therefore, it is hoped that these results can provide valuable insights and direction in the development of concrete teaching strategies for teaching and learning physics in an everyday context.

References

1. L. Keeves and G. Aikenhead, Science curricula in a changing world, In B.J. Fraser and H.J. Walberg (eds.), Improving Science Education (pp.13-45). IL: The University of Chicago Press, 1995.
2. J. Song and J. Choi, Students' preference on different contexts in learning basic concepts of mechanics, Physics Education, The Korean Physical Society, 12(2), 82-87, 1994.
3. D. Betty and L. Fred, Liked and disliked learning activities: Responses of swazi students to science materials with a technological approach, Research in Science & Technological Education, 14(2), 221-236, 1996.
4. F. Reif and J.H. Larkin, Cognition in scientific and everyday domains: Comparison and learning implications, Journal of Research in Science Teaching, 28(9), 733-760, 1991.
5. B. Schofield, J. Bell, S. Johnson and P. Murphy, A. Qualter and T. Russell, Science at age 13: Review of APU survey findings 1980-1984, London: HMSO, 1989.
6. W.L. Saunders and J. Jesunathadas, The effect of task content upon prepositional reasoning, Journal of Research in Science Teaching, 25(1), 59-67, 1998.
7. J. Song and P. Black, The effects of concept requirements and task contexts on pupils' performance in control of variables, International Journal of Science Education, 14(1), 83-93, 1992.
8. L.J. Renni and L.H. Parker, Placing physics problems in real-life context: Students' reactions and performance. Australian Science Teachers Journal, 42(1), 55.

Appendix 1 : E1 Problem

At an intersection without traffic signals, car A, turning left, collided with car B, which was going straight.
The following diagram and information were obtained by a policeman.

<Traffic Rules>
* Speed limit for the road was 19.4m/s (70km/h)

<Information obtained by policeman>
* As car B passed under the footbridge, car A started to turn left.
* After passing the footbridge, car B started to decrease its speed.
* 2 seconds after car A started to turn left, car A collided with car B.

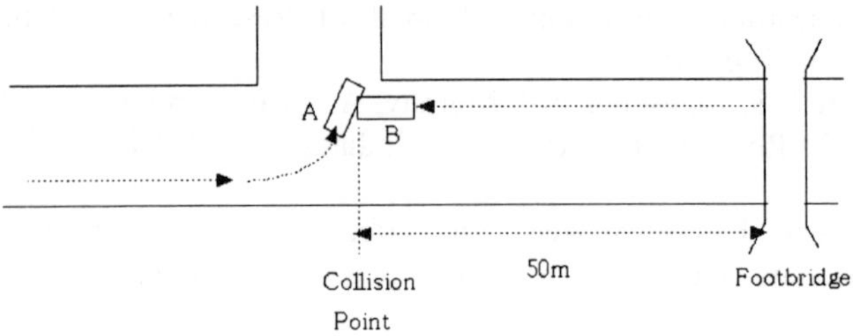

Choose all the correct answers among the following:
(1) The average velocity of car B from the footbridge to the collision point was 20m/s.
(2) The instantaneous velocity under the footbridge exceeded 20m/s.
(3) Car A caused the traffic accident.
(4) Car B caused the traffic accident.

Appendix 2 : D1 Problem

Object moves from point A to point B in 2 seconds, while decreasing its speed.

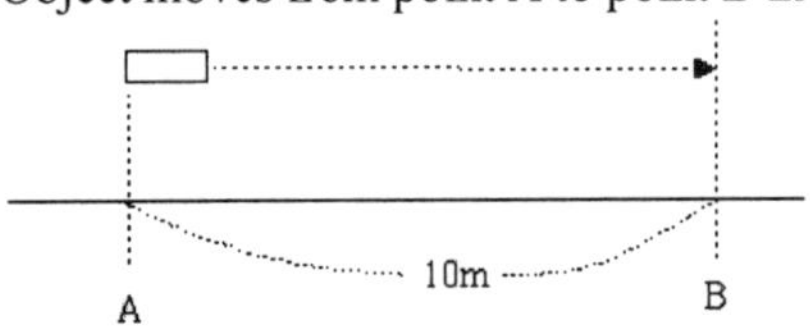

Choose all the correct answers among the followings:
(1) The average velocity of the object from A to B was 5m/s.
(2) The instantaneous velocity at point A exceeded 5m/s.

CRITICAL THINKING AS A STRATEGY TO PROMOTE CONCEPTUAL CHANGE AND TO ENHANCE BETTER PHYSICS PROBLEM SOLVERS

MAURÍCIA OLIVEIRA & ALICE RODRIGUES
Center for Educational Research, School of Sciences, University of Lisbon
Campo Grande, C4-Piso 1, 1749-016 Lisbon, Portugal

To enlighten the relation between problem solving, critical thinking and conceptual change, a study was outlined to find out if enhancing conceptual change by developing critical thinking can be a strategy to promote better physics problem solvers. The research model adopted was a quasi-experimental pretest/posttest, control group design, no random. The experimental and control group were each one constituted by 93 students from the 8th grade with successful academic achievement and six pre-service teachers of physics. The experimental group performed during one and a half months a set of physics activities. They were developed and implemented taking into account some strategies considered promoters of conceptual change by being facilitators of the development of critical thinking. During the same time, the control group performed physics activities not intentionally developed with the same purpose but based on the same contents. Several sources of data were used to collect data. The results seem to point out that the teacher's practices promoters of conceptual change by being facilitators of the development of critical thinking seem to be important factors that can influence if a pupil is a good or a week problem solver. Some implications of the study can be drawn for science education.

1. Theoretical Understandings and Significance of The Study

Problem solving and critical thinking are two areas of research that are a major concern of physics teaching in the 21st century. In fact, two of the major goals of science education are to develop the students' skills related with problem solving [1, 2, 3 and 4] and the critical thinking skills [5, 6 and 7].

The importance of these two areas of research arose from the need to prepare our students to become citizens able to live in a society dominated by science and technology. It cannot be denied that the information and new technologies increase in an exponential way in our society. So, it is impossible to teach in our schools all the information that it will be necessary in a life long. However, in school we can teach not only scientific facts and concepts, but also develop student's skills in order to make them able to think in a critical way about the world around them. Namely to be able to select the information needed to solve problems and make rational decisions about their own future and the future of mankind.

In Portugal, as in many other countries, critical thinking and problem solving are, besides areas of research, also two important science teaching concerns. The official programs of physics state it very clearly. Nevertheless, the results of the Third International Mathematics and Science Study (TIMSS) [8] reveal that the outcomes of Portuguese students in science, including physics are weak. The students reveal difficulties in solving problems, even mere exercises, especially those of physics and chemistry [9, 10]. For example, it is not unusual that students enunciate solutions without a physical mean: the mass of the moon bigger than the mass of the earth, negative values for mass and so on.

They have difficulties probably because they do not know the routines and mainly, how to use critical thinking skills when demanded. In spite of the current reforms in science education state as a very important need that students understand science conceptually, rather than memorise facts [11], the memorisation of facts is still privileged [12]. Even the laboratory activities are too often "cookbook" style, requiring little more of students than the ability to follow the steps in a certain order [13].

But, another source of difficulties in solving problems is the misconceptions that students have regarding the topic of the particular problem they are trying to solve.

Research results reveal that school children can proceed through their school careers retaining misconceptions about many science concepts [14, 15 and 16]. In fact, the children's ideas are stable and resistant to change [17, 18]. As a consequence, teachers cannot neglect the children's ideas or constructs. They must be aware of the student's misconceptions and try to conceive strategies for conceptual change. As Cachapuz [19], stated "the teacher must help students to change conceptual structures and doing so contribute for the reorganization of students concepts in a qualitatively different way". But, how can teachers promote conceptual change? According to Hewson, Beeth and Thorley [20] when teaching for conceptual change, it is necessary that the range of ideas held by different people in the class related to the topic are made explicit. Those ideas should be discussed in classroom leading students to choose between ideas. But, to choose and discuss involves critical thinking skills. To discuss, for example, demands critical thinking skills such as presenting a position (argumentation), interacting with others and asking for clarification.

Furthermore, Akerson et al. argue that "developing understanding presupposes that students are actively engaged with the ideas of science and that scientific understanding can be gained through inquiry instruction generated from student experiences" [14]. Again, inquiry instruction underlies the promotion of critical thinking skills from the different areas of critical thinking that is elementary clarification, basic support, inference, advanced clarification and strategy and tactics. As a consequence, we believe that the development of critical thinking can probably be a way to promote conceptual change.

Still, according to Ennis [21, 22, 23 and 5], the activity of problem solving also demands critical thinking skills, mainly those skills inherent to the strategic and tactics area (Appendix A). Consequently, it can be possible that the developing of critical thinking can not only promote conceptual change but also contributes to enhance better physics problem solvers.

Based on the previous theoretical framework, a research study was developed to find out if enhancing conceptual change by developing critical thinking could be a strategy to promote better physics problem solvers.

2. Methodology

2.1 *Sample*

The research model adopted was a quasi-experimental pretest/posttest, control-group design, no random. The research was carried out with 8th grade Portuguese students and their teachers of physics from four different schools. The experimental and control groups were constituted selecting randomly two of thosc schools.

The sample consisted of 277 students from the 8th grade of the four schools. They were enrolled in 12 different classes, each one with a teacher. Their teachers of physics were pre-service teachers. The teachers of the experimental group were selected because they were aware of the importance of critical thinking, conceptual change and problem solving in physics education. Also, they had previously developed physics or chemistry activities intended to promote critical thinking. As suggested by Finegold and Mass [24] only students with academic achievement were selected to guarantee that a week performance as problem solvers could not be explained by a lack of basic knowledge in physics. A sub sample of 186 students was constituted. Being so the experimental and control group were each one constituted by 93 students with successful academic achievement and six pre-service teachers of physics.

2.2 *Treatment*

The experimental group performed during one and a half months a set of physics activities based on electricity. The activities and strategies to be implemented were conceived by the teachers of the experimental group in collaboration with the researchers, in the two and a half months that precede the intervention program, becoming a part of the teacher training process. The process involved discussing and sharing ideas, making decisions together and sometimes negotiating points of view.

As the aim was to find out if promoting conceptual change by enhancing critical thinking could be a strategy to promote better physics problem solvers, the activities emphasised the use of the critical thinking skills underlying an inquiry instruction approach. Accordingly, they were implemented also taking into account some of the strategies potentially promoters of conceptual change by being facilitators of the development of critical thinking. Those strategies are: giving students time to think and answer the questions; students working in pairs or in groups of tree elements; asking students to communicate orally the work they were developing; discussing polemic questions, asking students to write summaries; asking students to think aloud; promoting students to ask questions and to elicit problem solving activities. The activities were created using the Ennis critical thinking taxonomy (appendix A) as a template, according to a methodology assessed previously by Tenrreiro-Vieira [25].

During the same time, the control group performed physics activities not intentionally developed with the same purpose, but based on the same contents.

2.3 *Procedures and Instruments*

To ensure the validity and reliability of the inferences, varied sources of data were used. The sources were: the Cornell Critical Thinking Test-Level X [26]; a set of problems selected according to some theoretical criteria; a checklist; audio and observation records; students' record sheets; final grades in science, a list identifying the good and week problem solvers and a criteria list for analysing the problems solved by the students (appendix B).

At the beginning and end of the study, the Cornell Critical Thinking Test (level X) was applied to the experimental and control groups. Also the student's misconceptions concerning each problem were identified. Yet, the teachers identified the week and good problem solvers. Besides, each teacher used a checklist to identify which were the phases of the problem solving process followed by each student. This checklist was developed taking into account some phases of process followed by the good problem solvers [27, 28, 29, 30 and 24]. Still, the students with academic achievement were identified.

In the end of the study, the students solved a set of three physics problems focusing curriculum contents of. Electricity. Those three problems were selected from a pool of problems. They were considered adequate to be solved by the students from the teacher's point of view. Yet, based on the student's record sheets, the student's performance as problem solvers was evaluated using the checklist mentioned before.

Nevertheless, the description of the phases was adapted to each problem. The evaluation took into account not only the process but also the final result presented by each student. Also the teachers solve the problems and the same procedures were made. All data collected were used to distinguish between week and good problem solvers at the end of the study.

3. Findings and Implications of The Study

The results point out that the students of the experimental group became better physics problem solvers regarding not only the process but also the product. Deduction and Induction seem to be two aspects of critical thinking that have a major importance in the promotion of better physics problem solvers.

Also, the results seem to point out that the teacher's practices promoters of conceptual change by being facilitators of the development of critical thinking seem to be important factors that can influence if a student is a good or a week problem solver. Furthermore, when teachers are not aware of the importance of promoting critical thinking or when they do not know how to do it, they tend to ignore students misconceptions and they do not promote conceptual change. However, the analysis of the student's record sheets reveals that student's have misconceptions related with the concepts of the problems. But, teachers also have misconceptions regarding the same concepts. Even, a comparative analysis shows that some of the student's misconceptions are those of the teachers.

Yet, the analysis of data shows that teacher's misconceptions influence the selection of physics problems to be solved by their students and also the activities they do.

The analysis of the curriculum materials selected by the teachers belonging to the control group and the observations carried out during their classes show that the teachers practices concerning the activities and the strategies that they implement are much more centered in the teacher then those of the experimental group. Even when it seems that the activities are students centered their purpose is not attained due to the strategies they implement. This happens even if accidentally the activities selected or conceived by the teachers of the control group are promoters of critical thinking.

Summarizing some of the implications for physics teaching of this study it seems that to enhance better physics problem solvers, the teachers should, namely: implement activities promoters of critical thinking; emphasize the use of activities promoters of skills from the strategies and tactics area of critical thinking, according to the Ennis critical thinking taxonomy [23]; give a major importance to Deduction and Induction in all activities; be aware of the students misconceptions and of their own; avoid to reinforce students misconceptions and to increase those with their own; implement strategies facilitators of the development of critical thinking as well as of conceptual change.

References

1. W. M. Roth, Short-Term Memory and Problem Solving in Physical Science, *School Science and Mathematics,90*(1.4), 271-282, 1990.
2. E. L Pizzini; D. S. Shepardson, and S. K. Abell, A Rationale for and the Development of a Problem Solving Model of Instruction in Science Education, *Science Education, 73*(5), 523-534, 1989.
3. J. Stewart, Two Aspects of Meaningful Problem Solving in Science, *Science Education, 66*(5), 731-749, 1982.

4.. A. B. Champagne and L. E. Klopher, A Sixty-Year Perspective on Three Issues in Science Education: I. Whose Ideas are Dominant?; II. Representation of Women; III. Reflexive Thinking and Problem Solving, *Science Education, 61*(4), 431-452, 1977.
5. R. H. Ennis, *Critical Thinking*, Prentice-Hall, New Jersey, 1996.
6. A. Zohar and P. Tamir, Incorporating Critical Thinking into a Regular High School Biology Curriculum, *School Science and Mathematics, 93*(3), 136-140, 1993.
7. J. G. Kurfiss, *Critical Thinking: Theory, Research, Practice, and Possibilities. ASHE-ERIC Higher Education Report No. 2,* (Report No. ISBN-O-913317-44-6), Association for the Study of Higher Education, Washington, D.C., 1988.
8. TIMSS, Highlights of results from TIMSS: Middle School Years, *Publications Page.* |on-line|, Available: http://wwwcsteep.bc.edu/timss, 1998.
9. N. Cruz and M. O. Valente, Estratégias Metacognitivas e Resolução de Problemas: Um Estudo com Alunos do 11° ano [Metacognitive Strategies and Problem Solving: A Study with 11 Grade Students], *Revista de Educação*, pp. 87-103, 1993.
10. J. B. Lopes, *Resolução de problemas em física e química: modelo para estratégias de ensino-aprendizagem [Problem Solving in Physics and Chimestry: A Model for Instructional Strategies]*, Texto Editora, Lisboa, 1994.
11. National Research Council, *National Education Standards*, National Academic Press, Washington, D. C., 1996.
12. K. Tobin; D. Tippins and A. Gallard, Research on Instructional Strategies for Science Teaching, in D. Gabel; P. Blosser & D. Butts (eds.), *Handbook of Research on Science Teaching and Learning,* pp. 45-93, Macmillan, New York, 1994.
13. J. B. Tugel, Pollution, pH, & Problem Solving. *The Science Teacher*, 21-25, 1994.
14. V. L. Akerson; L. B. Flick and N. G. Lederman, The Influence of Primary Children's Ideas in Science on Teaching Practice, *Journal of Research in Science Teaching, 37(*4), 363-385.
15. B. A. Bishop and C. W. Anderson, Student Conceptions of Natural Selection and its Role in Evolution, *Journal of Research in Science Teaching, 27,* 415-427, 1990.
16. M. Hashweh, Descritive studies of Students' Conceptions in Science, *Journal of Research in Science Teaching, 25,* 277-299, 1988.
17. R. Driver; E. Guesne and A. Tiberghin, *Children's Ideas in Science,* Open University Press, New York, 1985.

18. F. Vasconcelos and N. Costa, O Trabalho Experimental no Ensino de Tópicos de Electricidade (8° ano) [The experimental Work in the teaching of Electricity Topics (8th grade)], *22. Gazeta de Física,* vol. 3, pp. 11- 15, 1999.
19. A. F. Cachapuz, Ensino por Mudança Conceptual [Teaching for Conceptual Change]. In A. F. Cachapuz (Ed.), *Fomação de Professores - Ciências: Perspectivas de ensino [Teacher Training in Science: Views in Education],* pp.19-44, Centro de Estudos de Educação em Ciência, Porto, 2000.
20. P. W. Hewson; M. E. Beeth and N. R. Thorley, Teaching for Conceptual Change, in B. J. Fraser and Tobin (eds.), *International Handbook of Science Education,* vol. 1, PP. 199-218, Kluwer Academic Publishers, Boston, 1998.
21. R. H. Ennis, A Logical Basis for Measuring Critical Thinking Skills, *Educational Leadership, 43*(2), 44-48, 1985a.
22. R. H. Ennis, Goals for a Critical Thinking Curriculum, in A. L. Costa (ed.), *Developing minds: a resource book for teaching thinking,* pp. 54-57, Association for Supervision and Curriculum Development, Alexandria, 1985b.
23. R. H. Ennis. A Taxonomy of Critical Thinking Dispositions and Abilities, in J. B. Baron e R. J. Sternberg (eds.), *Teaching thinking skills: Theory and practice*, Freeman and Company, New York, 1987.
24. M. Finegold and R. Mass, Diferences in the Processes of Solving Physics Problems Between Good Physics Problem Solvers and Poor Physics Problem Solvers, *Research in Science & Technological Education, 3*(1), 59-67, 1985.
25. M. C. Tenrreiro-Vieira, *Pensamento crítico na educação científica: Proposta de uma metodologia para a elaboração de actividades curriculares.* [Critical Thinking in Science Education: a Frame Work to Develop Curriculum Activities]. Unpublished Master thesis, School of Sciences, University of Lisbon, Portugal, 1994.
26. R. H. Ennis and J. Millman, *Cornell Critical Thinking Test:Level X,* Midwest Publications, Pacific Grove, 1985.
27. S. L. Helgeson, Research on Problem Solving: Middle School, in D. L. Gabel (ed.), *Handbook of research on science teaching and learning,* pp. 327-354, National Science Teachers Association, Washington, D.C., 1994.
28. R. H. Barba, Problem Solving Pointers: Techniques that Can be Taught, *The Science Teacher, 57*(7), 32-35, 1990.
29. J. Stewart, and J. Van Kirk, Understanding and Problem-Solving in Classical Genetics, *International Journal of Science Education, 12*(5), 575-588, 1990.
30. D. R. Woods, *What research says to science teacher: problem solving.* National Science Teachers Association, Washington D.C., 1989.

Appendix A : Ennis Critical Thinking Taxonomy[1] [23]

WORKING DEFINITION: Critical thinking is reasonable reflective thinking that is focused on deciding what to believe or do. Critical thinking so defined invokes both dispositions and abilities:

A. DISPOSITIONS:

1. Seek a clear statement of the thesis or question
2. Seek reasons
3. Try to be well-informed
4. Use credible sources and mention them
5. Take into account the total situation
6. Try to remain relevant to the main point
7. Keep in mind the original and or basic concern
8. Look for alternatives
9. Be open-minded
   a. Consider seriously, other points of view than one's own ("dialogical thinking")
   b. Reason from premises with which one disagrees-without letting the disagreement interfere with one's reasoning ("suppositional thinking")
   c. Withhold judgment when the evidence and reasons are insufficient
10. Take a position (and change a position) when the evidence and reasons are sufficient to do so
11. Seek as much precision as the subject permits
12. Deal in an orderly manner with the parts of a complex whole
13. Be sensitive to the feelings, level of knowledge, and degree of sophistication of others (2)

B. ABILITIES: (Classified under these categories: Elementary Clarification, Basic Support, Inference, Advanced Clarification, and Strategy and Tactics)

Elementary Clarification

1. Focusing on a question
   a. Identifying or formulating a question
   b. Identifying or formulating criteria for judging possible answers
   c. keeping the situation in mind
2. Analyzing arguments
   a. Identifying conclusions, b. Identifying stated reasons
   c. Identifying unstated reasons d. Seeing similarities and differences
   e. Identifying and handling irrelevance
   f. Seeing the structure of an argument g. Summarizing

3. Asking and answering questions of clarification and/or challenge, for example:
   a. Why?
   b. What is your main point?
   c. What do you mean by
   d. What would be an example?
   e. What would not be an example (though close to being one)?
   f. How does that apply to this case (describe case, which might well appear to be a counter example)?
   g. What difference does it make?
   h. What are the facts?
   i. Is this what you are saying: _________?
   j. Would you say some more about that?

Basic Support

4. Judging the credibility of a source; criteria:
   a. Expertise
   b. Lack of conflict of interest
   c. Agreement among sources
   d. Reputation
   e. Use of established procedures
   f. Known risk to reputation
   g. Ability to give reasons
   h. Careful habits
5. Observing and judging observation reports; criteria:
   a. Minimal inferring involved
   b. Short time interval between observation and report
   c. Report by observer rather than someone else (i.e., not hearsay)
   d. Records are generally desirable. It report is based on a record, it is generally best that:
      1) The record was close in time to the observation
      2) The record was made by the observer
      3) The record was made by the reporter
      4) The statement was believed by the reporter, either because of a prior belief that the observer was habitually correct
   e. Corroboration
   f. Possibility of corroboration
   g. Conditions of good access
   h. Competent employment of technology, if technology is useful
   i. Satisfaction by observer (and reporter, if a different person) of credibility criteria *(#4* above)

Inference

6. Deducing, and judging deductions
   a. Class logic-Euler circles
   b. Conditional logic
   c. Interpretation of statements
      1. Double negation
      2. Necessary and sufficient conditions
      3. Other logical Words: "only'", "if and only if", "or", "some", "unless", "not" not both", etc.
7. Inducing and judging inductions

a. Generalizing
   1) Typically of data: limitation of coverage   2) Sampling
   3) Tables and graphs
b. Inferring explanatory conclusions and hypotheses
   1) Types of explanatory conclusion, and hypotheses
      a) Causal claims
      b) Claims about the beliefs and attitudes of the people
      c) Interpretation, of author, intended meanings
      d) Historical claims that certain things happened
      e) Reported definitions
      f) Claims that something is an unstated reason or unstated conclusion
   2) Investigating
      a) Designing experiments, including planning to control variables
      b) Seeking evidence and counterevidence
      c) Seeking other possible explanations,
   3) Criteria: Given reasonable assumptions
      a) The proposed conclusion would explain the evidence (essential)
      b) The proposed conclusion is consistent with know facts (essential)
      c) Competitive alternative conclusions are inconsistent with known facts (essential)
      d) The proposed conclusion seems plausible (desirable)

8. Making and judging value judgment
   a. Background tactics   b. Consequences
   c. *Prima facie* application of acceptable principles
   d. Considering alternatives   e. Balancing, weighing, and deciding

Advanced Clarification

9. Defining terms, and judging definitions: three dimensions:
   a. Form
      1) Synonym   2) Classification
      3) Range   4) Equivalent expression
      5) Operational   6) Example-nonexample
   b. Definitional strategy
      1) Acts
         a) Report a meaning ("reported" definition)
         b) Stipulate a meaning ("stipulative" definition)
         c) Express a position on an issue ("positional", including "programmatic" and "persuasive" definition)
      2) Identifying and handling equivocation
         a) Attention to the context
         b) Possible types of response:
            i) "The definition is just wrong" (the simplest response)

ii) Reduction to absurdity: "According to that definition, there is an outlandish result"
iii) Considering alternative interpretations: On this interpretation, there is this problem: on that interpretation, there is that problem
iv) Establishing that there are two meanings of key term, and a shift in meaning from one to the other

c. Content

10. Identifying assumptions
a. Unstated reasons
b. Needed assumptions: argument reconstruction

Strategy and Tactics

11. Deciding on a action
a. Define problem
b. Select criteria to judge possible solutions
c. Formulate alternative solutions
d. Tentatively decide what to do
e. Review, taking into account the total situation, and decide
f. Monitor the implementation

12. Interacting with others
a. Employing and reacting to "fallacy" labels (including)
1) Circularity
2) Appeal to authority
3) Bandwagon
4) Glittering term
5) Name calling
6) Slippery slope
7) Post hoc
8) Non sequitur
9) Ad hominem
10) Affirming the consequent
11) Denying the antecedent
12) Conversion
13) Begging the question
14) Either-or
15) Vagueness
16) Equivocation
17) Straw person
18) Appeal to tradition
19) Argument from analogy
20) Hypothetical question
21) Oversimplification
22) Irrelevance

b. Logical strategies
c. Rhetorical strategies
d. Presenting a position, oral or written (argumentation)
1) Aiming at a particular audience and keeping it in mind
2)Organizing (common type: main point, clarification, reasons, alternatives, attempt to rebut prospective challenges, summary - including repeat of main point)

APPENDIX B : Checklist

Phases of the Problem Solving Process Followed by the Students

1. Identifying easily the nature of the problem
2. Analysis the problem situation
3. Seeks to clarify the problem
4. State by his/her own words the problem
5. Plans procedures
6. Makes predictions
7. Changes easily from one solving method to another
8. Checks if the solutions fulfill the problem requirement

Notes

[1] This is only an overall content outline. It does not incorporate suggestion, for level, sequence, repetition in greater depth, emphasis or infusion in subject matter area (which might be either exclusive or overlapping).

2. Item 13 under "Dispositions" is not strictly Speaking a critical thinking disposition. Rather it is a social disposition that is desirable for a critical thinker to have.

EFFECTS OF ESTIMATION ACTIVITIES ON PHYSICS PROBLEM SOLVING PROCEDURES

JUNGAH SUH, JINWOONG SONG, KWANGHEE JO & SUNG-JAE PAK

Dept. of Physics Education, Seoul National University, Gwanakgu, Seoul

The current study investigated the effects of estimation activities, aimed at helping students compare and estimate, on problem-solving procedures. Previous research has reported that students solve physics problems without thinking about their physical meanings. Therefore, it was considered that if students could develop a better understanding of quantities and an ability to estimate them properly, this would help students solve problems while considering their physical implications. One hundred and ninety two 7th graders participated in the current study, and five students were individually interviewed. Before the estimation activities, most interviewees calculated the problems without understanding the related concepts. However, after the estimation activities, some of the students learned to relate the problems to the physical world. Furthermore, the estimation activities helped the students develop the ability to predict the results of the problems. Accordingly, estimation activities would appear to have the potential to enhance students' problem-solving procedures.

1. Introduction

Problem solving has already been emphasized by many science educators[1, 2, 3], and the real world is abundant in physics problems. However, students do not seem to be good at problem solving. For example, McMillan and Swadener reported that even if students solve a problem, this does not mean that they understand the problem qualitatively[4]. In addition, Crane reported that most problems used in schools are not closely related with the real world, which means that physics problems are apt to become mere exercises unrelated to the physical mind[5].

Science educators have studied the reasons why students have difficulties in solving such problems. Research on the differences between novice and expert problem-solving procedures can also be included in this area[6, 7, 8]. For example, Larkin and Reif revealed that the major difference between a novice and an expert is the expert's construction of a qualitative physical description(see also [6]). Whereas a novice usually proceeds directly from the original description to a mathematical description without any qualitative thinking. In addition, several researchers have found that the knowledge structure of the problem solver also causes differences[9, 10]. Thro reported 'Thus, the subject's ability in solving problems seemed to vary directly with the establishment of related concepts in their cognitive structure(see also [9], p.976). Furthermore, de Jong and Ferguson-Hessler argued that just having knowledge was insufficient; it must be organized in a useful manner[11].

According to such previous research, students who learn a new physics concept and then are confronted with a problem generally have difficulty in thinking about the problem qualitatively. Therefore, in a science classrom, students should have their knowledge constructed in a way that is easy to use in a problem solving situation.

There have been several studies that have examined the effect of pedagogical intervention on the development of novice problem-solving skills[12, 13]. However, these studies mainly emphasized the development of several kinds of problem-solving strategies, including qualitative problem analyses. Whereas, there have been relatively few studies that have emphasized changing student knowledge to enhance their problem-solving skills.

Accordingly, the current study attempted to change student knowledge from being useless in problem-solving situations into being useful through estimation activities. This assumption was based on the nature of estimation activity and suggestions by previous researchers.

An estimation activity can change an abstract meaningless physical concept into a meaningful realistic concept. For instance, the size of a proton, 10^{-15} m cannot be easily understood intuitively. However, 'if we regard the size of an atom as a playground and the size of a proton as a grain of sand', this is more easy to understand. Such a comparison is a kind of estimation activity, which allows the physical meaning of numbers written on a sheet of paper to come alive. In a Korean science class, more than 50% of the students were found to estimate the average density of an apple as 100 g/cm^3 or more, which is quite close to the mass or volume of the object[14]. However, if they learned that the density of an apple was less than 1g/cm^3, their concepts might be challenged. Schofield suggested that if pupils estimate that the volume of water filling a beaker is 10 cm, or 100 g or even 5 cm^3, this reveals quite important information on their concept of volume[15]. As such, since estimation is closely related to the estimator's concepts, the current study assumed that estimation activities may help students understand concepts from a more physical perspective, thereby allowing such concepts to be more practicable in problem-solving situations.

Moreover, several researchers have insisted that estimation is important, as it helps students deal with experiments with more confidence and inspiration[16], and facilitates a better experiential understanding of the physical quantities [17]. Morrison stated that those abilities needed to solve Fermi problems (a kind of estimation problems) are the same as those needed to become a scientist, therefore such skills should be taught in science classes(see also [16]). Fermi problems can also help students to generate concepts, theories, hypotheses, and so on. For example, 'how much does a clock slow down on the top of a mountain,', or 'how many piano tuners lived in Chicago' can improve students' abilities to think and hypothesize reasonably. Consequently, the current study assumed that estimation activities could improve students' physical thinking based on such previous studies.

Therefore, estimation activities about density were developed and used in a junior high school, then the effects on the students' problem-solving procedures were investigated.

2. Research Methods

2.1 *Subjects*

One hundred and ninety two 7th graders at a boys middle school located in a large Korean city participated in the estimation activities and five were individually interviewed.

The five interviewees participated voluntarily. Three of the students' estimation abilities were slightly higher than average, 1.79 had a perfect score of 10, while one student's ability in the experimental group and one student's ability in control group were lower than average. Seo, Park, and Jo belonged to the experimental group, which participated in the estimation activities, while Che and Hong belonged to the control group, which performed traditional experiments about density. The students' conversations during every activity and interviews about problem solving were all tape recorded.

2.2 *Activities*

In the experimental group, the students participated in estimation activities that consisted of three parts. The first part was a theoretical class, where the students were introduced to the concept and several formulas. The second part was a counter-estimation class, where students were given numerical information on the density, mass, and volume of various materials, then compared them with their experiences guided by worksheets. This was a kind of estimation activity, as defined by Bright[18]. The third part was a direct-estimation class, where the students estimated the physical quantities according to their own ideas. Each part took about forty five minutes.

In the control group, which was formed to compare the results with the group involved in the estimation activities, the students participated in the same theoretical class as the experimental group, then conducted traditional laboratory activities. The time taken by the control group was the same as that taken by the experimental group. The students were guided by worksheets to measure the density of coins and fluids.

In the control group, after the traditional laboratory activities, students performed supplementary estimation activities. Hence, in the control group, the changes in the students' perceptions due to both the traditional and estimation activities were analyzed and compared. Because the students in the control group began with the traditional experiments, the changes in their perceptions were investigated after the traditional experiments and after the estimation activities, thereby revealing the common effects of estimation in the two groups.

In both groups, the students had small group discussions during the activities, excluding the theoretical class. Table 1 and 2 shows the items dealt with in the classes.

Table 1. Activity items in estimation group

	Items
Counter-estimation	-To compare two 30 cm cubes where each side is made of gold and ice with the weight of a man, as regards the densities of the materials -To define a density -To compare the average densities of planets -To compare the sizes and weights of a golf ball and hummingbird and count hummingbirds equal to the weight of a golf ball -The weight of a man made of gold
Direct-Estimate	-Tables showing densities of various materials -To estimate a human's average density -To estimate the density of a bob -To estimate the density of wood -To estimate the range of densities on earth

Table 2. Activity items in control group

	Items
Measurement I	-To measure the mass, and volume of coins -To find out the density of coins
Measurement II	-To measure the mass and volume of a fluid in a cup -To find out the density of a fluid

2.3 *Instruments*

Tools for evaluating estimation abilities

In the current study, the students' estimation abilities were analyzed based on the following questions in Table 3. The questions were verified as suitable for evaluating students' estimation abilities by four researchers. The students' answers were analyzed by two researchers, then marked.

Tools for probing problem-solving procedures

Students solved the problems in Table 4 before and after the activities, then five students were individually interviewed.

Table 3. Questions for evaluating estimation ability

	Estimation Questions
First test (Before activities)	-Estimate the density of plastic -Choose the materials with a density that is within the defined range
Second test (After activities)	-Estimate the density of an apple -Choose the materials with a density that is within the defined range
Third test (After estimation activities in the control group)	-Estimate the density of glass -Choose the materials with a density that is within the defined range

Table 4. Questions for probing problem-solving procedures

	Problems
First test (Before activities)	-To compare and calculate the mass of cans made from aluminium and iron -To compare and calculate the volume of objects made from pure gold and gold mixed with silver
Second test (After activities)	-To compare and calculate the mass of rings with different percentages of gold -To compare and calculate the volume of objects made from pure cupper and an aluminium alloy
Third test for control group	-To compare and calculate the mass of copper and silver-plated wires -To compare and calculate the volume of objects made from pure iron and galvanized iron

Before the direct calculation, the students were also asked about their concepts of density using prepared questions. In addition, the students were requested to predict the results of the problems without calculation. For this the exact numbers of the problem were hidden. This procedure was referred to as 'prediction'. The prediction procedure was aimed at probing if the students could understand and explain the qualitative features of the problems, and determine if there was a relation between their predictions and solutions.

3. Results

3.1 *Student Discussions During Estimation Activities?*

During the counter-estimation activity, a student named Jo seemed to be interested in the density values. In this activity, the weights of two 30cm cubes made of gold and ice, respectively, were compared with the weight of a man. The gold cube weight was similar to the weight of seven adults, while the ice cube weight was similar to the weight of a small boy. Through the following conversations, Jo expressed surprise that the material densities did not vary as much as he had expected.

Jo: The densities of the gold and ice are not much different
Park: What?
Jo: You know the difference is only about ten.
Park: No, it's about nineteen.
Jo: At first, I thought it would differ by more than one hundred.

Generally, students are familiar with physical quantities larger than 100g or 100mL in their everyday lives, which may have been why Jo was surprised. These results were also reflected in the survey of the students' estimation abilities, where many of the students estimated the densities of the objects to be over $100g/cm^3$ (Suh, 2000). After the activity, the students came to know that there are very few materials that have a density of more than $20g/cm^3$.

Furthermore, there was information on the worksheet that the total mass of twenty humming birds was the same as a golf ball.

Seo: Why are the densities different….
Jo: (reads the sentences loudly)
Seo: How many humming birds are equal to one golf ball?
Park: How many? (count)
Park: Too many humming birds per one golf ball….
Seo: If the density is large, the volume can be smaller
Jo: Right.
Park: What?
Seo: The bigger the density, the smaller the volume

As such, this process helped the students understand the relation between density, mass, and volume. For example, Seo wrote on his worksheet, 'If the density is large, the volume is small.' This means the volume of objects with the same mass can differ depending on their densities. Furthermore, in this situation, volume is in inverse proportion to density.

Based on their everyday experiences, the students thought that if the volume of an object became larger, the mass also increased. However, in this process, the students were able to see that even if the volume of one object was larger than another, the mass could still be smaller if the density was small. The students also came to realize that the volume was inversely proportional to the density when the mass was the same. This type of understanding is different from the rote memorization of formula, which states that density is equal to mass divided by volume, or mass is density times volume, and so on. Rather, this kind of understanding is based on second hand experience. As a result, the concept of density was understood not only in the context of a textbook, but also in relation to their everyday lives.

In addition, the students in direct-estimation activity used a variety of methods to estimate the densities of various materials. In this activity, they were asked to estimate the average density of a human, scale weights, and wood. The methods used differed according to the subject. When estimating the average density of a human, the 'floating' phenomenon was used based on the idea that if one ninth of a body buoyed on the water, its density would be about one ninth g/cm^3. When estimating the density of a scale weight, which sank in water, the students attempted to determine the component material. The following are some of the conversations while estimating the density of a human body.

Park: Density of a human body? Can a man float on water?
Seo: No
Jo: Yes, it floats
Park: Ah, yes.
Jo: Sometimes it floats and sometimes it sinks
Seo: Yes, when a man is dead, it will float
Park: Babies can also float
Jo: Then, it floats
Seo: Then, it may be about one point...
Park: Yah, then one point...
Seo: About one point eight. One point eight.
Seo: Roughly, measure the weight and volume...
Seo: How can we measure the volume?
Park: We should put you into water

The following are some of the discussions while estimating the density of the scale weight.

Jo: a scale weight?
Seo: There is...
Jo: A scale weight...I know it.
Park: Is a scale weight made from iron?
Jo: No. the weight is a scale weight.
Seo: Ah, the scale weight is made from a metal
Jo: No, the scale weight is made from iron
Park: Is metal the same as iron?
Jo: No. The weight is made from the iron
Park: Yes.
Seo: Is it lighter than iron?
Park: Metal is iron?
Seo: Then 7.9 g/cm^3?
Park: Yes, 7.9.

The above discussions held during the estimation activities reveal that the students began to think qualitatively about the values of the physical quantities, and the activities gave the students the chance to apply their concepts in various physical contexts, resulting in a change of concepts, predictions, and finally problem solving.

3.2 *Change in Problem-Solving Procedures After Estimation Activities*

Before calculating the problems, Seo, Park, and Jo in the experimental group and Che and Hong in the control group were asked to explain their concepts and predict the results of the problems without calculations. Seo, Park, and Jo's estimates were higher than those of Che and Hong after their respective activities.

Table 5. Estimation abilities of interviewed students

	Group	Before activity	After activity	After estimation activity in control group
Park	Estimation activity (Exp.)	2.00	6.00	-
Seo		2.50	7.50	-
Jo		.00	7.50	-
Che	Traditional activity (Con.)	.00	3.50	6.00
Hong		2.00	2.50	9.00

The first finding was the students' misconceptions of density, especially the relation between volume, density, and mass. Before the estimation activities, two students believed that density was a kind of unique quantity, which did not change, while another two students had just memorized it as a formula, mass divided by volume. Furthermore, some students had misconceptions, such as volume is proportional to density.

For example, Park and Hong thought density was like a characteristic amount that did not change.

I: What do you think the density is?
Park: Density?
I: Um...
Park: Characteristic amount of an object

Hong also said,

Hong: I don't know exactly the meaning of density, The density is, if the quality or character is the same, whatever the size is, or whether it is small or larger, does not change.

In the theoretical class the students learned that materials can be discriminated based on their density, which can lead to the concept that the density of material will not change under any circumstances. However, density, especially that of gases, can be changed by pressure. Therefore, this kind of understanding can promote misconceptions.

Jo and Che thought of density as a formula. Yet, even though they remembered the formula and had the ability to calculate it, they had difficulty in explaining density to the interviewer. In addition, Jo's thought was from rote memorization of what that teacher had said, plus it also included the possible misconceptions mentioned above.

I: What do you think density is?
Jo: Volume over mass.
I: If we cut a small piece of metal from a large piece, what is the density of the small piece?
Jo: The density is not related with the size.
I: How do you know that?
Jo: The teacher said.

Che memorized the formula mechanically. He knew the equations, however, this did not mean he could solve the problems or qualitatively understand the physics of the problem.

I: Good, and what do you think the density is?
Che: Mass over volume, isn't it?
I: Can you explain it any other way?
Che: Density, density is mass over volume.

Other misconceptions were about the relation between volume and density. Four students, except for Park, stated that they thought volume was proportional to density when asked to compare the volume of two objects made of different materials.

For example, Seo thought that if the volume or mass of an object were large, the density would also be large.

I: What do you mean by density?
Seo: When we put something into water, if a heavy person gets into water, ...if a light person into water, less water overflows.
I: Then whose density is bigger between the two people?
Seo: The larger person, The density of the larger person is..
I: Um..
Seo: It...It is a kind of amount.
I: Um.
Seo: In this case, a smaller density is...is...a larger density is...the smaller one is small, so the density is small, while the larger one is large, then the density is higher...

Jo also wrote, "If there is a piece of pure gold and a gold crown containing silver, the volume of the piece of pure gold will be larger, because gold is heavier than silver."

Hong had another misconception about the reason why the volume of a gold crown containing silver would be heavier than a piece of pure gold of the same mass. He explained that if an object is mixed with different materials, the volume becomes smaller, because, when two different kinds of balls are mixed, the total volume becomes smaller than the summation.

Hong: If the atoms of gold, gold is small, and silver is smaller, then, what I am saying is, if we put just baseballs in one box and baseballs and golf balls in another box, the latter will be smaller in volume.

The second finding was that students' misconceptions often result in incorrect problem solving. Students' misconceptions that volume is proportional to density, or that mass, volume, and density are similar and proportional to each other resulted in incorrect predictions about the problems, mistaken qualitative thinking related to the problems, misunderstanding about the aim of the problem, and errors in writing the units.

For example, when Seo was asked to predict the result of problem no. 2, comparing the volumes of a piece of pure gold and a gold crown and containing silver without calculation, he only compared the mass of the two materials, not the volumes, i.e. 'Gold is heavier than silver'

This means he confused the concept of volume with mass, and although the problem asked to predict the volume, he predicted the mass, instead. Furthermore, when Seo was asked to solve problem no. 2 using the given exact values, he wrote 66g and 100g. The correct answer was 100mL and 66mL, indicating that he was unable to differentiate the units of volume as well as the concepts.

In addition, the lack of calculating ability also created difficulties in problem solving. For example, Seo wrote 1000/20 as 100, while Hong divided it incorrectly, which resulted in too small number decimal fractions. In this case, Hong reflected on his calculation result.

Hong: The volume of the crown is 0.0015mL. Very strange.
I: Why?
Hong: It is a decimal fraction
I: Um.
Hong: Uh... Why was that?
I: Would you show me your result and explain it?
Hong: This figure... is wrong.

The third finding was that the estimation activities caused a noticeable change in the students' concepts about density and the relation of mass, volume, and density. After the estimation activities, Seo developed concepts on density. For example, in answer to the question comparing the respective mass of pieces of aluminium and iron with the same volume, he wrote, 'if we cut one cubic centimeter of aluminium and iron, then measure their mass, iron will be heavier'.

For the prediction questions comparing the volumes of tools made of copper and an aluminium alloy with the same mass, he wrote, 'The density of copper is larger. So if the mass is the same, the volume of the aluminium alloy will be bigger'. In his calculation, Seo also said, "The density is 8, and the mass is 750. The smaller the density , the bigger the volume."

In the other problem, he did not write the formulas properly. However he did write the correct answer based on qualitative thinking. For example, when he confronted the question asking him to calculate the mass of two kinds of rings, one ring consisted of pure gold with a density of $19g/cm^3$, while the other was gold plated with a density of $17g/cm^3$, plus the volumes of both rings were $2cm^3$, the correct formulas were 19 times 2, and 17 times 2, respectively, and he wrote: 'Pure gold ring: 19/2 = 38g, Gold planting ring: 17/2=34g,'

Seo also wrote, 'The pure gold ring will be heavier. The density of pure gold is greater than that of gold plate'.

As such, Seo's thoughts were able to help him answer the problems. Through the interview, it was found that Seo thought the volume of an object was inversely proportional to its density. Before the estimation activities, Seo had thought density, mass, and volume were proportional to each other. However, after the estimation activities, the concept of density was generated, which led him to predict the correct results and helped him calculate the result corresponding to the predicted one, even though he did not remember the formulas exactly. The way in which he calculated the answer was incorrect, yet he still happened to write the correct answer. Although this is not the perfect result, when considering the thought process preceding the calculation, the outcome was still a good result. Despite Seo's inability to solve the equations and remember the formulas, he was still able to predict the answer and understand the physics behind the problem. Accordingly, this result demonstrates that learning to estimate can help students who are not good at mathematics to solve problems.

An interesting fact was that the students' calculations were closely related to their predictions. In the density problem, most of the students' calculation results were the same as their predictions($p<.01$).

Che's concepts about the relation between volume and density remained unchanged after the traditional activities, for example, 'Copper is lighter than an aluminium alloy, so if the mass is the same, the size of the copper will be smaller. Also he wrote: 'Volume is equal to mass times density'

Table 6. Students' predictions and calculations

calculate \ predict	Aluminium alloy	Copper	Total
Aluminium alloy	44	21	65
Copper	11	25	36
Total	55	46	101

(No answer: 72, Chi-square=12.88, $p<.01$)

After Che completed the estimation activities, his concepts changed slightly. Although not completely correct, he offered a different explanation from his previous one. In relation to the problem asking which volume would be larger between galvanized iron and pure iron with the same mass, he said,

Che: At first, the galvanized iron object will be larger
I: Why?
Che: Because, galvanized iron is iron covered with zinc, it will be larger.

When the interviewer probed Che's thinking more deeply, his scientific concept was revealed.

I: If the iron and galvanized iron have the same weight, which will have the larger volume?
Che: Zinc will be heavier, I think.
I: Zinc is heavier?
Che: lighter?
I: The density of galvanized iron is smaller.
Che: Then, the density will be smaller
I: If the volume is smaller, will the density be smaller?
Che: If the volume is larger, the density will be smaller, isn't it? If the mass is the same, the volume will be larger and the density smaller.

Of course, sometimes the estimation activities did not help change the students' conceptions. This happened when the student had a tendency to calculate mechanically or strong misconceptions. Jo was such a case. After the estimation activities, he failed to understand the relation between density and volume, so he wrote, 'The volume of the aluminium alloy will be larger than the copper. Aluminium is heavier than copper'

His concept was that a heavier object has a larger volume, even though he had the ability to calculate the equations. He calculated and wrote his answer as 'The volume of the aluminium alloy was 100mL, the volume of the copper was 93mL', which agreed with his own prediction based on his misconceptions. Therefore, the interviewer decided to probe further about his concept.

I: How did you solve the problem?
Jo: The volume is density over mass
I: Does it agree with your prediction?
Jo: Yes
I: You predicted that this (the volume of aluminium alloy) is larger than that (the volume of copper), didn't you?
Jo: I thought the aluminium alloy was heavier than the copper
I: Look at the problem, the density of the aluminium alloy is smaller than that of the copper.
Jo: I was wrong. It did not agree with my prediction.

This dialogue showed he did not really read the problem. He only calculated it. After the traditional activities, Hong's concepts did not change, nor did his predictions. When he compared the volume of objects made from an aluminium alloy and pure copper, he said the pure copper would be larger. Also, even though he remembered the equations for producing the volume using the mass and density, he failed to calculate it properly. "Aluminium alloy was 750/7.5 = 1 cm^3, copper was 750 / 8 = 93.7 cm^3"

After completing the estimation activities, Hong's misconceptions still remained. However, they came into conflict with the correct concepts.

I: What do you mean by that?
Hong: Well, if there are basket balls and base balls, if the baseballs are in the space of the basket balls, ..I thought.
I: If there is galvanized iron with a small density and iron with a large density, which will be larger in volume?
Hong: Galvanized iron.
I: Why?
Hong: ...the smaller the density is, the larger the volume is when the mass is the same.

The estimation activities did not seem to be effective in remembering the equations. Before the estimation activities, Park seemed to know the relations between density, volume, and mass and was able to solve problems using equations. Before the estimation activities, for the second question about predicting which volume would be larger, a gold crown containing silver or a piece of pure gold, he explained 'So, here, the equation for producing the density is mass over volume. If we put a mass with an unknown quantity, volume over mass, then it is.'

However, after the estimation activities, he solved the problem incorrectly as he did not remember the equations properly. In Park's case, the estimation activities influenced his concepts and prediction abilities in a positive way, yet his calculation abilities in a negative way. Because estimation activities put an emphasis on qualitative thinking, not remembering formula, this is an understandable result.

5. Conclusions and Discussion

Most interviewees did not seem to understand the concept sufficiently, and some had misconceptions, which caused difficulties in problem solving, such as misunderstanding the problem, misuse of units. In addition, a lack of calculation ability resulted in calculation errors. Before the estimation activities, none of the interviewees were able to conceive the problems qualitatively.

However, after the activities, some students learned to think of the problems in a qualitative manner. This kind of understanding then helped them to reasonably predict the results of the problems even without any specific values. Although the students' tended not to see the physical implications of the problems and the students' misconceptions were not always influenced, the estimation activities were found to potentially enhance the students' problem-solving procedures.

References

1. Helgeson, Problem Solving in Middle Level Science, What Research Says to the Science Teacher, vol. 5, Problem Solving, pp. 13-34, in D. Gabel (Ed.), Washington, NSTA, 1988
2. B. B. Barr, Research on Problem Solving : Elementary school, Handbook of Research on Science Teaching and Learning, in D. Gabel (Ed.), pp. 237-247. Macmillan Pub., 1994.
3. D. P. Maloney, Research on Problem Solving : Physics, Handbook of Research on Science Teaching and Learning, in D. Gabel (Ed.), pp. 661-670. Macmillan Pub., 1994.
4. C.III McMillan and M. Swadener, Novice Use of Qualitative Versus Quantitative Problem Solving in Electrostatics, Journal of Research in Science Teaching, 28(8), 661-670, 1991
5. H. R. Crane, Problems for Introductory Physics, The Physics Teacher, 7, 371-378, 1969.
6. F. H. Larkin and F. Reif, Understanding and Teaching Problem Solving in Physics, European Journal of Science Education, 1(2), 191-203, 1979.
7. F. H. Larkin, Skilled Problem Solving in Physics : A Hierarchical Planning Model, Journal of Structured Learning, 1, 271-297, 1980.
8. M. W. Hackling and P. J. Garnett, Expert-Novice Differences in Science Investigation Skills, Research in Science Education, 22, 170-177, 1992.
9. M. P. Thro, Relationships between associative and content structure of physics concepts, Journal of Educational Psychology, 70(6), 971-978, 1978
10. M. T. H. Chi, P. S. Feltovich and R. Glaser, Categorization and Representation of Physics Problems by Experts and Novices, Cognitive Science, 5, 121-152, 1981.
11. T. De Jong and M. G. M. Ferguson-Hessler, Cognitive Structures of Good and Novice Problem Solvers in Physics, Journal of Education Psychology, 78(4), 279-288, 1986.

12. J. Bascones and J. D. Novak, Alternative Instructional Systems and the Development of Problem-Solving Skills in Physics, European Journal of Science Education, 7(3), 253-261, 1985.
13. D. S. Wright and C. D. Williams, A wise strategy for introductory Physics. Physics Teacher, 24(4), 211-216, 1986.
14. J. A. Suh. The effect of Estimation and Measurement Activities related to Quantitative Concepts of Physics on Problem Solving Procedure. Ph.D. thesis, Seoul National University, Seoul, 2000.
15. B. Schofield, Use of Apparatus and Measuring Instruments, Assessment of performance Unit, Science at age 13: A review of APU Survey findings 1980-1984, London: Her Majesty's Stationary Office, 1989.
16. P. Morrison, Fermi Questions, American Journal of Physics, 31, p. 626, 1963.
17. K. S. Kunz, Visualizing Large Numbers, American Journal of Physics, 39(4). p. 452.
18. G. W. Bright, Estimating Physical Measurement, 79(8). 581-5

ROLE OF EMBODYING LEARNER'S THINKING IN COMPUTER SIMULATION

JAEHYEOK CHOI
Seoul National University, Seoul, Korea

Current research says that computer simulations help students create a bridge between physics and the real world. However, it is not yet understood which factors influence the effect of computer simulations. In a computer simulation, the learner, as the player, can control and create situations. Whereas in the real world, students can only engage in rule-based activities and practice their predictions. Furthermore, computer simulations allow the learner to visualize and test their thinking. In the current study, a student was asked to solve real-world problems by creating and practicing their own explanatory hypothesis in a computer simulation. Finally, the role of embodying the learner's thinking in a computer simulation is discussed in the context of scientific inquiry.

1. Introduction

A computer simulation as a method of rule-discovery is a promising area for scientific reasoning [1, 2]. de Jong et al have already described different types of assignment using computer simulations. Investigating assignments prompt learners to identify relationships, while specification assignments ask learners to predict the value of a certain variable, and explanation assignments ask learners to explain a certain phenomena in the simulation environment [3]. As such, the first type of assignment is based on a descriptive hypothesis, the second is a predictive hypothesis, and the third is an explanatory hypothesis (see also [4]).

Wenham argued that the creation of hypotheses is an essential part of any scientific investigation [4]. Meanwhile, Klahr and Dunbar accentuated the role of interaction between experimental design and hypothesis formation in scientific reasoning [3]. To explain phenomena scientifically, learners need their own hypothesis and then need to test this hypothesis, i.e. the interaction between two spaces, the hypothesis space & the experiment space [4, 5]. However, when students encounter counterevidence, they do not usually abandon their hypothesis but rather create an auxiliary hypothesis or doubt the results of the experiment (see also [1], [6], [7], [8]). Yet, according to Kim & Park, a high achieving student can easily abandon their hypothesis for an alternative hypothesis [8]. With a reliable computer simulation, learners can control the situation to test their hypothesis (see also [2, 9]). This is a strong point in computer simulations, especially for high achieving students. Accordingly, the current study compared the role of real world experiments and computer simulations in modifying an explanatory hypothesis. Plus the role of computer simulations in explanatory hypotheses is discussed. As a result, a model bridging real-world experiments and computer simulations is proposed for scientific explanation.

2. Outline of Research Methods

The current study investigated the hypothesis formation, modification, and transformation process of a gifted eighth grade student as regards a Newtonian cradle using the real object and a computer simulation.

The task was to explain the phenomena related to a Newtonian cradle. First, the student was allowed to play freely with a Newtonian cradle for twenty minutes. Second, the student was asked to explain the phenomena related to the Newtonian cradle for ninety minutes. Third, the student was allowed to play with a computer simulation of Newtonian cradle made by Interactive Physics 2000 for thirty minutes. Finally, the student was asked to explain the phenomena related to the computer simulation for one hundred and twenty minutes.

3. Results

When using the real Newtonian cradle, the student developed his own explanatory hypothesis, then tested his explanatory hypothesis by conducting experiments using the Newtonian cradle. In the case of encountering phenomena beyond his explanatory hypothesis, the student modified his hypothesis or created an ad-hoc hypothesis.

When using the computer simulation, the student spent most of his time testing his explanatory hypothesis, especially to find confirmation. However, in the case of encountering phenomena beyond his explanatory hypothesis, in addition to modifying his hypothesis or creating an ad-hoc hypothesis, the student also abandoned his hypothesis, and made a new hypothesis.

The details of the investigation process are given below.

Stage PR. Playing in real world

The student was allowed to play with the Newtonian cradle for twenty minutes. He lifted one ball up and released it, then did the same with two, three, and four balls. Next, the student lifted one or two balls from each side. Each activity was repeated two or three times. The student then commented "Now I can explain all the phenomena"

Stage IR Inquiry in real world

Explanatory hypothesis for related motion of balls : The student was asked to explain the phenomena by writing down his explanatory hypothesis. His explanatory hypothesis was that one force (he was unable to discriminate between force and energy and perceived them to be the same) moved one ball, two balls, and three balls. Suddenly, he wanted to play more because he realized that his explanatory hypothesis could not explain one phenomenon, i.e. when lifting one ball on each side, after colliding, all the balls should stop, however, the balls did not stop.

To solve his conflict, the student performed various experiments, including varying the height of the lifted ball and swinging balls. Then the student revised his hypothesis, whereby the motion of the ball was not related to its height or velocity, but rather to its mass, however, an explanation of the phenomenon was still unclear. Finally, an auxiliary hypothesis was created to explain this phenomenon, which stated that "Energy must be conserved."

Explanatory hypothesis for cease of motion :The student discovered that the Newtonian cradle finally stopped. His explanatory hypothesis was that "Gravity made the rope remain vertical, so gravity made the cradle stop."

Stage PS. Playing with computer simulation

The student experienced the same phenomena using the computer simulation as he witnessed in the real world, and commented that the computer simulation was very similar to the real world.

Stage IS. Inquiry using computer simulation

After the student felt satisfied with the computer simulation, he conducted various experiments that were impossible in the real world. First, he observed the motion with zero gravity. Second, he changed the material of the ball from steel to wood. This was done once. Then, he experimented how the elasticity of the ball affected its motion. His prediction hypothesis was "If the elasticity of the ball is 1, the ball will not stop". However, the student observed that the total energy of the ball decreased even though the elasticity was one. (He also experimented in a world where the friction was not zero) Then, the student recalled his explanatory hypothesis (g); the reason why the ball stopped was gravity and said, "Enough, I understand"

When the student was asked how he could confirm his hypothesis (g), he answered, "All forces make zero, except gravity. If the ball does not stop, my thought is correct, and since I did it, I know my thinking is right". The teacher then asked the student why he did not make the friction zero. He answered, "Friction cannot affect the motion of the balls, as there are no slopping points. Even if the friction is zero, if the gravity is not zero, the ball must stop." The student then conducted an experiment and observed counterevidence to his explanatory hypothesis (g). As a result, he abandoned his explanatory hypothesis (g) and made an alternative explanatory hypothesis, "The reason why the ball stopped is because the elasticity is not zero, air-resistance, and friction between the balls, rod, and bar exists in the real world"

A summary of the results is given below:

- When playing in the real world, the student did not experience any conflicts. Yet when asked to explain the phenomena, the student was forced to create an explanatory hypothesis, then when he observed counterevidence, this pushed him to inquire further.
- When inquiring in the real world, the student conducted experiments to resolve his conflict, which resulted in an auxiliary hypothesis.
- When playing with the computer simulation, the student created various phenomena and compared them with his real world experiences.

- When inquiring using the computer simulation, the student was able to experiment in a way that was impossible in the real world. When he observed confirmation of his hypothesis, he did not conduct any more experiments. When he observed counterevidence this made him abandon his explanatory hypothesis and make an alternative hypothesis.

4. Conclusions

In scientific explanation, a hypothesis is made based on real world experience, however, computer simulations provide the opportunity to test a hypothesis through embodying the learner's thought. The reliability of a computer simulation and the embodiment of the learner's thought to test their hypothesis allows a student to elaborate or change their hypothesis. Accordingly, the role of play and inquiry in the real world involves making hypotheses and elaborating on them, whereas the role of computer simulations involves elaborating on hypotheses and making alternative ones.

Consequently, for gifted children, the opportunity to make, elaborate, and change explanatory hypotheses as a 'model for bridging real-world experiments and computer simulations' would appear to be effective in solving conflicts resulting from experimental results.

References

1. W. van Joolingen and T. de Jong, Exploring a Domain with a Computer Simulation: Traversing Variable and Relation Space with the Help of a Hypothesis Scratchpad, *Simulation-based experiential learning,* Springer-Verlag, pp. 191-206, 1993.
2. T. de Jong, E. Martin, J. Zamarro, F. Esquembre, J. Swaak, W. R. van Joolingen, The Integration of Computer Simulation and Learning Support: An Example from the Physics Domain of Collisions, *Journal of Research in Science Teaching,* vol. 36, pp. 567-615, 1999.
3. D. Klahr and K. Dunbar, Dual Space Search During Scientific Reasoning, *Cognitive Science,* vol. 12, pp. 1-48, 1988.
4. M. Wenham, The nature and role of hypotheses in school science investigations. *International Journal of Science Education,* vol. 15, pp. 231-240, 1993.
5. J. Park, Analysis of Students' Processes of Generating Scientific Explanatory Hypothesis – Focused on the Definition and the Characteristics of Scientific Hypothesis, *Journal of the Korean Association for Research in Science Education,* vol, 20, pp. 667-679. 2000.
6. J. Park, The Role of Deductive Reasoning in Scientific Activity, *Journal of the Korean Association for Research in Science Education,* vol. 18, pp. 1-18, 1998.
7. J. Park & S. Pak, Students' responses to experimental evidence based on perceptions of causality and availability of evidence, *Journal of Research in Science Teaching,* vol. 34, 57-67.

8. I. Kim & J. Park, The students responses on the conflict observation, data and result in electricity experiment, *paper presented at the annual meeting of the Australasian Science Education Research Association (ASERA),* Bendingo, Australia, 1995.
9. J. Choi & S. Pak, The Analysis of Effects of openness in simulation on the middle school students' ability to make interpreation about mechanics, *Proceeding in International conference Physics Teacher Education beyond 2000*, Barcelona, 2000.

DEVELOPMENT OF TEACHING MATERIALS FOCUSED ON SEQUENTIAL CONCEPTS: CASE OF ELECTROMOTIVE FORCE AND VOLTAGE DROP

D. KIM, H. LEE [a]

[a] *School of Science Education, College of Education, Chungbuk National University, Cheongju, Korea.*

K. CHOI, J. KIM & H. JUN [b]

[b] *Dept. of Science Education, Ewha Womans University, Seoul, 120-750, Korea.*

As an academic discipline, physics has a basic structure that can be expressed and represented by basic and sequential concepts. Therefore, understanding these concepts in an appropriate sequential order is important for effective physics education. Accordingly, the current study examines the basic concepts and sequential order of electromotive force and voltage drop in high-school level electricity. Using the analytical results and simple experiments, sequential teaching/learning materials focusing on electromotive force and voltage drop are developed and tested on sophomore students majoring in physics education. The effectiveness of the new teaching-learning materials was demonstrated as they provided procedural steps for experiments and data collection.

1. Introduction

Bruner(1978) suggested that concepts or knowledge bases have structures, and that these structures can be optimized to enable students to understand the concepts more easily. He claimed that the structure of knowledge should be emphasized for easier understanding and a more effective transition between learning and a longer memory of the study subjects. Bruner also emphasized the importance of the serialization of teaching/learning materials as part of a strategy for teaching the structure of knowledge.

As a field of study, physics has a basic structure. From a previous analytical study of the structure of dynamics concepts in physics textbooks used in Korean high schools, Choi and Kim(1997) found that most dynamics concepts are arranged in a logical order. They also found that there was a significant commonality in the structure of the contents among textbooks. However, they pointed out that many teachers only understood some of the sequential relationships among dynamics concepts, whereas others tended to be overlooked. Furthermore, one of the problems in physics education is that most physics concepts are taught using textbooks containing very abstract examples(Song, 1997). For this reason, the opportunity of learning physics through common phenomena related to real life or situations easily understood by students is almost nonexistent.

Kim and Choi(1999) analyzed the basic concepts in high school and college mechanics and then developed teaching materials on the motion of a dynamic cart. The results of their study demonstrated that the use of teaching materials focused on sequential concepts was effective in producing a systematic and accurate understanding of dynamics concepts.

Accordingly, the current study examined the basic concepts and sequential order of electricity in high-school level electricity and then developed sequential teaching/learning materials focusing on electromotive force and voltage drop.

2. Development of Teaching Materials for Electromotive Force and Voltage Drop.

2.1 *Teaching Materials for Electromotive Force and Voltage Drop.*

The voltage decreases when an electric current flows through a resistance. This phenomenon is called voltage drop. As such, the terminal voltage of a battery becomes lower when it is connected to an electric load. The voltage in a battery drops when the electric cell is connected to an electric load and the current flows through an internal resistance in a closed circuit. Therefore, the purpose of this experiment is to study the voltage drop of a battery with an electric load.

In Figure 1, in an open circuit, when the voltage is measured from 0 to 6 points in turn, the voltage in the battery increases, while the electric load remains unchanged. Yet, in a closed circuit, when the voltage in the battery is increased, the electric load decreases. Thus the voltage is increased by electromotive force and reduced by resistance. This phenomenon clearly demonstrates the concept of potential difference and Ohm`s law.

Apparatus

· voltmeter(DC 3, 15, 30V)
· ammeter(DC 0.05, 0.5, 5A)
· switch
· three types of electric bulb with different resistances
· electric bulb sockets (three connected in series)
· 6 1.5 V batteries
· 6 battery holders
· 3 connecting wires

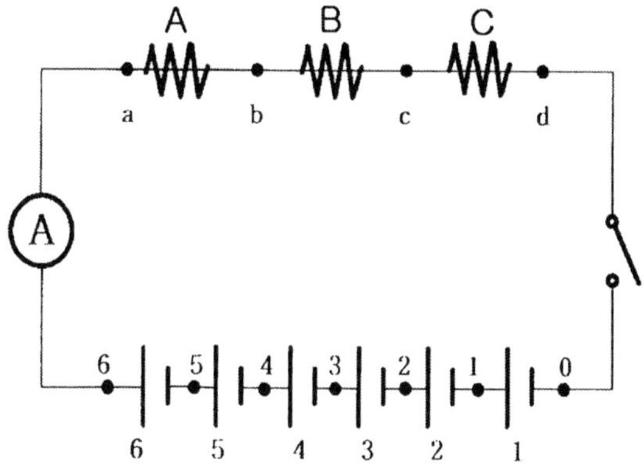

Figure 1 Electric Circuit

Procedure

1) In sequence, connect the bulb sockets (three connected in series) using three 1.5V bulbs and the ammeter using six 1.5V batteries, then examine all the illuminated bulbs. Measure the voltage between two points and the current in the circuit when the switch is off, then when the switch is on. Insert the data in Table 1-1.

Table 1-1. Voltages between two points.

Two points		00	01	02	03	04	05	06	0a	0b	0c	0d
SW	OFF											
	ON											

Type of bulb: 3 x 1.5 V Current: mA

2) a) Explain why the voltage between two points in the circuit changes when the switch is on and when it is off.
   ① in the case of the batteries
   ② in the case of the ammeter
   ③ in the case of the bulbs

   b) What is the term used to describe the difference in the voltage between two points in the circuit when the switch is on and when it is off ?

   c) Is it correct to define the voltage between two batteries as an electromotive force when the switch is off?

3) Referring to Table 1-1, complete Table 1-2 and calculate the internal resistances of the batteries and ammeter, plus the electric resistance of the

bulbs. The potential difference in Table 1-2 is the difference between the two voltages when the switch is on and when it is off.

Table 1-2. Electric resistances of batteries, ammeter, and bulbs.

battery		1	2	3	4	5	6	V_{56}	Am meter V_{6a}	battery		a	b	c
		V_{01}	V_{12}	V_{23}	V_{34}	V_{45}	V_{56}					V_{ab}	V_{bc}	V_{cd}
SW	OFF									SW	OFF			
	ON										ON			
potential difference										potential difference				
internal resistance										resistance				

Type of bulb: 3 x 1.5 V Current: mA

4) Is the electromotive force of the battery equal to the reduced voltages in the internal resistance of the batteries, ammeter, and bulbs?
5) Repeat the above experiments using 3V and 4.5V bulbs, and complete Tables 2-1, 2-2, 3-1, and 3-2.

Table 2-1. Voltages between two points.

Two points		00	01	02	03	04	05	06	0a	0b	0c	0d
SW	OFF											
	ON											

Type of bulb: 3 x 3.0 V Current: mA

Table 2-2. Electric resistances of batteries, ammeter, and bulbs.

battery		1	2	3	4	5	6	V_{56}	Am meter V_{6a}	battery		a	b	c
		V_{01}	V_{12}	V_{23}	V_{34}	V_{45}	V_{56}					V_{ab}	V_{bc}	V_{cd}
SW	OFF									SW	OFF			
	ON										ON			
potential difference										potential difference				
internal resistance										resistance				

Type of bulb: 3 x 3.0 V Current: mA

Table 3-1. Voltages between two points.

Two points		00	01	02	03	04	05	06	0a	0b	0c	0d
SW	OFF											
	ON											

Type of bulb: 3 x 4.5 V Current: mA

Table 3-2. Electric resistances of batteries, ammeter, and bulbs.

battery		1	2	3	4	5	6	V_{56}	Am meter V_{6a}	battery		a	b	c
		V_{01}	V_{12}	V_{23}	V_{34}	V_{45}	V_{56}					V_{ab}	V_{bc}	V_{cd}
SW	OFF									SW	OFF			
	ON										ON			
potential difference										potential difference				
internal resistance										resistance				

Type of bulb: 3 x 4.5 V Current: mA

6) Repeat the above procedures using three different types of electric bulbs and complete Tables 4-1 and 4-2.

Table 4-1. Voltages between two points.

Two points		00	01	02	03	04	05	06	0a	0b	0c	0d
SW	OFF											
	ON											

Types of bulb: 1.5, 3.0, and 4.5 V Current: mA

Table 4-2. Electric resistance of batteries, ammeter, and bulbs.

battery		1	2	3	4	5	6	V_{56}	Am meter V_{6a}	battery		a	b	c
		V_{01}	V_{12}	V_{23}	V_{34}	V_{45}	V_{56}					V_{ab}	V_{bc}	V_{cd}
SW	OFF									SW	OFF			
	ON										ON			
potential difference										potential difference				
internal resistance										resistance				

Types of bulb: 1.5, 3.0, and 4.5V Current: mA

7) Draw graphs points for the measured voltage vs. voltages for the experimental results in Tables 1-1, 2-1, 3-1, and 4-1. Answer the following questions:

① How does one determine the current(I=V/R) flowing in an electric circuit?

② Explain the electromotive force, terminal voltage, and internal resistance of each battery. Are the electromotive force, terminal voltage, and internal resistance of each battery always equal?

③ Is the internal resistance of the ammeter always the same? Explain.

④ What are the resistance values for each electric bulb? Is the resistance for each bulb always the same?

8) Is there a current flow in a battery in a closed circuit? What is the direction of the current flow?

9) Explain Ohm`s law.

10) Write what was interesting about these experiments.

2.2 *Sample Data and Explanation of Electromotive Force and Voltage Drop*

1) Data (See from Table 5-1 to Table 8-2)

Table 5-1 Voltages between two points.

Two points		00	01	02	03	04	05	06	0a	0b	0c	0d
SW	OFF	0.00	1.50	2.95	4.50	6.20	7.78	9.10	9.10	9.10	9.10	9.10
	ON	0.00	1.30	2.20	3.58	5.00	6.60	7.98	7.95	5.20	2.50	0

Type of bulb: 3 x 1.5 V Current: 95 mA

When the circuit is open, no voltage drop is observed. When the switch is on, electric current flows in the circuit and a voltage drop is observed at each battery. This means that each battery has an internal resistance, and when the electric current flows through a battery, the voltage drops. Plus, the voltage drop varies according to the internal resistance of the battery.

Although there is no internal resistance in an ideal ammeter, in reality, an ammeter has internal resistance. Thus, when an electric current flows through an ammeter, a voltage drop appears. Yet, since the internal resistance in an ammeter is small, a voltage drop seldom appears if the electric current is weak.

A voltage drop in an electric bulb occurs when an electric current flows through the filament of a bulb.

All electric conductors have two terminals. Therefore, the voltage between the two terminals is called 'the terminal voltage'. However, a distinction has to be made between a voltage charged with an electric current and not charged.

The terminal voltage is 0.00 volts if there is no current in the circuit. Yet, the voltage of a battery is not zero. The voltage only increases within a range from V_{00} to V_{06}, because each battery has a different electromotive force.

The voltage drop in the internal resistance of a battery is due to the current in the voltmeter, as shown by measuring the terminal voltage of the battery with a voltmeter. Thus, the measured terminal voltage of a battery is less than the actual electromotive force when the switch is off. Since the electromotive force of a battery cannot be measured directly, this is done by measuring the terminal voltage of the battery when it is not charged.

Table 5-2. Electric resistances of batteries, ammeter, and bulbs.

battery		1 V_{01}	2 V_{12}	3 V_{23}	4 V_{34}	5 V_{45}	6 V_{56}	V_{56}	Ammeter V_{6a}	battery		a V_{ab}	b V_{bc}	c V_{cd}
SW	OFF	1.50	1.45	1.55	1.70	1.58	1.32	9.10	0.00	SW	OFF	0.00	0.00	0.00
	ON	1.30	0.90	1.38	1.42	1.60	1.38	7.98	0.03		ON	2.75	2.70	2.50
potential difference		0.20	0.55	0.17	0.28	0.02	0.06	1.12	0.03	potential difference		2.75	2.70	2.50
Internal resistance		2.11	5.79	1.79	2.95	0.21	0.63	11.8	0.32	resistance		29.0	28.4	26.3

Type of bulb: 3 x 1.5 V Current: 95 mA

The internal resistance of the battery V_{06} was 11.8Ω. The internal resistance of the ammeter was 0.32Ω. The total resistance of the 3 electric bulbs was 83.7Ω. The electric current was 95 mA. The voltage drop through the whole closed circuit was: [11.8+0.32+(29.0+28.4+26.3=83.7)]Ω x 95 mA = 9.10 V

Table 6-1. Voltages between two points.

Two points		00	01	02	03	04	05	06	0a	0b	0c	0d
SW	OFF	0.00	1.50	2.95	4.50	6.20	7.78	9.10	9.10	9.10	9.10	9.10
	ON	0.00	1.38	2.25	3.70	5.20	6.80	8.10	8.10	5.50	2.60	0.00

Type of bulb: 3 x 3.0 V Current: 75 mA

Table 6-2. Electric resistances of batteries, ammeter, and bulbs.

Battery		1	2	3	4	5	6	V_{56}	Am meter V_{6a}	battery		a	b	c
		V_{01}	V_{12}	V_{23}	V_{34}	V_{45}	V_{56}					V_{ab}	V_{bc}	V_{cd}
SW	OFF	1.50	1.45	1.55	1.70	1.58	1.32	9.10	0.00	SW	OFF	0.00	0.00	0.00
	ON	1.38	0.87	1.45	1.50	1.60	1.32	8.12	0.02		ON	2.60	2.90	2.60
potential difference		0.12	0.58	0.10	0.20	0.02	0.00	0.98	0.02	potential difference		2.60	2.90	2.60
internal resistance		1.60	7.73	1.33	2.67	0.27	0.00	13.1	0.27	resistance		34.7	38.7	34.7

Type of bulb: 3 x 3.0 V Current: 75 mA

Table 7-1. Voltages between two points.

Two points		00	01	02	03	04	05	06	0a	0b	0c	0d
SW	OFF	0.00	1.50	2.95	4.50	6.20	7.78	9.10	9.10	9.10	9.10	9.10
	ON	0.00	1.38	2.60	4.20	5.80	7.40	8.65	8.65	5.80	2.80	0.00

Type of bulb: 3 x 4.5 V Current: 14.5 mA

Table 7-2. Electric resistances of batteries, ammeter, and bulbs.

battery		1	2	3	4	5	6	V_{56}	Am meter V_{6a}	battery		a	b	c
		V_{01}	V_{12}	V_{23}	V_{34}	V_{45}	V_{56}					V_{ab}	V_{bc}	V_{cd}
SW	OFF	1.50	1.45	1.55	1.70	1.58	1.32	9.10	0.00	SW	OFF	0.00	0.00	0.00
	ON	1.38	1.22	1.60	1.60	1.60	1.25	8.65	0.00		ON	0.00	0.00	0.00
potential difference		0.12	0.23	0.05	0.10	0.02	0.07	0.45	0.00	potential difference		2.85	3.00	2.80
internal resistance		8.28	15.36	3.45	6.90	1.38	4.83	31.0	0.00	resistance		197	207	193

Type of bulb: 3 x 4.5 V Current: 14.5 mA

Table 8-1. Voltages between two points.

Two points		00	01	02	03	04	05	06	0a	0b	0c	0d
SW	OFF	0.00	1.50	2.95	4.50	6.20	7.78	9.10	9.10	9.10	9.10	9.10
	ON	0.00	1.32	2.55	4.00	5.55	7.10	8.40	8.40	8.30	7.98	0.00

Type of bulb: 1.5, 3.0, 4.5 V Current: 25 mA

Table 8-2. Electric resistances of batteries, ammeter, and bulbs.

Battery		1	2	3	4	5	6	V_{56}	Am meter V_{6a}	battery		a	b	c
		V_{01}	V_{12}	V_{23}	V_{34}	V_{45}	V_{56}					V_{ab}	V_{bc}	V_{cd}
SW	OFF	1.50	1.45	1.55	1.70	1.58	1.32	9.10	0.00	SW	OFF	0.00	0.00	0.00
	ON	1.32	1.23	1.45	1.55	1.55	1.30	8.40	0.00		ON	0.10	0.32	7.98
potential difference		0.18	0.22	0.10	0.15	0.03	0.02	0.70	0.00	potential difference		0.10	0.32	7.98
internal resistance		7.20	8.80	4.00	4.00	1.20	0.80	28.0	0.00	resistance		4.00	12.8	319

Types of bulb: 1.5, 3.0, and 4.5 V Current: 25 mA

2) Explanation of electromotive force and voltage drop

There are two ways to determine the current flowing in a circuit. One is based on the total resistance and electromotive force(V_{06}) in a closed circuit consisting of a battery, ammeter, and electric bulbs, while the second is based on the resistance of the bulbs and terminal voltage of the batteries(V_{06}) when the switch is on.

The properties of a battery are represented by its electromotive force, terminal voltage, internal resistance, and capacity.

- The electromotive force is the ability to change a certain amount of energy into electric energy:
- The terminal voltage in an open circuit = the electromotive force of the battery. The terminal voltage in a closed circuit = the electromotive force of the battery - the reduced voltage due to the internal resistance of the battery
- The internal resistance sometimes changes like an electric generator and/or sometimes changes like a chemical cell. The internal resistance of an alkaline cell is reduced when the current is increased.
- The electromotive force, internal resistance, and terminal voltage differ for each battery, as they are not manufactured identically or may have already been used.

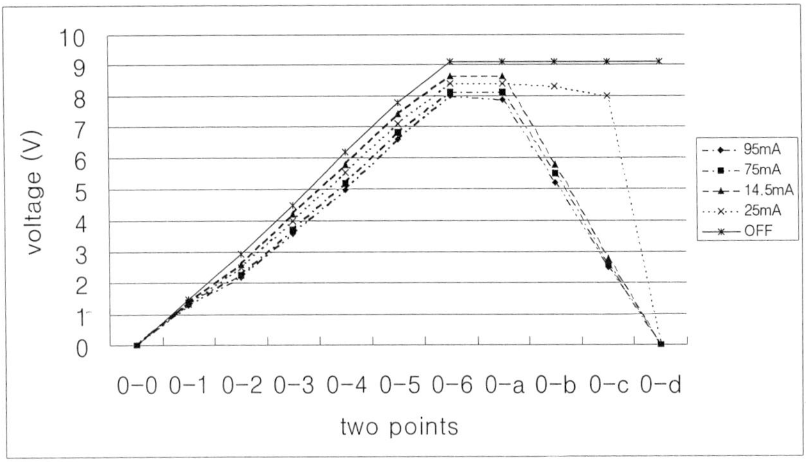

Figure 2. Voltages between two points

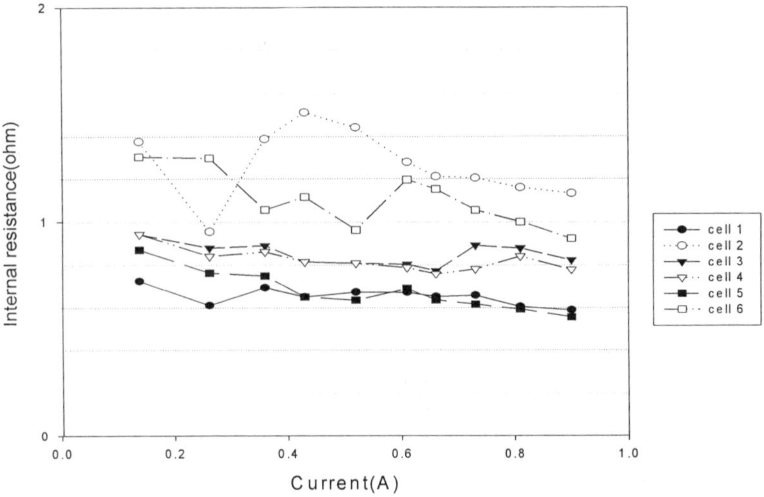

Figure 3. Internal resistance vs current

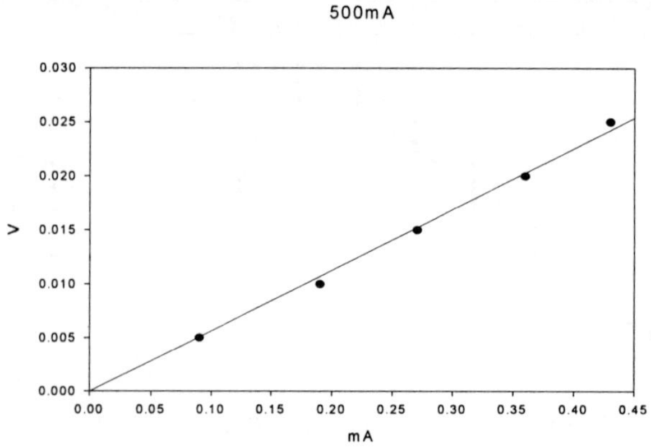

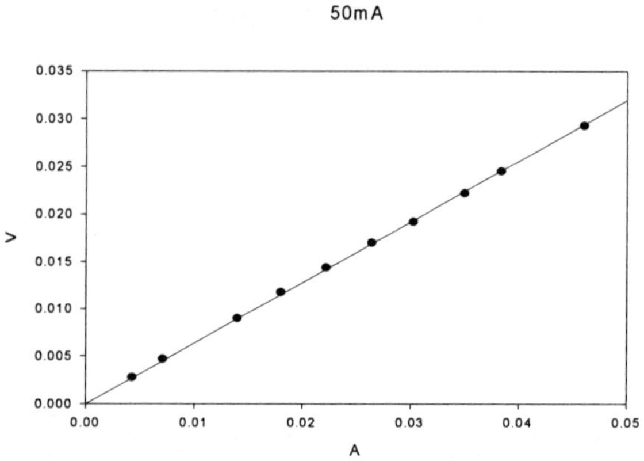

Figure 4. Voltage drops through ammeter

The internal resistance of an ammeter can be measured differently according to its measurement range, yet in the current experiment the data value was always the same with three measurement ranges: 0 ~ 50mA, ~500mA, ~5A. The two graphs in Fig. 4 show the results of the voltage drop when the current was changed within each measurement range: 50mA and 500mA.

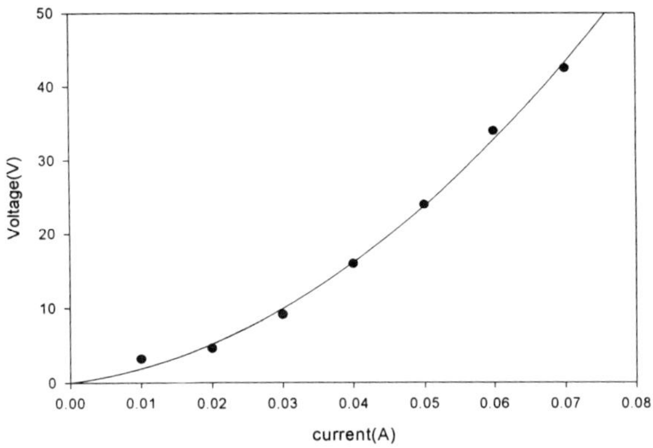

Figure 5. Current vs. voltage in bulb

Measuring the resistance of a bulb is very difficult, because the resistance increases according to the high temperature of the filament even with a small current. As such, the resistances of the three bulbs used in Table 8-2 were all different, because the filament temperatures were also different.

The circuit components, battery, socket and holder, and wire are all conductors and thus full of charge carriers. The electro motive force of the electric cell makes the charge carriers move to one side, thereby making the current flow through the circuit. Namely, the current flows from (-) to (+) in a battery.

Ohm`s law can be expressed by three numerical formulae.

① $V = I \times R$ --- used in Question 4

② $I = V / R$ --- used in Question 7 - ①

③ $R = V / I$ --- used in Question 7 - ③

3. Summary and Conclusion

Physics as an academic discipline has a basic structure that can be elaborated in sequential concepts. For effective physics education, it is necessary to understand these concepts in the appropriate sequential order. Therefore, the current study examined the basic concepts and sequential order of electromotive force and voltage drop for high-school level electricity. The main concepts are structured based on current, resistance, internal resistance, voltage, terminal voltage, voltage drop, electromotive force, Ohm's law, and so on. As such, sequenced teaching materials using a few simple experiments were developed according to the analysis of the main concepts.

The serialized and structured learning materials were tested using sophomore students majoring in physics education. The results indicated that the new materials were able to correct and confirm students' pre-conceptions related to basic electricity. Plus, the students clearly understood Ohm's law through the simple experiments.

Consequently, the current results indicate that sequentially structured concepts can improve student learning and provide teachers with an opportunity to enhance their teaching efficiency.

References

1. Bruner, J. S.(1978). The Process of Education. Harvard University Press.
2. Choi, K. & Kim, B.(1997). Comparative study of mechanics concepts in textbooks and teachers' conceptual structure using interpretative structural modeling. Korean Physics Education. 15(2). 79-87.
3. Kim. D. & Choi, K.(1999). Development of teaching materials focused on sequential concepts 1: The motion of a dynamic cart. Korean Physics Education. 17(1), 64-71.
4. Song, J.(1997). Study of university students' understanding of major ideal conditions used in physics. Korean Physics Education. 15(1). 1-7.

EFFECTS OF EXPLANATIONS OF PHYSICAL PHENOMENA GIVEN IN NON-PHYSICS TEXTBOOKS ON FORMATION OF STUDENT PHYSICAL CONCEPTIONS

M.J. PARK & Y.M. KIM

Dept. of Physics Education, Pusan National University, Pusan, 609-735, Korea

The purpose of the current study was to investigate the effects of explanations about physical phenomena given in non-physics textbooks on the formation of student physical conceptions. Two classes, 39 students in each, were sampled from two middle schools in Pusan, Korea, and two kinds of test tools for investigating student conceptions were developed for the study. The first test tool(a) investigated student conceptions after reading explanations about physical phenomena in non-physics textbooks, while the second(b) investigated student conceptions after reading explanations revised by physics education experts about the same physical phenomena. The two test tools were applied to each class, and for a fair invetigation, test(a) followed by test(b) was applied to one class, while test(b) followed by test(a) was applied to the other class. The results were as follows: In both classes, the students' level of understanding from explanations revised by physics education experts was significantly ($p < .01$) higher than that from explanations in non-physics textbooks. As such, it is feasible that false or inappropriate explanations in non-physics textbooks can cause student misconceptions. Moreover, the improper expression of physical science concepts, improper choice of scientific terms, and incorrect grammatical structures, along with the use of unsuitable examples and improper model pictures can make it difficult for students to understand physics concepts. Furthermore, differences in the terms used in physics textbook and those used in other textbooks can also confuse students.

1. Introduction

According to constructivism, students have their own conceptions related to physical phenomena before instruction, and they acquire scientific concepts through the interaction between a new concept and their own conceptions. As such, learning can be defined as a conceptual change[1].

The characteristics of a misconception are as follows:

1) Students have their own plausible explanation for physical phenomena.
2) The concept is not easily changed even after instruction[2][3].
3) Students may be confused based on holding a scientific concept separately from their own misconception within their cognitive structure[3].

The causes of student misconceptions[4][5][6][7] are classified into innate factors and environmental factors. The environmental factors are divided into physical, cultural, and the school environment. David Maloney[8] also supposed that 'loose' statements in textbooks induce students to have misconceptions. In addition, physical science explanations included in non-physics textbooks may be another important source of student misconceptions.

Accordingly, the purpose of the current study was to investigate the effects of explanations about physical phenomena given in non-physics textbooks on the formation of student physical conceptions.

2. Research Problems

The research questions used in this study were as follows :

1) What physical science concepts are explained in non-physics textbooks currently used in secondary schools in Korea?
2) What do students understand from the explanations of the physical science concepts provided in the above textbooks?
3) What do students understand from the explanations revised by physics education experts about the same physical science concepts provided in the above textbooks?

3. Research Methods and Procedures

The null hypothesis in the current study was "the explanation of physical science concepts given in non-physics textbooks do not cause student misconceptions." The procedures used for the research are shown in Figure 1.

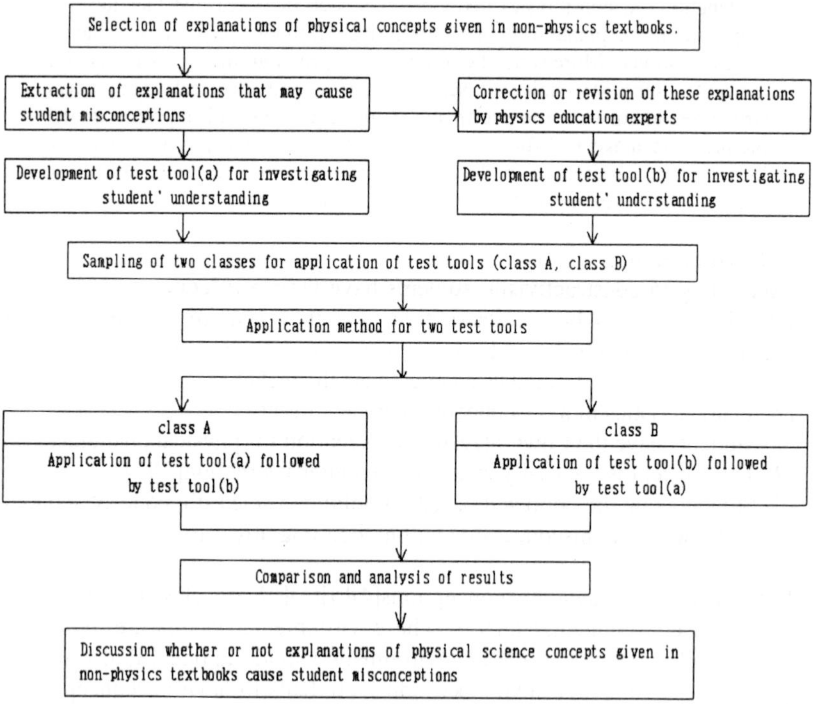

Figure 1. Research procedures

Two classes, 39 students in each, were sampled from two middle schools in Pusan, Korea. The classes were named class A and class B. Two kinds of test tools were also developed to investigate the students' conceptions. The first test tool (a) investigated the students' conceptions after reading explanations about physical phenomena from non-physics textbooks, while the second(b) investigated the students' conceptions after reading explanations revised by physics education experts about the same physical phenomena.

The two test tools were applied to each class, however, for a fair investigation, test tool (a) followed by test tool (b) were applied to class A, while test tool (b) followed by test tool (a) were applied to class B.

The textbooks analyzed for the current research are shown in Table 1. They contents were related to mechanics in physical education(PE) textbooks for the 7th and 10th grade, energy in technology & industry textbooks for the 7th grade, and electricity in technology, living & technology, technology & industry, and technology & domestic science textbooks for the 8th grade.

4. Results of Tests

The correct answer rates for the problems in test (a) and test (b) are shown in Table 2. In both classes, the students' understanding of the explanations revised by physics education experts was significantly ($p< .01$) higher than that of the explanations in the non-physics textbooks.

Figure 2 shows a comparison of the correct answer rates for test (a) and test (b) from class A. From most problems, the students' understanding from explanations revised by physics education experts was higher than that from explanations in non-physics textbooks. Figure 3 shows a comparison of the correct answer rates for test (a) and test (b) from class B. The results were almost the same as from class A.

Table 1. Textbooks used for analysis

Name of textbook	Publication year	Publishing company
PE for 7th grade	1997	Po Chin Chai Co,. LtD.
PE for 7th grade	1995	Kyo Hak Sa Co,. LtD.
PE for 7th grade	1996	Ji Hak Sa Co,. LtD.
Technology & Industry for 7th grade	1997	Dong-A Publising company
Technology for 8th grade	1990	Chungnam National University - Industrial Education Research Center
Living & Technology for 8th grade	1984	Chungnam National University - Industrial Education Research Center
Technology & Industry for 8th grade	1997	Tam Gu Won
Technology & Domestic Science for 8th grade	1994	Korea Educational Development Institute
PE for 10th grade	1997	Po Chin Chai Co,. LtD.
PE for 10th grade	1998	Kyo Hak Sa Co,. LtD.

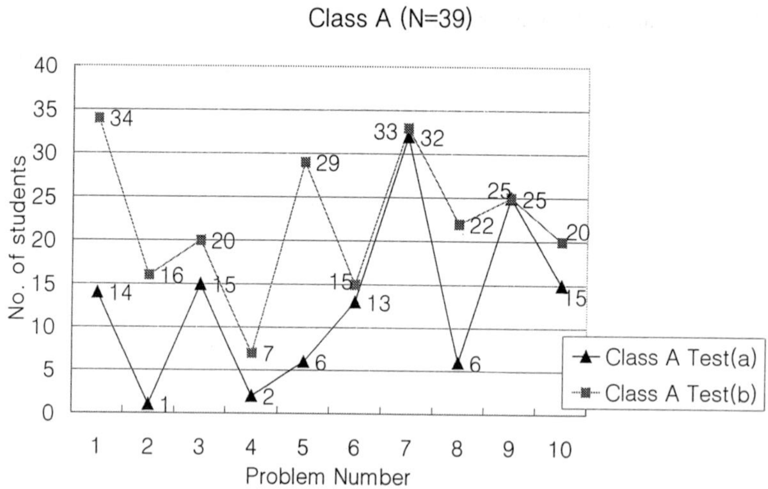

Figure 2. Comparison of correct answer rates for test (a) and test (b) from class A.

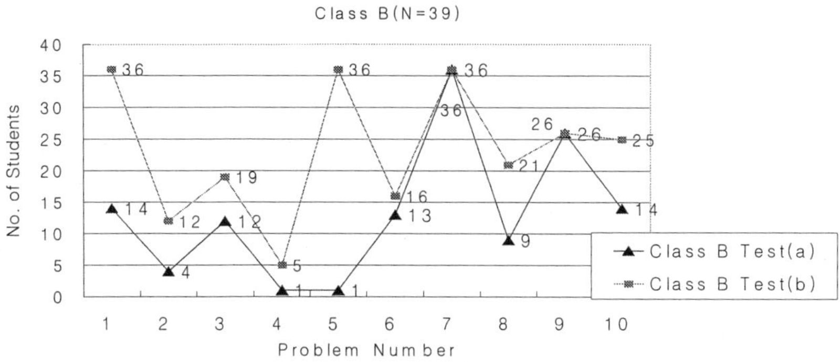

Figure 3. Comparison of correct answer rates for test (a) and test (b) from class B.

Table 2. Correct answer rates for problems in test (a) and test (b).

Problem	Class A(N=39)				Class B(N=39)			
	test(a)		test(b)		test(b)		test(a)	
	freq.	Correct answerrates	freq.	Correct answer rates	freq.	Correct answer rates	freq.	Correct answer rates
1	14	35.9	34	87.2	36	92.3	14	35.9
2	1	2.6	16	41.0	12	30.8	4	10.3
3	15	38.5	20	51.3	19	48.7	12	30.8
4	2	5.1	7	19.7	5	12.8	1	2.6
5	6	15.4	29	74.4	36	92.3	1	2.6
6	13	33.3	15	38.5	16	41.0	13	33.3
7	32	82.1	33	84.6	36	92.3	36	2.3
8	6	15.4	22	56.4	21	53.8	9	23.1
9	25	67.6	25	71.4	26	66.7	26	66.7
10	15	38.5	20	51.3	25	64.1	14	35.9
Average		33.1		57.2		59.5		33.3
T-test	T = 7.01*				T = 8.79*			

*p < .01

5. Conclusions and Suggestions

The conclusions of this study are as follows :

1) It is feasible that false explanations or inappropriate explanations in non physics textbooks can cause student misconceptions.
2) The improper expression of physical science concepts, improper choice of terms, and incorrect grammatical structures can make it difficult for students to understand physics concepts.
3) The presentation of unsuitable examples and improper model pictures can cause student misconceptions.
4) The use of different terms in non-physics textbooks and physics textbooks about the same phenomena can confuse students.

Accordingly, this implies that curricula development should involve much more communication between physics and subjects related with physics, plus unified terms should be used.

References

1. R. Driver, The Pupils as Scientist?. Milton Keynes : The Open University Press, 1983
2. J. K. Gilbert and D. J. Swift, Towards a Lakatosian Analysis of the Piagetian and Alternative Conceptions Research Programs, Science Education, vol. 69, no. 5, pp. 681-696, 1985.
3. J. K. Gilbert, R. J. Osborne and P. J. Fensham, Children's Science and its Consequences for Teaching, Science Education, vol. 66, no. 4, pp. 623-633, 1982.
4. K. M. Fisher, a Misconception in Biology : Amino Acids and Translation, Journal of Research in Science Teaching, vol. 22, no. 1, pp. 53-62, 1985.
5. U. M. O. Ivowi and J. S. O. Oludotu, an Investigation of Sources of Misconception in Physics. Proceedings of the second International Seminar: Misconceptions & Educational strategies in Science and Mathematics, Cornell Uni. Ithaca, NY, USA, vol. Ⅲ, pp. 252-259, 1987.
6. J. W. Renne and M. R. Abraham, Understandings and Misunderstandings of Eighth Graders of Four Physics Concepts Found in Textbooks, Journal of Research in Science Teaching, vol. 27, no 1. pp. 35-54, 1990.
7. D. Kim, Effect of Teaching Model to Decrease Misconception in Learning of Mole Concept. Ph D. Dissertation, Seoul National University, 1991.
8. M. David, Charged Poles? Physics Education, vol. 20, pp. 310-317, 1985.

Appendices

Examples from test tool (a)

※ Answer the question after you have read the following explanation carefully.

All objects thrown up fall down because of gravity. For example, when a man throws a cannonball it moves up parabolically. When it reaches a certain point, it moves down parabolically and its speed increases due to gravitational force.

2. According to the above explanation, when does gravitational force act on the cannon ball.

① When the cannonball is going up.
② When the cannonball reaches the highest point.
③ When the cannonball is going down.
④ If there are other cases, write them below

Underline the phrases or statements from the above explanation that made you choose your answer.

※Answer the question after you have read the following explanation carefully.

A glow lamp creates lights when electrical current flows through the filament, because the filament becomes heated.

Most electrical energy is consumed as heat and less than 10% of electric energy is changed into light energy.

3. From the above explanation, do you agree with the statement 'energy disappears eventually'?

① Yes. ② No.
③ If you have a different answer, write it below

Underline the phrases or statements in the above explanation that made you choose your answer.

Examples from test tool (b)

※ Answer the question after you have read the following explanation carefully.

All objects thrown up fall down because of gravity. For example, when a man throws a cannonball, for a while it goes up parabolically. However, the upward speed decreases due to the downward gravitational force. Therefore, the cannonball finally starts to fall and the downward speed increases.

2. According to the above explanation, when does gravitational force act on the cannon ball.

① When the cannonball is going up.
② When the cannonball reaches the highest point.

③ When the cannonball is going down.
④ If there are other cases, write them below

Underline the phrases or statements in the above explanation that made you choose your answer.

※ Answer the question after you have read the following explanation carefully.

A glow lamp creates lights when an electrical current flows through the filament, because the filament becomes heated.

In this case most electrical energy is changed into heat and less than 10% is changed into light energy.

3. According to the above explanation, do you agree with the statement 'energy disappears eventually'?

① Yes. ② No.
③ If you have a different answer, write it below

Underline the phrases or statements in the above explanation that made you choose your answer.

Part III

Media

TAKING THE PHYSICS CLASSROOM INTO THE WORLD

CHRISTOPHER J. CHIAVERINA
New Trier High School, Winnetka, IL 60093, USA
Fizzforfun@aol.com

1. Introduction

Physics may be one of the easiest subjects to teach because it touches virtually every aspect of our students' lives! The world outside the classroom provides endless possibilities for discovering and understanding nature's laws. Through experiments and exploratory activities performed in non-traditional settings, students of all ages can be made aware of the beauty and wonders of physics. Furthermore, such experiences make evident the myriad applications of physics in everyday life. This paper discusses how the use of auditoriums, parking lots, amusement parks, interactive science centers, the home and other unique learning environments can make physics come alive!

2. Out of The Classroom Experiences

Although a school's physics laboratory is the traditional arena for hands-on exploration and experimentation, a host of alternative venues exist. We will first examine laboratory settings that lie outside the classroom, but in or near the school.

Our first stop will be the school auditorium. In an attempt to illustrate acoustical phenomena on a large scale, we have an annual "sound and light" show in our school's auditorium.[1] The program consists of demonstrations of interference and reflection of sound waves and additive color mixing.

We begin by reviewing the principle of superposition of waves. This is accomplished by projecting a ripple tank interference pattern on a screen. The large image allows students to observe changes in the nodal pattern as the spacing of the sources and the frequency of the waves are changed.

A transition from water waves to sound waves is made by using two loudspeakers to approximate point sources. While these two sources (typically 3-m apart) are being driven in phase at a fixed frequency (for example, 500 Hz), students are asked to move around the auditorium and locate points where the sound level is low. By using a large number of students, an easily discernible nodal pattern emerges. While students remain in their seats where the sound level was determined to be low, the frequency of the sound and then the separation of the sources are changed. The resulting changes in the nodal pattern are observed as nodes and anti-nodes sweeping across the auditorium. Finally, music is played to demonstrate how a large number of frequencies played simultaneously give rise to a "washing out" of the nodal pattern.

The fact that the nodal lines in the interference patterns are not totally "dead" leads naturally to a discussion of virtual sound sources.[2] These virtual sources result from multiple reflections from the walls, ceiling and floor of the auditorium. The plane-mirror analog is used to predict the location of the virtual sources of sound.

To reinforce the notion of virtual sound sources, a single speaker and a large sheet of plywood are used to set up an interference pattern. The sheet of wood serves as a reflector and hence a virtual source of sound. The interference pattern formed by a real source and its image is easily observed.

Before leaving the auditorium, we allow students to observe how the color of an object depends on the color of the incident light. We invite the students to join us on the stage and observe the color of their clothing under white light. Then the white light is turned off and each primary color is used in turn to illuminate the stage and its occupants. The students' reaction to the changes in the color of their clothing is something to behold!

While many physics activities can be performed in non-classroom settings inside the school, some must be done outdoors. The following experiment takes place in our school's parking lot. To get a feel for Newton's Laws, students push a car. After using simple equipment to gather data, they apply Newton's 2nd Law of Motion to determine the mass of the automobile.

First, three students are selected to push the car with the bathroom scales. The sum of the three bathroom scale readings provides the applied force. With stopwatch in hand, a student in the backseat of the car says "go" and the driver releases the brake. Every two seconds thereafter the timer chants "drop." While the student-powered car accelerates, a person in the passenger seat drops markers onto the pavement through an open door. The markers provide a record of the position of the car as a function of time. The students' reaction to the increasing distance between successive markers alone makes the experiment worth doing.

Using the measured distances between markers, the average velocity during each interval is calculated. From the slope of a plot of velocity versus time, students determine car's acceleration.

The frictional force acting on the car is found by determining the force needed to keep the car moving at a constant velocity. The net force is then calculated by subtracting the frictional force from the applied force. The mass, found by applying Newton's 2nd is compared with data found in the car owner's manual.

Opportunities for doing on-campus physics activities outside the classroom are almost limitless. The Physics Teacher magazine is an excellent source of ideas for such activities.[3]

3. Physics to Go: Doing Physics in The Home

For some time now we have been taking advantage of yet another non-traditional laboratory setting: the home.[4] Using simple materials, students are encouraged to do physics experiments with family and friends. Sometimes the materials needed to investigate physical phenomena may be found in the kitchen or workshop. When more specialized apparatus is called for, we create a "lab in a bag" by packaging the required equipment in a plastic food-storage bag.

The materials used in our "lab in the bag" experiments are safe, simple and generally easily obtainable. Because of the low-cost nature of the equipment, students are usually allowed to keep the bag and its contents. Return is requested only when higher priced items, such as polarizing filters, are included in the kit. Still the nominal cost of even these materials eliminates any concern about loss or breakage.

Though certainly not the focus of our physics course, these take-home experiments have greatly enhanced our program. The activities have allowed us to somewhat surreptitiously extend the time our students are thinking about and doing physics. Since many of the explorations focus on counter-intuitive phenomena, students delight in sharing unexpected outcomes with others. Equally important is the obvious satisfaction they derive from demonstrating their knowledge of physics.

Needless to say, parents love seeing what their children are doing in physics. We have received numerous written, email and phone messages from parents who have been intrigued by the experiments and delighted by their son or daughter's involvement in science.

Each year, prior to the first at-home experiment assignment, we send a letter to parents explaining the purpose and nature of the upcoming activities. The letter also informs parents that their son or daughter will receive credit upon the return of a signed sheet indicating the student's successful completion of and the parents' involvement in the activity.

Sometimes the experiments are qualitative. Students may only be required to make observations and form hypothesis regarding the phenomena they witness. When appropriate, students may be instructed to make measurements, record data, construct graphs and perhaps draw vector diagrams.

" Physics to Go" experiments may be designed for virtually any topic in physics. The following optics activities are meant to be illustrative examples.

Identifying Sources of Ultraviolet Light Inexpensive UV-sensitive plastic beads[5] are used to identify sources of ultraviolet light. The reusable, chemically treated beads undergo a dramatic change in color when exposed to UV. Students expose their beads to radiation from as many sources as they can find in their environment. After identifying strong UV sources, they shield their beads from these sources with a variety of materials in an attempt to find the best absorber. Glass, plastic, water, suntan lotion are among the materials tested.

Camera Physics Students learn about the workings of a camera by taking one apart. With the popularity of single-use cameras, it is possible to obtain a class set of used disposable cameras from virtually any camera store.

Students examine the camera's optics (these inexpensive cameras sometimes have up to three lenses!), flash electronics and film transport mechanism.

They form images with the camera's principal lens and measure its focal length and f-number. Dissecting and analyzing a camera is one of our students' favorite take-home experiments. (Caution: shock hazard! Care must be taken in preparing flash cameras for student use. Batteries should be removed from cameras and capacitors discharged prior to dissemination of cameras to class.)

Exploring Color This lab allows students to explore the principles of additive and subtractive color mixing. Along the way, they are made aware of examples of color mixing going on all around them.

Each student is given six color filters (red, green, blue, cyan, yellow, and magenta)[6] and a pair of inexpensive diffraction glasses.[7] Students examine the makeup of white light by looking an incandescent bulb through the diffraction glasses. They record what they observe with crayons or colored pencils. They then place each colored filter over the glasses and see that each filter removes a different portion of the spectrum. They again record their observations with colored markers.

To observe the effect of overlapping filters, they view white light through various combinations of the filters. The hope is that students will "discover" the rules of subtractive color mixing.

Placing a drop of water on the screen of a television or computer monitor reveals the wonders of additive color mixing. With close inspection of the drop, red, blue and green dots or rectangles are visible. Students realize that the myriad colors seen on the screen result from the additive mixing of these three primary colors.

CD-Spectroscope Using a shoebox and any compact disc, students may construct a no-cost spectroscope.[8] Functioning as a diffraction grating, the disc's pitted surface separates incident light into its component colors. Both emission and absorption spectra are observable using this easily constructed optical instrument.

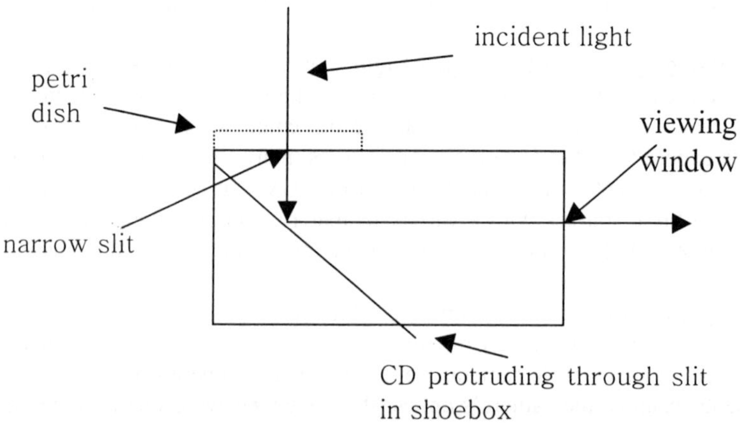

4. Interactive Science Centers

Interactive science centers have become extremely popular worldwide. Following the lead of San Francisco's Exploratorium,[9] cities around the globe have created hands-on museums where the watchword is "please touch" and visitors are encouraged to experiment and explore.

Dr. Frank Oppenheimer, physicist-educator and founder of the Exploratorium, was acutely aware of the importance of engaging the learner. According to Oppenheimer, "Students have to have the opportunity to ask 'what will happen if....' They just can't read. They have to have props to see, handle, and understand what's happening. Otherwise, teaching science is like teaching swimming but not allowing anyone near the water."

Inspired by the Exploratorium, art, mathematics and science teachers at New Trier initiated the Connections Project.[10] The project is an ongoing multidisciplinary endeavor whose mission is to create interactive, museum-type exhibits that illustrate connections between seemingly disparate disciplines. Teachers then use these exhibits to enhance concept development and demonstrate interdisciplinary linkages. Since the inception of the Connections Project, thousands of students have used the exhibits as extensions of the classroom. Through hands-on experiences, they have seen how science is intertwined with the arts, humanities, engineering and mathematics.

The teachers and students involved in this initiative have thus far created more than 125 museum-quality, interactive exhibits. The multi-disciplinary exhibits are grouped into thematic clusters that include "curves", "optics", "visual perception", "motion and energy", "tessellation", "iteration and fractals", and "symmetry".

An exciting aspect of The Connections Project is that it brings students and teachers together as co-learners. Students and teachers learn by producing exhibits that, in turn, are used to teach other students and teachers. The project also encourages collegiality and renews teacher enthusiasm.

The Connections Project displays are being used in a variety of venues. In addition to being presented in exhibitions at New Trier, the exhibits have been used in a wide range of classes at New Trier, in local and Chicago elementary schools, at local, state and national professional meetings, and in university classes. Recently we were asked to display our exhibits at the Explorit Museum in Nebraska and at "Science on the National Mall," an event sponsored by the AAAS and Smithsonian Institution.

5. The Amusement Park: An Unlikely Laboratory

In addition to thrill rides and cotton candy, what else does an amusement park offer? Would you be surprised if you were told that it's also the perfect place to study the laws of physics?[11, 12] In fact, an amusement park just may be the ultimate teaching laboratory.

At an amusement park, virtually all the topics included in the study of mechanics can be observed operating on a grand scale. Furthermore, phenomena, such as weightlessness, which can only be talked about in the classroom, may be experienced by anyone with sufficient courage.

When we take our students go Six Flags Great America, our local amusement park, they must quantify what they see and feel. Unlike textbook problems, no data is given. Therefore, students must start from scratch. Heights of rides, radii, periods of rotation, lengths of roller coaster trains and other quantities must be obtained before applying equations learned in the classroom.

Fortunately, only simple equipment is required to make such measurements. A stopwatch, meter stick and protractor are all that is needed to obtain data that will allow the calculation of such diverse quantities as a person's potential energy at the top of a roller coaster, the centripetal acceleration experienced by a rider in the English Rotor, or the speed of a passenger after 50 meters of free fall.

In recent years, a new type of recreational facility has hit the scene: the aquatic amusement park. Consisting of water slides, swimming pools and large wave pools, these parks allow the study of a variety of large-scale wave phenomena including traveling and standing waves. The centerpiece of most aquatic parks is an enormous wave pool. Waves in such pools may be produced by a large blade at one end of the pool that rhythmically pushes on the water or by blowers that pneumatically generate waves. At Magic Waters, a local pool we visit, pneumatic wave generators are employed.[13]

When the water is excited pneumatically at the proper frequency, intricate interference patterns may be produced on the surface of the water. In the case of Magic Waters, not only are nodes and anti-nodes clearly visible on the surface of the water, but a beautiful standing wave is seen along one end of the wave pool while traveling waves move along the sides of the pool. Among others things, our students analyze the interference patterns, measure the frequency and wavelength of the water waves, and calculate wave speed using three different methods: 1) $v = f\lambda$; 2) v = distance/time; 3) $v = (gh)^{1/2}$. All three methods agree to within 1 m/s.

Like conventional amusement parks, aquatic amusement parks offer wonderful kinesthetic learning opportunities. Where else can you ride an anti-node or relax while standing at a node?

The interest in amusement physics continues to grow. This past May, "Physics Day" at Six Flags Great America alone attracted over 20,000 physics students from four states. These students, like thousands like them around the world, would probably tell you that amusement parks provide the ultimate vehicles for learning physics.

6. Concluding Thoughts

The activities outlined in this paper are intended to provide students with an awareness, appreciation and understanding of the natural phenomena that surround them. Such consciousness-raising can enrich their lives just as an appreciation of art and music can expand their horizons. Seeing the physics involved in everything form a rainbow to a rock concert can be rewarding and enjoyable; having an understanding of the underlying principles behind these phenomena serves to heighten these experiences even more.

References

1. C. Chiaverina and J. Hicks, "People Demos," Phys. Teach. **21**, 169 (1983).
2. Thomas D. Rossing, "Acoustic Demonstrations in Lecture Halls: A Note of Caution," Am. J. Phys. **44**, 1220 (1976).
3. The Physics Teacher magazine is published by the American Association of Physics Teacher, One Physics Ellipse, College Park, MD 20740-3845; 301-209-3333.
4. M.B. Barrett and C. Chiaverina, "Optics Experiments to Go," Forum on Education, Fall 1999.
5. Ultraviolet detecting beads are available from Educational Innovations, 362 Main Avenue, Norwalk, CT 06851.
6. Sheets of colored filters may be obtained from Arbor Scientific, P.O. Box 2750, Ann Arbor, MI 48106, or any stage lighting outlet.
7. Inexpensive diffraction glasses are available through Arbor Scientific or Kipp Brothers, P.O. Box 781080, Indianapolis Indiana 46278. When ordering from Kipp, Request "Laser Amazers."
8. The plan for the spectroscope is based on a design by F. Wakabayashi, K. Hamada, and K. Sone in J. Chem. Educ. **75**, 1569 (1998).
9. Information on the Exploratorium may be found at <http://www.exploratorium.edu>.
10. J. Roeder, "Physics and the Amusement Park," Phys. Teach. **13**, 327-332 (1975).
11. J. Walker, "Thinking About Physics Whiled Scared to Death on a Falling Roller Coaster," Sci. Am. **249,** 162-169 (1983).
12. D. Rathjen, "Physics for the Adventurous," Exploratorium Quarterly, **1**, 12-17 (1987).
13. J. Hicks and C. Chiaverina, "Catch the Wave," Phys. Teach. **28**, 167-170 (1990).

TIME-DEPENDENT PERMEABLE INTERFACE AND IT-BASED PHYSICS EDUCATION*

JIN S. KIM & KEUM H. LEE
School of Science and Technology & Institute of Photonics and Information Technology
Chonbuk National University, Jeonju 561-756, Korea
E-mail: jin@moak.chonbuk.ac.kr and khl@moak.chonbuk.ac.kr

Education prepares people for the future and hence should have a feedback mechanism reflecting societal changes/needs, and physics education is no exception. Along with the ongoing transition to knowledge-based economies, equal access to scientific knowledge is a fundamental prerequisite for sustainable development and world peace. The use of current information and communication technology can significantly contribute to improving the quality of education by overcoming barriers, such as time and space, and promoting an active-learning mode of education, known to be superior to the traditional passive format. Accordingly, this paper briefly discusses the use of IT for improving physics education with an emphasis on resource sharing.

1. Introduction

The current paper focuses on the use of modern information technology to improve physics education for the future. For a general and encyclopedic reference on this subject, the reader is referred to the proceedings of the Conference on Undergraduate Education [1], particularly, the three plenary articles on the application of modern technologies [2,3,4] and articles by the current authors [5,6]. Meanwhile, the current paper is limited to various structural points associated with IT application, plus several examples with productive results are listed with relevant references.

2. Education: Interface & Feedback

Any system of interest is usually part of a larger system with an interface between the interesting part and the rest, plus it can also have a sub-system structure. Since there is no perfectly insulating interface, a sub-system will develop under the influence of the larger system, as such, the two interacting parts change together as one feedback system, which in turn leads to an interface change.

An educational system or activity, surrounded and divided by interfaces, is often characterized by space (location, course, school, district, country, etc.) and time (class period, semester, grade level, academic year, etc.) variables, along with more complex variables, such as academic discipline, educational unit, ethnicity, culture, etc., therefore, the time-dependency and permeability of the interfaces must be taken into account for a better result. Consequently, any education that prepares people for the future should have a feedback mechanism reflecting societal changes and needs, and physics education is no exception to this.

Education is an interactive process involving knowledge-information content between the educators and the educated, thereby combining teaching and learning. Developing societies emphasize the quantitative expansion of the educated population and productive teaching with fewer streamlined courses. However, the demand for quality improvement is bound to follow with a paradigm shift in education from teaching to learning as well as diversification.

3. Paradigm of Physics Education

The current wave of science education reform is driven in part by a post cold war restructuring of the global economy and its associated focus on the education of a more scientifically literate society. Since physics is the foundation of modern science and technology, physics educators have a unique responsibility to educate students in the basic concepts of modern science. Engineers and technology-related professionals need better physics education and industry needs more better-trained physicists. However, there is good reason to believe that current practices are not measuring up. Research data clearly indicates that changes are needed in the methods of physics education [1]. Effective solutions have already been developed, yet go unnoticed by large segments of the education community. Physics education can be made more effective, however, it requires the will to adapt.

Physics education research [7] reveals a wide gap between what is taught and what is actually learnt [8] and that active-learning (interactive-engagement) methods are the key [9] where current IT can be a major support.

4. IT-Based & Active-Learning Solutions

Along with the ongoing transition to knowledge-based economies, equal access to scientific knowledge is a fundamental prerequisite for sustainable development and world peace. Therefore, the use of new information technology (IT), particularly through networking, can significantly contribute to improving the quality of education for all by overcoming barriers, such as time and space, disparity of available funds, and availability of experts among institutions/countries, and promoting an active-learning (interactive-engagement) mode of education, known to be superior to the traditional passive format [9]. It is no wonder that the Science Agenda – Framework for Action adopted at the World Conference on Science (Budapest, 1999) stresses the leading role UNESCO should play in spreading IT applications for science education.

For a well-balanced review, the authors recommend the three plenary articles [2-4] in the encyclopedic reference [1]. The first article discusses the pedagogical, expository, and apprentice modes, the second examines a regular studio class and distance education through a virtual studio, and the third provides various examples of possible uses for computer technology: promoting learning, assessing understanding, providing information, delivering instruction, promoting communication, learning about computer technology, and supporting a classroom learning environment.

The curricular solutions given below for introductory physics are distinguished from other pedagogical tools in that they are research-based. They are developed via an iterative process of research, design, implementation, and assessment, and make use of results from cognitive studies as well as (in some cases) state-of-the-art interactive computer-based technology. For the most part they are constructivist: they guide the student through a highly-structured sequence of activities which enable them to directly confront their misconceptions so that a correct and consistent conceptual structure is developed. Therefore, casual editing of the materials is not recommended, making the implementation quite straightforward. The following list (U.S. source only in alphabetical order) is not meant to be exhaustive, merely representative of this class of materials.

•**Just-in-Time Teaching** [10] enhances interactivity and responsiveness among teachers and students via web-based assignments turned in "just in time" before the next lecture so the teacher can adjust their teaching materials as appropriate.

•**Peer Instruction** [11] actively involves students in large lecture courses by interspersing brief mini-lectures with conceptual questions.

•**Physics by Inquiry** [12] is an inquiry-based course that can also be used with a lecture-based course.

•**Real-Time Physics** [13] is a complete set of interactive microcomputer-based labs.

•**Tools for Scientific Thinking** [14] consists of a small set of interactive microcomputer-based labs.

•**Tutorials in Physics** [15] is a complete set of carefully designed tutorials that can also be used as labs/recitations.

•**Workshop Physics** [16] is an activity-based course without lectures.

5. Educational Resource Sharing

Several points should be noted when utilizing IT for education. A well organized concerted effort is required by more than a few people. Continual updating and improvements are needed, thereby operating as a feedback system. Help is also required from other experts, such as computer scientists, artists, etc.

A depository and clearing house are needed for all the materials for resource sharing.

In resource sharing among different educational units, be it inter-institutional or inter-national, dedicated human effort is essential for success since the paradigm of education is position and time dependent. A one-model-fits-all approach is not appropriate and diversity has to be accepted. The Asian Physics Education Network (established in 1981) [17] has been working to promote resource sharing to improve university physics education in the Asia-Pacific region, with a recent emphasis on active-learning. It should also be noted that the Korean Physics Society has recently placed a strong emphasis on education and educational resource sharing on a national as well as international level [18].

References

1. E. F. Redish and J. S. Rigden (eds.), The Changing Role of Physics Departments in Modern University-Proc. Int'l Conf. on Undergrad. Physics Education (College Park, Maryland, August 1996): AIP Conf. Proc., vol. 399, 1997.
2. I. Johnston, The Place of Information Technology in the Teaching of Physics Majors, Ibid., pp. 343-356, 1997.
3. J. M. Wilson, How Computing and Communications are Changing Physics Education, Ibid., pp. 357-373, 1997.
4. F. Goldberg, How Can Computer Technology Be Used to Promote Learning and Communication among Physics Teachers, Ibid., pp. 375-392, 1997.
5. M. Johnson, D. Johnson and K. Lee, A Pedestrian Guide To Reforming The Introductory Physics Course, Proc. 1998 Summer Seminar : Teaching Introd. Physics using Interactive Teaching Methods and Computers (NSF-Univ. of Oregon).
6. J. S. Kim and K. H. Lee, Information Technology Based Physics Education, in Y. Yao, H. Cheng, C. Chang and S. Lee (eds.), Proc. the 8th Asia-Pacific Physics Conf. (World Scientific, Singapore), pp. 621-624, 2001.
7. L. C. McDermott and E. F. Redish, Resource Letter: PER1: Physics Education Research, Am. J. Phys., vol. 67, pp. 755-767, 1999.
8. L. C. McDermott, Millikan Lecture 1990: What We Teach and What Is Learned -- Closing the Gap, Am. J. Phys., vol. 59, pp. 301-315, 1991, and Guest Comment: How We Teach and How Students Learn – A Mismatch? Ibid., vol. 60, 295.
9. R. R. Hake, Interactive-Engagement vs. Traditional Methods: A Six-thousand-student Survey of Mechanics Test Data for Introductory Physics Courses, Am. J. Phys., vol. 66, pp. 64-74, 1998.
10. G. M. Novak, E. T. Patterson, A. D. Gavrin, and W. Christian, Just-in-Time Teaching: Blending Active Leaning with Web Technology, Prentice Hall, 1999.

11. E. Mazur, Peer Instruction, A User's Manual, Prentice Hall, 1997.
12. L. C. McDermott and the Physics Education Group at the University of Washington, Physics by Inquiry, Vols. I and II, John Wiley & Sons, 1996.
13. D. Sokoloff, P. Laws and R. Thornton, RealTime Physics, Vernier Software, 1995.
14. D. Sokoloff and R. Thornton, Tools for Scientific Thinking, Vernier Software, 1995.
15. L. C. McDermott, P. S. Shaffer, and the Physics Education Group at the University of Washington, Tutorials in Introductory Physics, preliminary edition, Prentice Hall, 1998.
16. P. Laws, Workshop Physics Activity Guide, John Wiley & Sons, 1997. http://www.swin.edu.au/physics/aspen/
17. Korean Physical Society and AAPT Sign Agreement, Announcer, vol. 31, p. 10, AAPT, Summer 2001.

*Supported by the Korea Science and Engineering Foundation

BORDERLESS PHYSICS EDUCATION: CONSTRAINTS AND CHALLENGES

VIVIEN M. TALISAYON
University of the Philippines
National Institute for Science and
Mathematics Education Development

Borderless physics education faces constants and challenges in the 21st century. Teaching and learning borderless physics can mean universal physics concepts, skills and values, physics in daily life, outdoor physics, physics, society, and Internet-based physics. Physics education constants like poverty, untrained teachers, rote learning, negative attitude towards physics are continuing challenges requiring creative solutions. Digital developments such as global laboratory, online courses, and rapid, multiple mass media inputs vying for students' attention pose new challenges to teachers and all sectors involved in physics education. The paper examines issues and explores solutions.

1. Introduction

Almost everywhere physics is still taught within the confines of a classroom. However, the horizon of borderless physics teaching looms. Internet access has expanded the reach of both teacher and students beyond the classroom. Lecture notes and assignments via email and bulletin board, electronic chat and discussion groups, and Internet surfing on worldwide physics education resources have fostered teacher-student interactions and widened their perspectives.

In this global village, the same goals in physics education remain. Physics teachers and educators strive for one or more the the following (Fig. 1): meaningful learning, e.g., constructivist approaches for concept learning; practical work, inquiry, and development of higher thinking skills; physics is fun; development of scientific attitude and values; physics teaching-learning in different contexts, that is, daily life community, physics-technology-society; and physics interfacing with other sciences/disciplines.

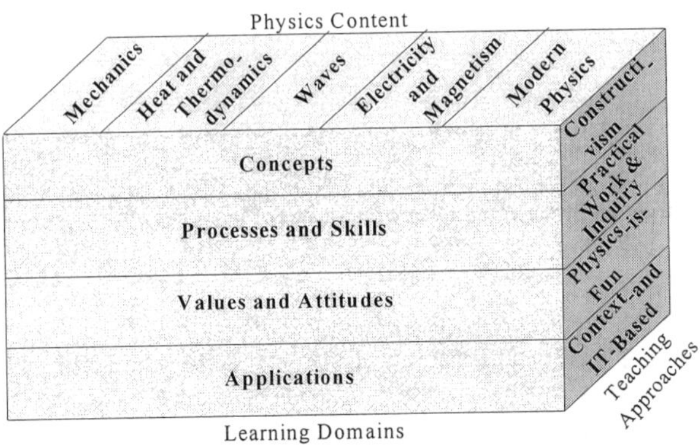

Figure 1. A Model for Physics Education for the 21st Century

This paper presents physics education constants and challenges in an increasingly IT-based society and implications for decision-makers, educators and teachers.

2. Physics Education Constants

Poverty. The lack of school and home educational resources for many poor countries is still a hard reality. Large classes are a direct result of lack of classrooms and teachers. TIMSS 1999 data indicate a class size average of 30 students, whereas poor countries like the Philippines have an average of 50 students. Classroom management of practical work remains a real challenge in high-density classrooms where teachers and students can hardly move around. Low-cost or no-cost materials continue to be viable learning aids in the face of poverty.

Untrained Teachers. At the heart of an effective teaching-learning process is the teacher. Poverty, low salary, and increasing population exacerbate the lack of qualified teachers. A digital world puts additional demands on the teaching skills of teachers. Under these conditions, pre-service education and in-service training face problems in producing physics teachers who can attain goals in physics education despite lack of teaching resources in high-density classrooms in an increasingly IT-based society.

Rote Learning. It takes a trained teacher for meaningful learning to take place in a physics classroom. In the Philippines, poverty, untrained teachers and rote learning are related problems that have persisted despite decades of educational reforms.

Creativity, (e.g., using low-cost measures and materials), dedication and perseverance on the part of school authorities and teachers to implement school-based mentoring for less effective teachers without deloading or additional compensation are possible solutions. Reducing teaching load or additional compensation is difficult to implement with lack of financial and human resources.

External grants and loans seem to have had only short-term or limited effects on the teaching-learning process in a recipient country like the Philippines. Creativity and determination are required of all implementing parties to spread and sustain the gains of an externally- funded project after its completion when lack of resources, again, prevails. _

Gender Differences. In student achievement at the secondary level, physics is still male-dominated. The TIMSS (Third International Mathematics and Science Study) 1995 results show that in 64% of 39 countries, Grade 8 male students significantly ($p<.05$) did better than the female students in physics (Mullis et al. 2000). In TIMSS 1999 (Fig. 2), the boys had, again, a significantly higher achievement in physics than the girls (Martin et al. 2000). Chemistry and earth science had similar results; in biology, the gender difference was not significant.

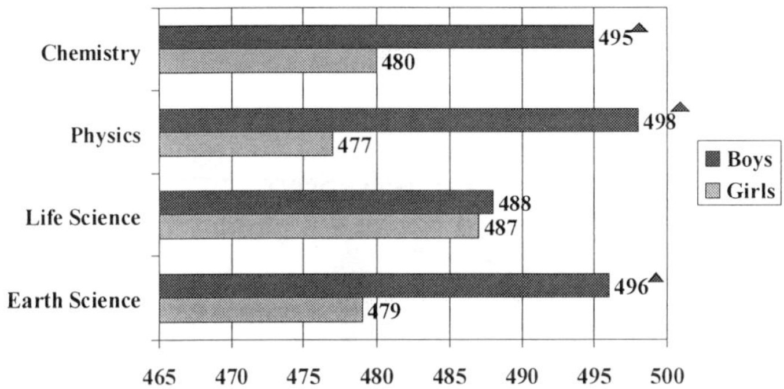

Figure 2. TIMSS 1999 Science Achievement Test Mean Scores*

*Source: Martin *et al.* 2000

▲ Gender mean difference significant at .05 level

Could this be due to mathematics performance, mathematics being a language of science, particularly, physics? In TIMSS 1999, the boys significantly ($p < .05$) performed higher than the girls. In a study conducted by UP NISMED on TIMSS 1999 inter-country data, mathematics score was the single most important significant predictor of science achievement. Using Philippine data for about 4,000 students, mathematics score accounted for nearly 40% of the variance in physics score (Fig. 3).

It will be interesting to determine the reasons for the significantly lower physics achievement of girls in some countries and not in others, including the Philippines. Nevertheless, interventions can be introduced in countries where girls have lower physics achievement. For example, enrichment or remediation classes can address learning difficulties unique to girls.

3. New Challenges

Students' Negative Attitude. Among the sciences in TIMSS 1999, physics had the lowest percentage of students with high positive attitude towards science and the highest percentage of students with low positive attitude (Fig. 4). General Science had the highest percentage of students with high positive attitude. For gender differences, a significantly ($p < .05$) higher percentage of boys taking the physics test had a high positive attitude towards science; for medium and low attitude, the percent of girls was significantly higher (Fig. 5). This was the pattern, too, for all the other science areas, except biology where results were reversed in gender.

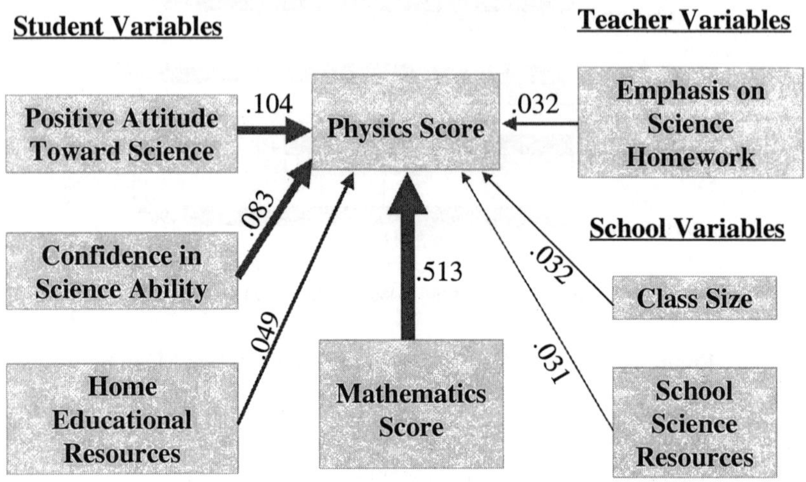

Figure 3. Significant Predictors of Physics Score, TIMSS 1999 Philippines

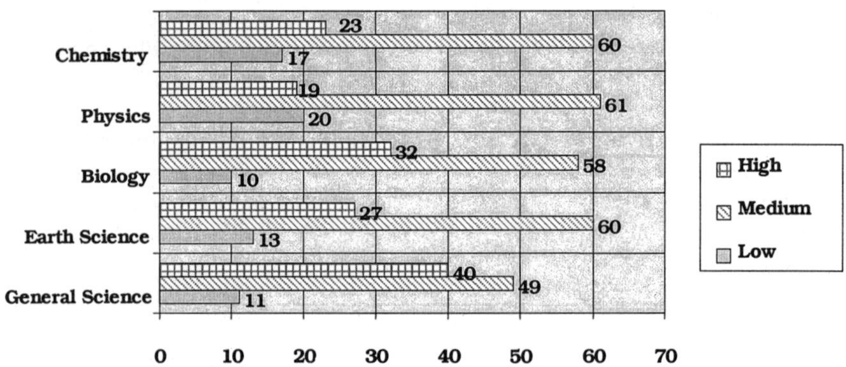

Figure 4. TIMMS 1999 Percent of Students with Positive Attitude Towards Science*

*Source:Martin et al. 2000

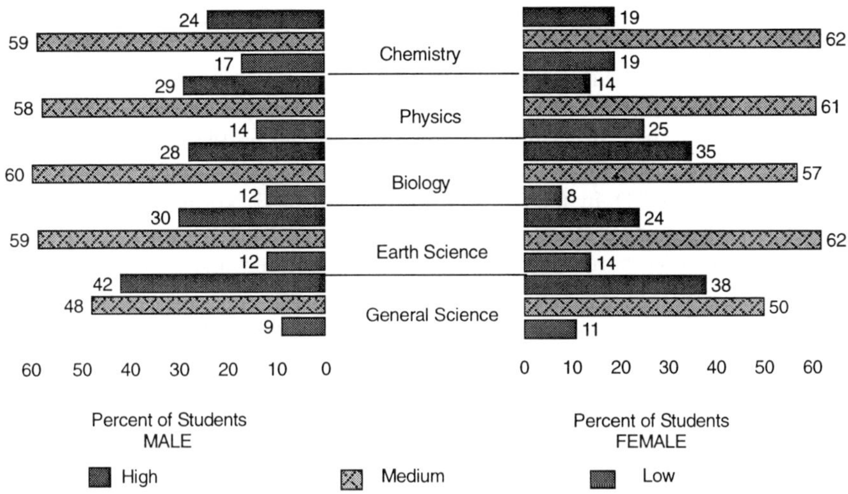

Figure 5. TIMSS 1999 Percent of Students with High, Medium and Low Positive Attitude Towards Science

A study of UP NISMED on TIMMS 1999 data using country as the unit of analysis showed attitude towards science as a negative predictor of science score. High-performing countries had lower attitude mean ratings. Further studies on underlying reasons and appropriate interventions are needed to improve students' attitude towards science, particularly, physics.

Digital and Global Laboratory. Since the eighties, a number of collaborative science research projects have been undertaken, linking students and their teachers within and outside the United States. These include Earth Lab, WaterNet, NORSTAR, Kids Network, MIX, TERC Star Schools Project (Roberts et al. 1990). The Philippine Science High School participates in the Global Lab (based in the United States) where students in several countries interact with scientists, including physicists.

Microcomputer-based physics laboratory software and hardware and simulation, interactive physics software in CD-ROMS for students at all levels are commercially produced in countries like the United States, England, and Netherlands. Truly, ICT has facilitated communication among physics teachers, students and physicists, enabled students to have insights and data analysis that would be difficult to do in conventional laboratory setups, and provided enrichment and remediation even at the individual level.

Online Courses. The online mode is increasingly used, especially for undergraduate and graduate education. One example is the I.MO. PHY Net Course for teacher education in seven universities in Italy (ICPE 2000). An online precursor is courses electronically linked by means of chat, email, and homepage. A project of the International Commission on Physics Education in November 2000 was the e-linked course of the Seoul National University in Korea, UP NISMED in the Philippines and Guangxi Normal University in China, which demonstrated the enhancement of the teaching-learning process by the simultaneous participation of inter-country lecturers and students.

Competing Multimedia Inputs. Television via cable or satellite, cell phones, video games, compact discs, and Internet websites easily attract the youth and provide instant access to vast, rapidly changing information to and from many points in the world. Traditional non-ICT modes or methods of teaching suffer in comparison due to slower information flow and less mechanisms to get and hold students' attention. Therefore, competing multimedia inputs are new challenges to a teacher's ability to hold students' attention in physics, a subject negatively perceived even before the information age.

4. Conclusion

Poverty in the midst of ICT poses a dilemma for countries that have to prioritize scarce resources. Will limited budget be used for basic science equipment and supplies, or will ICT hardware and software be purchased for teaching purposes? The Philippines adopted a two-pronged approach: it gives importance to ICT in education while striving to provide for basic educational requirements.

ICT adds dimension to the resources and learning interfaces (Fig. 6). Policy-makers in educational systems with rich and poor schools face the responsibility of providing high ICT access to poor students. The private sector, including parents' organizations, can narrow the digital divide within and between countries.

ICT can be harnessed, for example, for global learning to achieve the same goals more efficiently for a greater number of people. Some have argued that ICT in a distance education or online mode may be cost-effective and, therefore, appropriate even for schools in remote areas that have little access to human and material resources.

With or without ICT, meaningful learning remains a goal for both poor and rich students (Fig. 6). Thus, approaches promoting meaningful learning need to be considered in designing ICT-based instructional materials.

The young have easily embraced ICT, making it a potent means of transforming students' negative attitude towards physics. Unique capabilities of ICT in the hands of a qualified teacher can facilitate the teaching-learning process, strengthen teacher-student interactions, and maximize the knowledge resource base for both teachers and students. However, continuing vigilance needs to be exercised on possible ICT misuse, overuse or overdependence, and abuse.

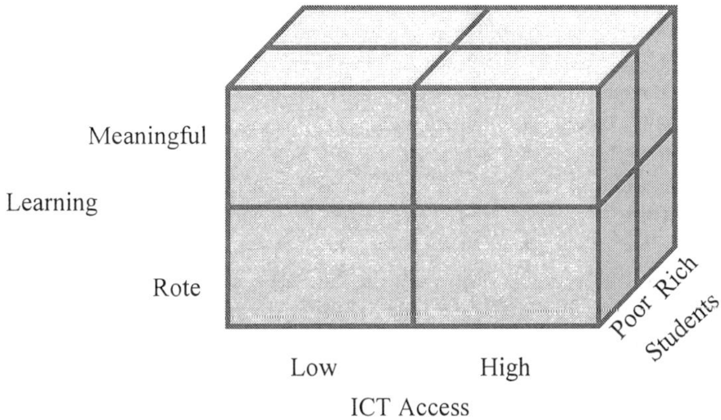

Figure 6. ICT, Resources and Learning Interfaces

While taking up the new challenges, countries affected by physics education constants like low educational resources, untrained teachers, role learning, gender differences need to address these problems creatively, systematically, and aggressively. Sharing of resources and expertise at the individual, community, country and international level may be the key, and ICT may be able to mediate the transformation process.

References

1. Roberts, N., Blakesless, G., Brown, M., and Lenk, C. (1990). Integrating Telecommunications Into Education. New Jersey: Prentice Hall.
2. Martin, M., Mullis, I., Gonzales, E., Gregory, K., Smith, T., Chrostowski, S., Garden, R., and O'Connor K. (2000). TIMSS 1999 International Science Report. Massachusetts: TIMSS International Study Center, Boston College.
3. Mullis, I., Martin, M. Fierros, E., Goldberg, and Stemler, S. (2000). TIMSS Gender Differences in Achievement. Massachusetts: TIMSS International Study Center, Boston College.
4. International Commission Physics Education Newsletter, October 2000, p.5-8; April 2001, p.1.

TEACHING PHYSICS AND THE ARTS

THOMAS D. ROSSING
Physics Department, Northern Illinois University, DeKalb, IL 60115 USA

Physics has much in common with the fine arts. Learning physics will greatly increase one's understanding of art, and learning about applications of physics to the arts will enhance one's understanding of physics. We discuss courses in physics and the arts, including textbooks, laboratory experiments, videos, and Internet sites

1. Introduction

Physics has much in common with the fine arts. Both are humanistic; physics can be described as the way physicists--human beings--view nature. Both have aesthetic appeal; there is great beauty in physics just as there is great beauty in the arts. Learning physics will greatly increase our understanding of art. It is possible for a painter to learn, by endless trial and error, how to combine pigments in a way that is pleasing to the eye, but how much better to learn about additive and subtractive color mixing and how the eye perceives and interprets color.

Likewise, it is possible to learn to play a musical instrument by endless practice alone, preferably with the help of a skilled teacher, but how much better to learn about how musical instruments produce sound. Similarly, learning about applications of physics to the arts will greatly enhance understanding of physics. Few of us have the ability to understand physics as a pure abstraction. In my own case, I understand a law much better if I can tie it to an interesting application. And to most people, sound and light are interesting.

This paper discusses courses in physics and the arts at both the high school and college levels. We focus especially on courses in light and color for students interested in the visual arts and courses in musical acoustics for students interested in music. Textbooks, laboratory experiments, videos, and Internet sites will be reviewed.

2. Sound and Light

Son et lumiere. Sound and light form the basis for some very popular presentations at historical monuments all over the world. These appealing media, which catch the attention of tourists, can also help us catch the attention of our physics-wary students. Although the study of sound and light was once included in most introductory physics courses, these subjects are generally short-changed these days.

Physics and physical science courses have traditionally begun with Newton's laws; yet, current research in physics education reminds us that many students retain misconceptions and are unable to apply mechanical principles to everyday life. This appears to apply to students who have had calculus-based courses, algebra-based courses, courses in high school, courses in college, lecture courses, hands-on workshop courses and almost any combination of the above. Yet we insist that students must "understand" Newtonian mechanics before going on to something else, such as wave motion. At many universities, the general physics course has essentially become a course in mechanics and electric fields.

On the other hand, many high school teachers, who are probably closer in touch with their students, are now beginning the study of physics with light, which the students find understandable and attractive. Certainly the thing I remember best from studying physics in college many years ago, is the optics laboratory. Experimenting with light, albeit with primitive equipment by today's standards, was fun. Here was an area where one could literally see what was going on and obtain precise results. Light and color have a strong aesthetic appeal to me; I love art museums.

3. Physics of Music

Similarly, I love music, although it wasn't until I was a teacher myself that I discovered the physics of music. Being asked to teach a course in the physics of music at St. Olaf College 40 years ago did much to shape my career as a teacher. It drew me into doing research in musical acoustics, and I began to write acoustics books (10 of them, at last count), because other physicists were not doing so. I became active in the Acoustical Society of America, and I met some very interesting musicians.

Over the years, we have had the opportunity to teach courses in sound to many high school teachers, as well as to future high school and college teachers in our physics teacher preparation program. Teachers who have taken our courses find that their students thoroughly enjoy learning about the physics of musical instruments and high fidelity sound reproduction, for example. A series of articles on the physics of high fidelity sound reproduction [The Physics Teacher **17**, 563 (1979); **18**, 278 (1980); **18**, 426 (1980); **19**, 293 (1981); **22**, 84 (1984); **25**, 556 (1987); and **32**, 458 (1994)] has been widely read by students as well as teachers.

Acoustics is the science of sound, and Musical Acoustics is the science that deals with musical sound. The term musical acoustics has become almost synonymous with physics of music, although it is a little broader, perhaps, because it includes the psychoacoustics of musical sound as well. In fact, musical acoustics deals with three main topics:

- The way musical instruments vibrate and produce sound (the physics of musical instruments);

• The propagation of musical sound from source to listener (via concert hall or recording);

• The perception of musical sound (psychoacoustics of musical sound). In my course Acoustics, Music, and Hearing, I try to cover all three. I begin with a unit on mechanics, physics of vibration, and an introduction to sound. This is followed by a unit on hearing and perception. Then we consider musical instruments, speech and singing, and finally room acoustics. In a second course Audio and Electroacoustics, we consider analog and digital circuits, microphones, loudspeakers, room acoustics, sound recording, digital audio, and computer music. The textbook for both of these courses is my Science of Sound [1]. There are a number of other good textbooks available for courses in musical acoustics [2--4]. In 1987 we produced a compact disc of auditory demonstrations that correlate very well with Science of Sound. The 37 demonstrations illustrate different aspects of hearing and auditory perception, especially the perception of loudness, pitch, and timbre [5].

To complement my textbook The Science of Sound, we have also written a laboratory manual Acoustics Laboratory Experiments, with 52 experiments at the introductory to advanced level and a manual for teachers, Teaching the Science of Sound, which includes many demonstration experiments on sound.

4. Physics of Light and Color

More recently, I decided to create a course on the physics of light for visual artists. This led me to team up with Christopher Chiaverina, an extraordinarily talented and enthusiastic high school physics teacher, in writing a textbook Light Science, which is published by Springer-Verlag [6]. We have tried to write this book so that it can be used to teach physics in high school as well as in college. Each chapter in this book includes experiments for home, laboratory, and classroom demonstration. We assign our students homework based on our "lab in a bag" that they do in their dormitory rooms or on the kitchen table, where they (hopefully) arouse the curiosity of roommates, parents, and friends.

The course begins with the fundamentals of light and color, presents some modern topics, such as holography, optical recording, and photonics, and finally discusses symmetry and optical illusions. The chapters, each of which represents approximately a week's work are: Our World of Light and Color; The Mystery of Light; Reflection, Mirrors, and Kaleidoscopes; Refraction of Light; Interference and Diffraction; Light Sources; Polarized Light; Color Vision; Holography; Photography; Computer Images; Optical Recording, Communication, and Photonics; Symmetry in Art and Nature; Visual Perception and Illusions.

The appendix discusses Materials in Paintings, Analysis of Art Using Physics and Chemistry, and Conservation and Restoration of Art.

Before each class period, while I am setting up demonstration experiments, we show a video. Needless to say, we have quite a collection, which I would be happy to share with anyone else who teaches such a course. Videos and other materials are listed in a Resource Letter on Teaching Light and Color that we published in the American Journal of Physics [7]. Of course there are other good texts for teaching light and color [8-10].

Sound and light have great appeal to many students, especially those with an interest in music and/or the visual arts. Although sound and light may never draw them to physics in the numbers that sex draws them to biology, I am convinced that they can provide attractive routes to science literacy for many students in middle school, high school, and middle school. Whenever I attend local teacher alliances, I hear of interesting experiences in teaching light and sound to high school students as well as to elementary and middle school students.

5. Light and Sound in Courses for Engineers

What role should sound and light play in physics courses for scientists and engineers? An important one, I would argue, especially if concepts are emphasized over theory. Yes, they will eventually have to master mechanics, but if two or three passes through Newton's laws have left them with so many misconceptions, can't that wait another semester or two? The unified approach to wave motion used to be an important foundation of the physics curriculum. Witness the beautiful book on Sound and Vibration by Philip Morse [11], the classic Physics of Waves by Elmore and Heald [12], or more recent books on wave motion by Anthony French [13] and by Frank Crawford [14]. How many physicists understand water waves?

6. Websites for Sound and Light Teaching

6.1 *Demonstration experiments in physics*

Many physics departments list their demonstration experiments on a website. Some sites have detailed descriptions and photographs so that these demonstrations can easily be reproduced. The Physics Instructional Resource Association (PIRA) maintains an excellent bibliography of demonstration experiments.

www.physics.umd.edu/deptinfo/facilities/lecdem/phys102.htm--U. of Maryland
www.physics.brown.edu/studies/demo--Brown U.
www.mip.berkeley.edu/physics/physics.html--U. of California, Berkeley
hendrix.uoregon.edu/~demo/Demo/demo.html--U. of Oregon
www.physics.ncsu.edu/demoroom--North Carolina State U.
www.ph.utexas.edu/~phy-demo--U. Texas
www.hpcc.astro.washington.edu/scied/physics/physdemo.html--U. Washington
www.physcis.unc.edu/~labs/demo/demo.html--U. North Carolina

http://demo2.physics.wisc.edu--U. Wisconsin
http://physics.bu.edu/~duffy--Boston U.
www.physics.uoguelph.ca/demo/demo.html--U. of Guelph
www.phys.vt.edu:80/teaching/demos.html--Virginia Tech
www.wfu.edu/academic-departments/physics/demolabs/demos--Wake Forest U.
www.pa.msu.edu:80/educ/lecdemo.htm--Michigan State U.
www.missouri.edu/~physic--U. of Missouri
www.physics.gatech.edu/demopage--Georgia Tech.
www.phys.ufl.edu/demoprep/data/view/index_win.html--U. of Florida
www.vmi.edu/physics/minnx.htm--VMI
http://astro.physics.sc.edu/S&ST/S&ST.html--"String and sticky tape" demos (Ron Edge, U. South Carolina)

6.2 *Acoustics demonstration experiments*

www.physics.auburn.edu/demo/waves/waves.htm--Acoustics demonstrations (Auburn U.)
physics.indiana.edu/~demos/sounddemo.htm—Acoustics demonstrations (Indiana U.)
http://www.physics.gatech.edu/demopage/3/index3.htm--Georgia Tech.
http://info.phys.uvic.ca/dbr/resman/acoustics.html--U. Victoria
http://www.physics.ncsu.edu/pira/oandw.html--PIRA bibliography of demonstrations on oscillations and waves
www.psrc-online.org--AAPT's Physical Science Resource Center. Click on "demos & labs."

6.3 *Websites for teaching light and color*

www.exploratorium.edu/exhibits/index.html--The Exploratorium, San Francisco
www.artscienceresearchlab.org--Art Science Research Lab.
www.pickover.com--Cliff Pickover's Graphics in Art and Science
www.nationalgallery.org.uk--The National Gallery, London
www.media.mit.edu/groups/spi/hrdcpyimages.html--Spatial Imaging Group, MIT
www.cnam.fr/fractals/mandel1.html
www.sprott.physics.wisc.edu
www.sciam.com/1998/0198issue/0198working.html
www.spie.org/web/abstracts/1600/1667.html
www.lfc.edu/physics/holography
www.holoworld.com/index.htm
www.scottkim.com
www.chaos.umd.edu/dsweet/spheres
www.grand-illusions.com/index.htm

www.lhup.edu/~dsimanek/3dillus1.htm
nths.newtrier.k12.il.us/academics/math/Connections/connections.htm
tqd.advanced.org/12740/netscape/index.html

7. Conclusion

I have had lots of fun teaching the physics of sound over the years, and now I am having a lot of fun learning to teach the physics of light. I enjoy teaching these subjects to physics and engineering students as well as to students majoring in other fields. I enjoy teaching them to advanced students, and I enjoy teaching them to teachers who (hopefully) will teach them to their students!

References

1. T. D. Rossing, Science of Sound, 2nd ed., Addison-Wesley, Reading, MA, 1900.
2. M. Campbell and C. Greated, The Musician's Guide to Acoustics, Schirmer Books, New York, 1987.
3. D. E. Hall, Musical Acoustics, 2nd ed., Brooks/Cole, Pacific Grove, CA, 1991.
4. W. J. Strong and G. R. Plitnik, Music Speech Audio, Soundprint, Provo, UT, 1992.
5. A. J. M. Houtsma, T. D. Rossing, and W. M. Wagenaars, Auditory Demonstrations (Philips Compact Disc 1126-061 and text), Acoustical Society of America, Melville, NY, 1987.
6. T. D. Rossing and C. J. Chiaverina, Light Science, Springer-Verlag, New York, 1999.
7. T. D. Rossing and C. J. Chiaverina, Resource Letter TLC-1: Teaching light and color, Am. J. Phys., vol. 68, pp. 881-887, 2000.
8. D. Falk, D. Brill, and D. Stork, Seeing the Light, J. Wiley & Sons, New York, 1986.
9. S. J. Williamson and H. Z. Cummins, Light and Color in Nature and Art, J. Wiley & Sons, New York, 1983.
10. R. D. Overheim and D. L. Wagner, Light and Color, J. Wiley & Sons, New York, 1982.
11. P. M. Morse, Vibration and Sound, McGraw-Hill, New York, 1948.
12. W. C. Elmore and M. A. Heald, Physics of Waves, Dover, New York, 1985.
13. A. P. French, Vibrations and Waves, Norton, New York, 1966.
14. F. S. Crawford, Waves, McGraw-Hill, New York, 1968.

IT-based Physics Education in Japan

KOBAYASHI AKIZO
Faculty of Education and Human Science, Niigata University, Niigata 950-2181, Japan

The current study presents a brief survey of the current status of IT-based Physics Education in Japan. As such, some of the activities and collaborative projects of the NEP (Network for Education of Physics) group and JAVA mailing list group in Japan intended to make university physics education a more attractive and active learning experience are introduced, resource sharing to make interesting educational materials using IT is dicussed, the modularized materials produced by collaboration between the NEP and JAVA mailing list groups are investigated, and some examples of IT-based education on the world of Aristoteles-Ptolemaios and Copernicus-Kepler, Galileo-Newtonian Mechanics, and Galileo-Einstein Relativity are presented.

1. Introduction

Recently, the total number of the Internet users has reached almost 400 million worldwide, and the increasing speeds, especially in Asian countries, are remarkable.

Almost all Japanese universities (649) have extensive Internet connections (10~100Mbps LAN), i.e. all national universities (99), all local public universities (72), and 97% of private universities (478).

Almost all students attending Japanese universities (95%) have an Internet ID [1], plus some large universities have about 10,000 LAN sockets, while others have wireless LAN systems.

Based on a recent survey, about 40% of all educators at Japanese universities use the Internet and multimedia in their lectures [1]. In the Academic Year (AY) 2000, 57.4% of Japanese K12 schools were connected to the Internet, and the current plan is to connect 100% of K12 schools by AY2001 [2]. However, these are mainly by ISDN (64kbps) connections (3.2 lines/school), therefore, the plan is to switch to T1 (1.5Mbps) connections by AY2005. In addition, rapid changes to ADSL LAN or 10~100Mbps LAN have started for home and school networks in city regions.

2. IT-based Physics Education in Japan

In AY2000, the "New Teacher's Certification Law" was initiated in Japan, which includes 2 credits (minimum) of IT- based education. Accordingly, to prepare for this IT-based education, one Japanese university has constructed two wired LAN rooms (with a total of 200 LAN sockets), where each room has about 100 LAN sockets and 100 electric power sockets under each desk. Plus, many rooms are being equipped with wireless LAN systems, as shown in Fig.1.

Figure 1. Wireless LAN system (Air Station™ and Card)

From April AY2000, the same Japanese university started two new IT education courses using student-owned portable computers connected to the above-mentioned 200 LAN sockets, thereby enabling the students total access anytime and anyplace (Fig. 2).

Figure 2. IT education class using portable computers

Before taking the IT course, 34% of the students had no previous IT experience, 58% had a little (could use the mouse), and 8% had reasonable skills (could use the keyboard for word-processing). However, after completing this active learning course, most students were capable IT users.

Further attempts are also being made to create active learning-based education for training courses in science education, based on creating web pages including interesting experiments demonstrated by students, refer to the Quick Time(QT) movie 1,2,3,and 4 web pages.

3. NEP Project and JAVA Mailing List

3.1 *Problems with science literacy Education in Japan*

The serious problems related to physics education in Japan are listed below. An increasing number of students do not learn any physics in high school, even those with scientific majors. Plus, the ratio of physics classes in high school has decreased to 20~30%. Many investigations have shown a significant decrease in scientific abilities and science literacy in Japanese K12 schools.

Such difficulties originate from a mismatch in physics education between high school and university, i.e., physics is not required for the entrance examination of many Japanese universities. In response to this serious educational situation, the following initiatives have been implemented in Japanese universities, as shown in Fig. 3.

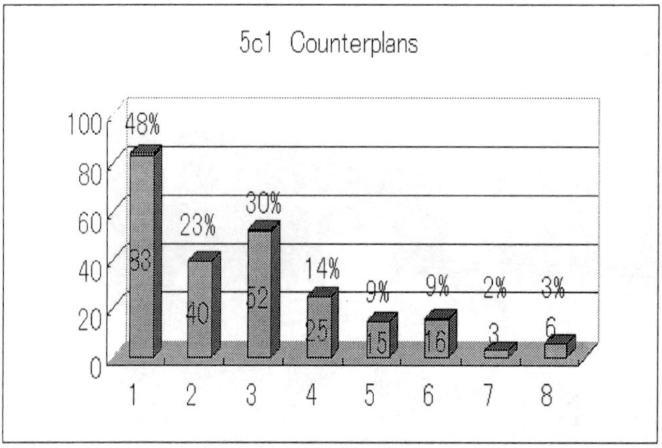

1 - physics courses for beginners, 2 - skills-based courses,
3 - introductory courses, 4 - remedial courses, 5 – distance learning courses,
6 - variety of suitable courses, 7 - only standard courses, 8 - no response [4].

Figure 3. Initiatives implemented in Japanese universities

Therefore, it has become very important to make physics education more attractive and interactive, plus to offer "various types of physics courses" to suit various types of students.

3.2 *NEP and JAVA mailing collaboration*

Japan is now encouraging collaborative activities based on the NEP (Network for Education of Physics) group [4] and JAVA-mailing-list group for the purpose of developing new types of networking systems and sharing basic resources for physics education, e.g. Internet, multimedia, and JAVA simulation.

Many resources including JAVA simulations and multimedia have already been produced by collaborative projects between the NEP and JAVA mailing-list groups [5], as demonstrated by the following web sites :

http://www.nep.chubu.ac.jp/~nepjava/applet.htm
Sun-Earth-Moon System by Kato,
Movement of constellations by Uozu
NEP member's PAGE (http://www.nep.chubu.ac.jp/pvt/index.html).
Web Page for Evaluation
(http://www.nep.chubu.ac.jp/pvt/NEPTop.jp.html) [4].
Modules (http://www.nep.chubu.ac.jp/module/index.jp.html).
http://www.nep.chubu.ac.jp/v17/v17-147.html

4. New Attempts to Teach Historical Development of Physics World [6].

Through the use of JAVA simulations on Web, all kinds of phenomena can be realistically and creatively simulated, even on a huge scale (e.g. macroscopic scale of astronomical world) or minute scale (microscopic scale as atomic world) for display in a teaching environment.

Furthermore, the observer's reference frame can be freely changed, e.g. from a center of mass frame to a laboratory frame, moving frame, or rotating frame etc., using a click button or scroll bar etc..

Here, the usefulness of JAVA technology is demonstrated in the context of changing the reference frame and connecting to different reference frames using a click button and scroll bar.

The following are web pages for teaching the historical development of Galileo-Newtonian Mechanics and Galileo-Einstein's Relativity, and the world of Aristoteles-Ptolemaios and Copernicus-Kepler using JAVA simulation technologies.

- **Historical development of Galileo-Newtonian Mechanics and Galilean Relativity. [6]**

Ball falling from top of mast fixed on moving ship (Figure 4)
Monkey and Canon. [6]

- **JAVA Demonstration of Galilean Relativity by changing inertial frames of reference [6].**

JAVA; Showing Momentum Conservation Law for Collision of Two- Balls by Changing Observation Frame from Center of Mass to Laboratory Frame.(Figure 5) [6]

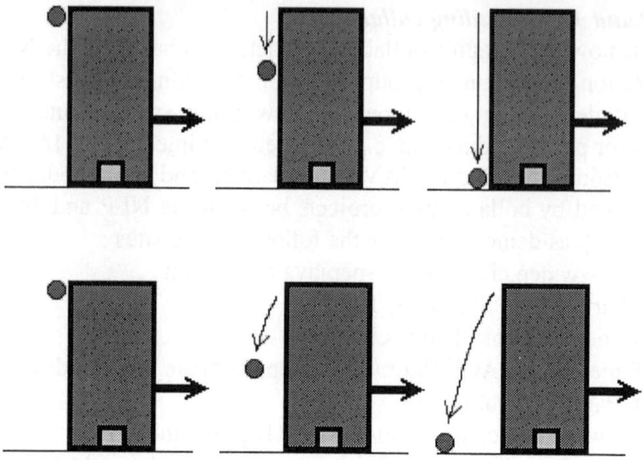

Figure 4. Ball falling from top of mast fixed on moving ship[6]

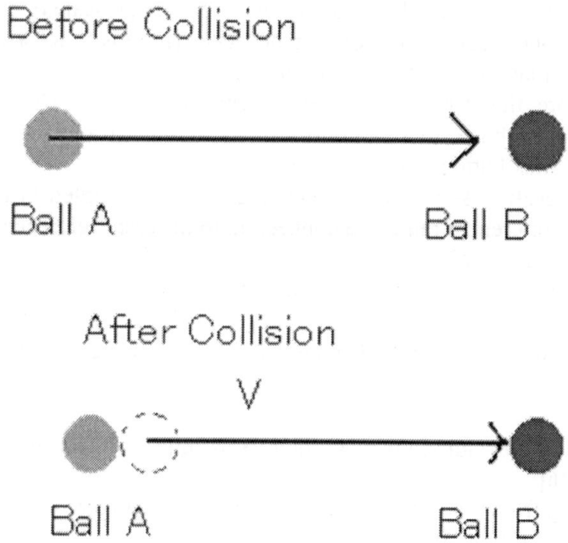

Figure 5. Conservation law in the collision phenomena of two-balls

• **JAVA Demonstration of Einstein's Special Relativity:**

Moving "Light clock" and "**time delay"; JAVA for Case 1.**
Moving "Light clock" and "**Lorentz Contraction"; JAVA for Case 2.** [6]

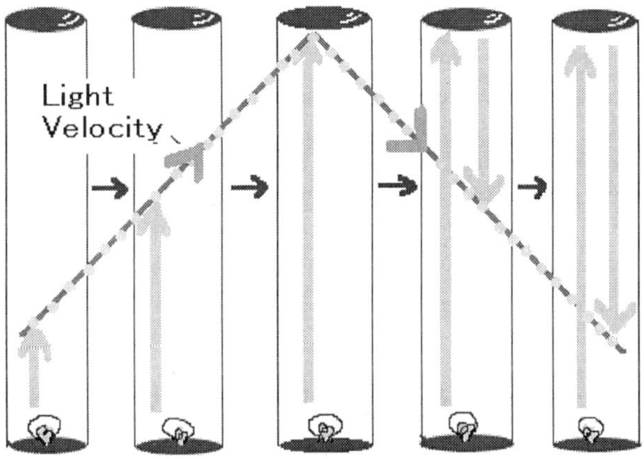

Figure 6. Moving Light clock with invariant Light Velocity

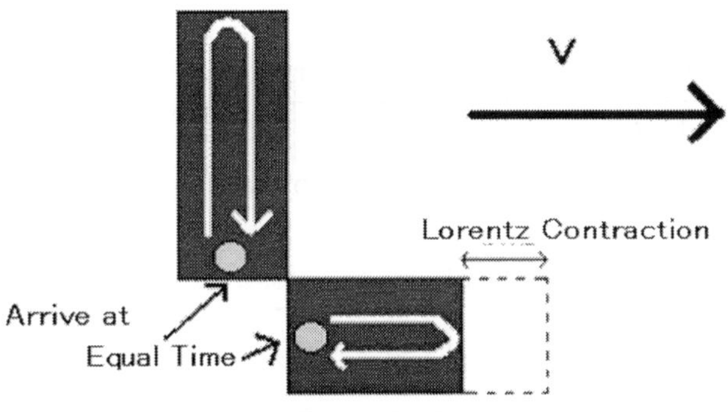

Figure 7. Relationship between Lorentz Contraction and time delay

JAVA Demonstration of Einstein's General Relativity.
JAVA on "Equivalence Principle" for Falling Photon (Curving Light Beam). [6]

The gravitational forces and inertial forces acting on a body, like a ball or photon etc., are equivalent and indistinguishable from each other. [6]

• JAVA simulation of world of Aristoteles-Ptolemaios and Copernicus-Kepler connected by rotating frames of reference; explanation of reverse direction motion in Ptolemaic system. [6]

Geocentric motion of Mars and Venus and heliocentric motion of Mars and Venus, JAVA simulation connecting geocentric motions to heliocentric ones
Jean Bernard Leon Foucault; JAVA on "A Foucault Pendulum at the North Pole".

5. New Attempts for Teaching Newtonian Mechanics

New teaching methods for Newton's' First and Second Laws of Motion are as follows : Using video movie and excel, Using digital movie camera and VideoPoint software, and Using distance sensor and force sensor, etc.

The important problems are 'How to change F = k v to F = ma'.

• Newton's First Law of Motion

First, the motion of a dynamic cart pushed by hand is investigated(Fig. 8). Predictions of the related x-t and v-t graphs (as shown in Fig.9) are useful for "Active Learning".

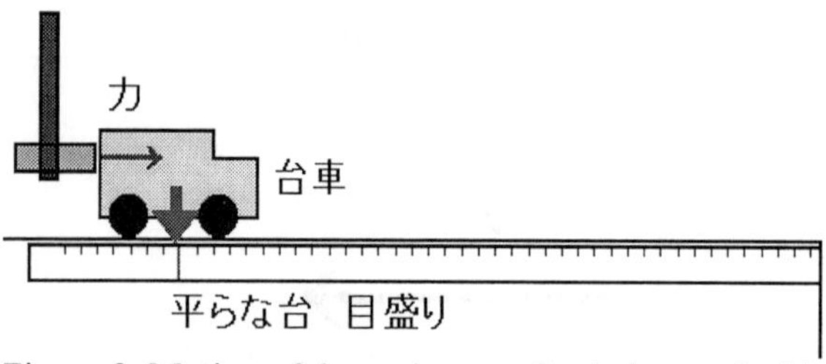

Figure 8. Motion of dynamic cart after being pushed by hand

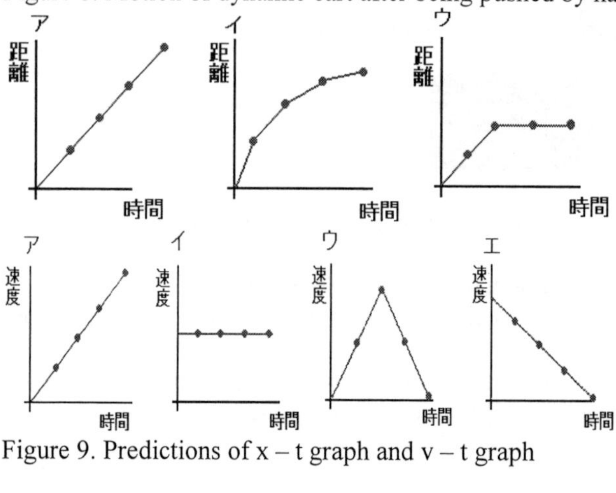

Figure 9. Predictions of x – t graph and v – t graph

• **Newton's Second Law of Motion**

As shown in Fig. 10 and Figs. 11,12, and 13 in section 6, the 4 movements are represented by 4 blocks with different heights, and, after touching one, there are no other forces acting on the cart system.

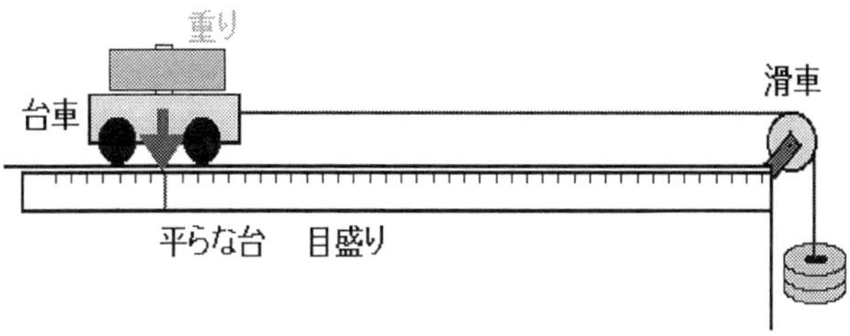

Figure 10. Seven coins with a hole (5 Japanese yen) are used as weights

The predictions for the x-t and v-t graphs(as shown in Fig.9) are down for movements 1-4. After disscussing these predictions, and showing real-time experiments of these movements, the relation between force and acceleration are investigated.

Relationship between force and acceleration

x – t graph and v – t graph

Relationship between mass and acceleration

6. Video-Point Analyses of Newton's Law

The relationship between force and acceleration can be quickly and convieniently analyzed with the use of a digital movie camera and VideoPoint software, as shown by the following sample VideoPoint analyses : (1), (2), (3), (4) and (5).

• The 4 movements of the cart can be investigated by a VideoPoint analysis as follows :

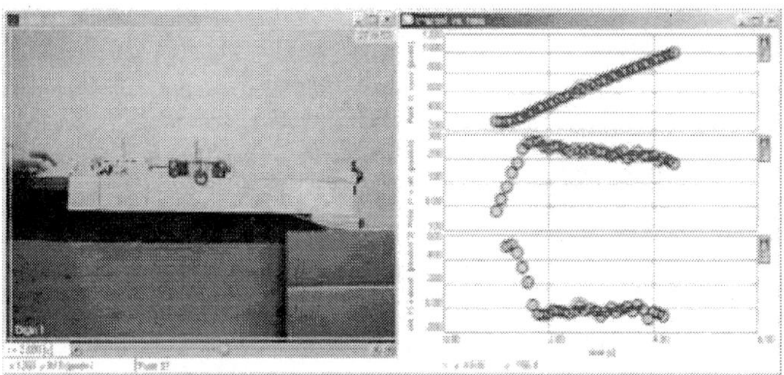

Figure 11. Movement 1

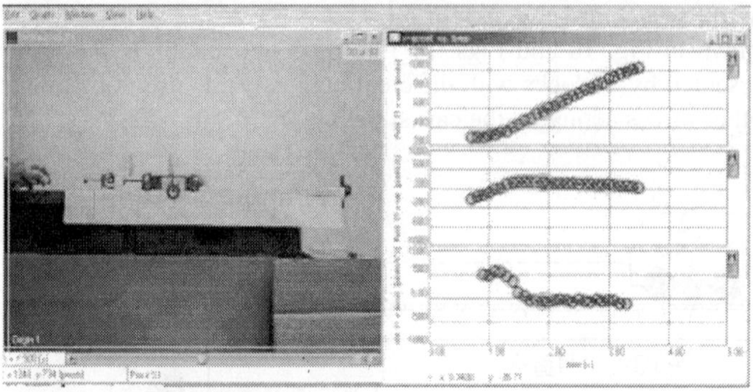

Figure 12. Movement 2

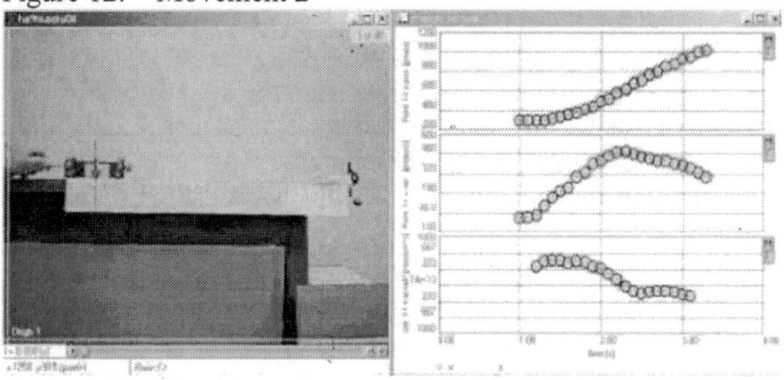

Figure 13. Movement 4

• Movement of bowling ball (Figure 14)

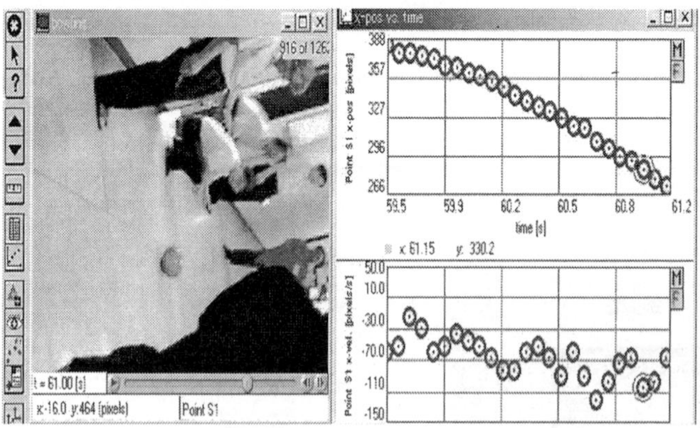

Figure 14. Movement of bowling ball

• Dry ice movement with constant speed(Fig. 15)

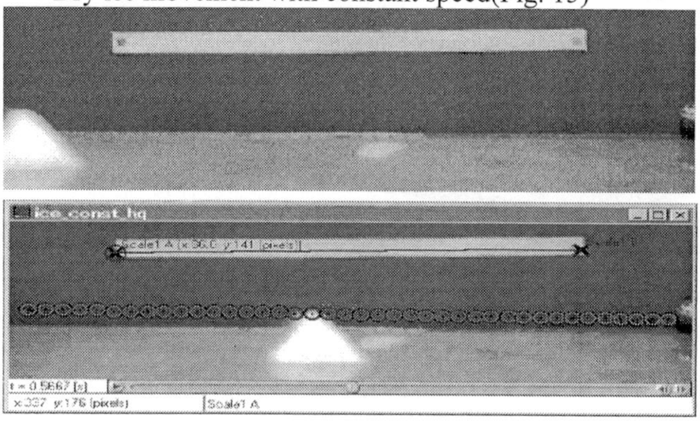

Figure 15. Movement of dry ice

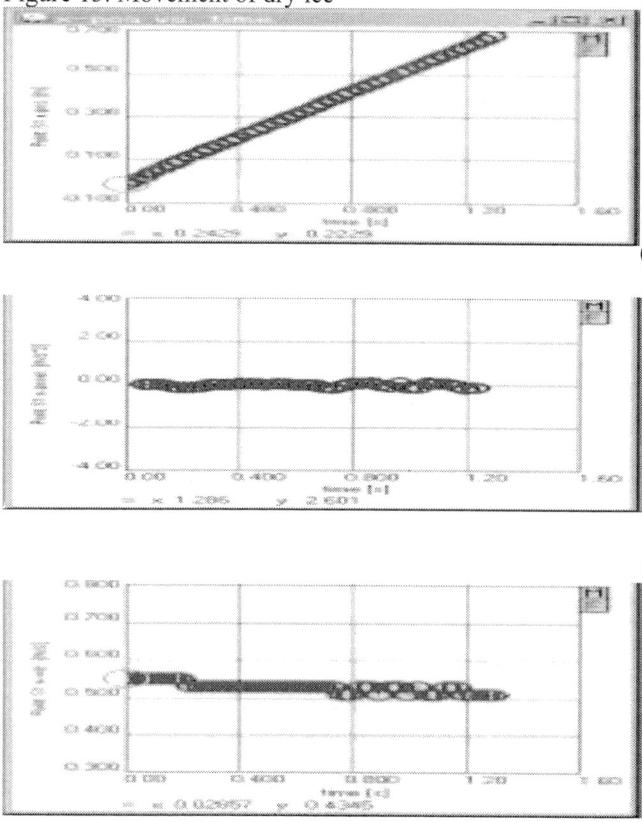

(a) x-t graphs

(b) v-t graph

(c) a-t graphs

Figure 16. VideoPoint analysis of dry ice movements

• **Cart with fun movement and constant acceleration**.

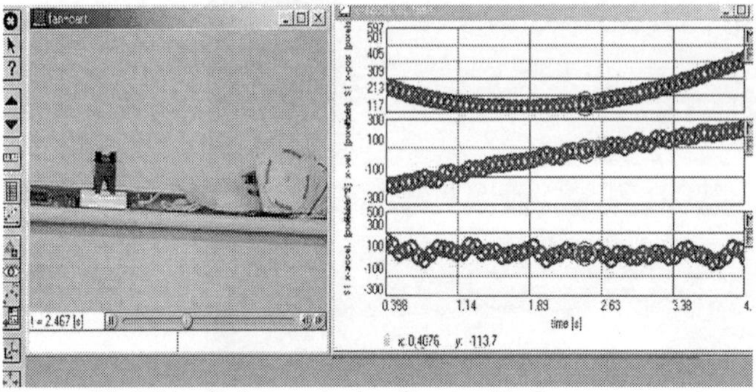

Figure 17. VideoPoint analysis of cart with fun movement

• **Basket ball movement with constant acceleration**.

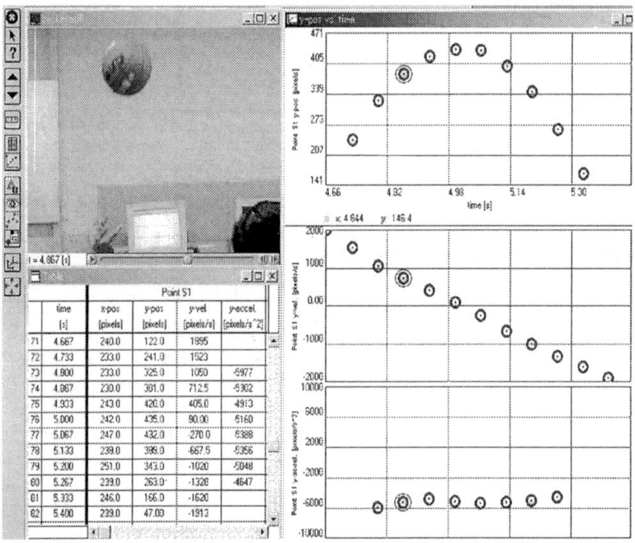

Figure18. VideoPoint analysis of basket ball movement

• **Three typical cases of falling motions**: (a) No friction, (b) Moderate friction, (c) Large friction.

(a) No friction (constant acceleration of 980m/s^2)

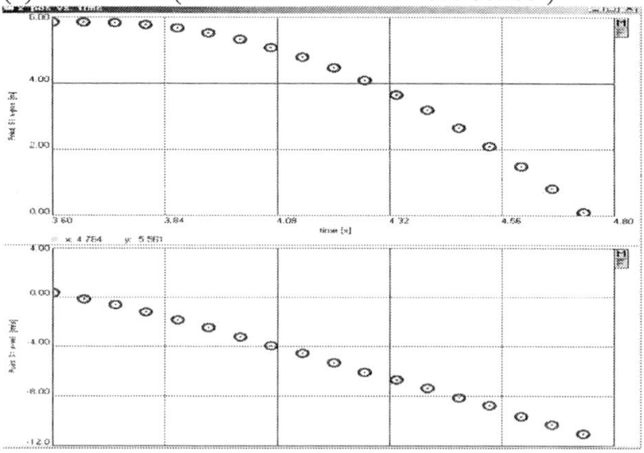

(b) Moderate friction (constant last-velocity after increasing friction with increasing velocity)

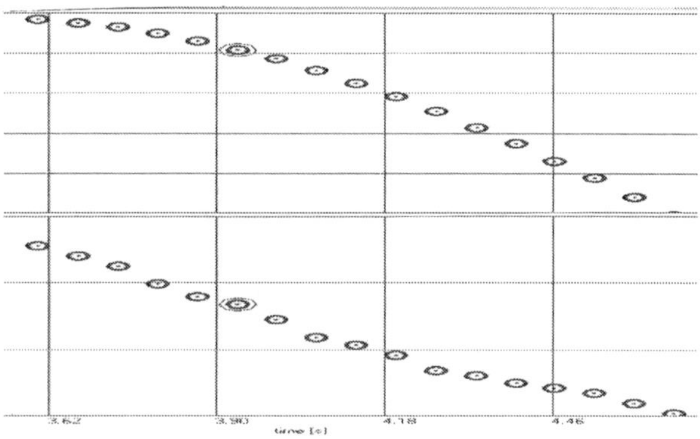

(c) Large friction (quickly reach constant last-velocity)

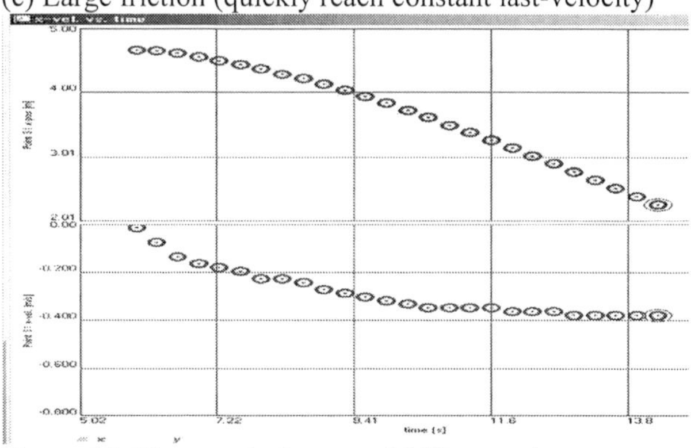

Figure 19. Three typical cases of falling motion

It should be noted that case (c) corresponds to the motion laws of Aristoteles-Ptolemaios, where the frictional force is just balancing gravity, then it become a constant motion in the last-velocity due to a totally zero net force.

Therefore, today there are many possible ways of investigating quick motions, for example, a digital-movie-camera and VideoPoint software, distance sensors with force sensors etc., all of which were unavailable in the Galileo-Newtonian era.

Consequently, there are many ways to significantly improve the teaching methods used to explain Newton's' First and Second Laws of Motion, as shown above.

In particular, VideoPoint annalyses are especially suited for very quick motions, such as falling balls, while distance sensors analyses are more suitable for very slow motions, such as a cart.

7. Concluding Remarks

The use of information technology enables a wide variety of materials to be presented, such as graphics, movies, and sounds etc., plus many complicated phenomena can be simply visualized through JAVA simulations.

Recently, it has become very important to offer various types of physics course to suit various types of student. As a result, many collaborative projects have been initiated by groups such as NEP and JAVA-ML etc..

The quality of physics education can be drastically improved through the use of IT, such as digital movie cameras, Java simulations, VideoPoint analyses, distance sensor analyses, and force sensor analyses etc.

The use of IT makes it possible to implement rapid innovations in physics education and drastic changes in science education, thereby offering bright prospects for physics education in the future.

References

1.Internet White Paper 2001, Internet Association of Japan, Impress Corporation, July 1, 2001.
2. News of Ministry of Education, Science, Sports and Culture, January 1999; http://www.mext.go.jp/
3. http://kakuda.ed.niigata-u.ac.jp/semi/semi.html.
4. NEP(Network for Physics Education)、 http://www.nep.chubu.ac.jp/
5. JAVA Physics ML (buturijava@kaku.ed.niigata-u.ac.jp, http://www.nep. chubu.ac.jp/~nepjava da /)
6. JAVA Collection and various enjoyable experiments, http://kakuda.ed.niigata-u.ac.jp/semi/java/inde

PHYSICS EDUCATION COURSE IN VIRTUAL UNIVERSITY PROJECT AT KYUSHU INSTITUTE OF TECHNOLOGY

Y. CHIKAURA*, Y. SUZUKI*, T. MATSUSHITA°, E. S. OTABE°, K. KAKIZAKI° & F. KOBAYASHI#

Dept. of Materials Science, Fac. of Eng., Kyushu Institute of Technology, Kitakyushu, 804-8550, Japan,

°Dept. of Computer Science and Electronics, Fac. of Computer Science and Systems Eng., Kyushu Institute of Technology, Iizuka, 820-8502, Japan

#Dept. of cntrol Eng. and Science, Fac. of Computer Science and Systems Eng., Kyushu Institute of Technology, Iizuka, 820-8502,

The Virtual University (VU) project was conducted by Kyushu Institute of Technology (KIT) as a joint research and development effort with the Japanese Ministry of Education, Culture, Sports, Science and Technology (MEXT), National Institute of Multimedia Education (NIME), and Japan Advanced Institute of Science and Technology (JAIST) from 1999 to 2001. The role of KIT in this project was the production of educational materials for undergraduate science and engineering. The subjects are as follows: a) Physics; b) Introduction to LSI; c) Information Technology. The VU physics course includes introductory mechanics and electromagnetism. For the mechanics section, a digital textbook was prepared using *Mathematica* with the contents transcribed from the recently published book, "Mechanics Exercises with *Mathematica*" (1999, Baifukan Co. Ltd., Tokyo). Plus, significant emphasis is placed on the student conducting physics exercises using a computer. For the electromagnetism section, special attention was paid to providing a unified theory for application to practical problems involving various engineering situations. Accordingly, the current report gives an overview of the process used to develop the VU physics program, along with insights on the significance of a VU based on computers and communication networks as a new education system.

1. Introduction

Two of the current authors (Y.C. & Y.S.) have already published several textbooks [1-4] on computer-aided physics education intended for entering students at the Kyushu Institute of Technology (KIT). These textbooks require students to work on exercises and handle computer source-codes. Recently, application software for algebraic computation and graphics, such as *Mathematica*, have also become available. Plus, the current authors previously released a textbook entitled "Elementary Mechanics Exercises Using *Mathematica*"[5] for students in basic science and engineering courses. In addition, the NEP (Network for Education of Physics) project, a joint effort involving several Japanese universities, has transformed a lot of physics education materials into digital media, which were presented at a recent NEP project conference [6]. Consequently, all the above efforts are contributing towards the construction of a new physics educational system in the next decade.

At KIT, various materials were initiated as part of the virtual university (VU) program. The objective of the VU program is to provide a higher education service to the public. Figure 1 shows a schematic diagram of the VU services. The present authors were involved in creating the materials for the physics component of the program, namely, elementary mechanics and electromagnetism. Accordingly, the current paper reports on the process used to make the materials and the experience gained while making them.

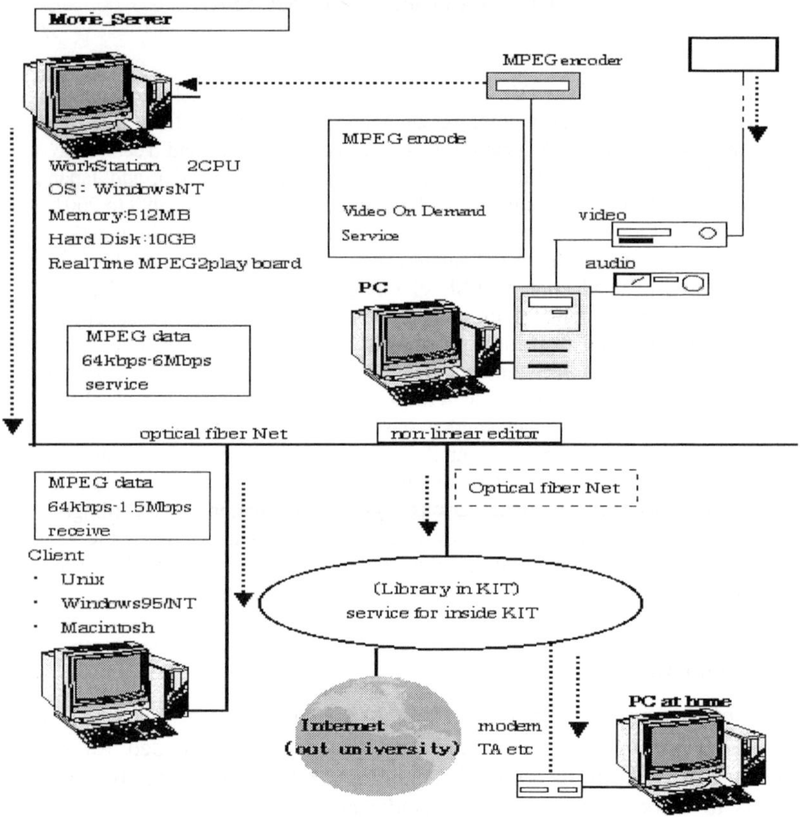

Figure 1. A schematic diagram of VU broad service

2. Production of Educational Materials

The process used to make the educational materials was comprised of the following steps:

Step 1 : Taking video clips;

Step 2 : Creating animated segments and pictures.

Step 3 : Making slides, including equations and figures.

Step 4 : Editing.

Step 5 : Making home pages.

In step 1, assistance was provided by professional cameramen. For steps 2 – 5, the current authors incorporated their own lectures using a personal computer (PC). In step 4, due to the amazing recent development of PC technology, i.e., non-linear editor systems, the editing was quite easy, which allowed more flexibility to improve the educational materials. As for step 2, the creation of animated segments was straightforward when using *Mathematica* or a PC. Consequently, the generation of multimedia educational materials using this procedure was relatively easy.

The figures and equations were made based on the draft of the textbook, "Mechanics Exercises using *Mathematica*", which was created by *Tex*. The use of *Tex* also made the equations appear neat and clear. The VU project committee required that the digital materials were made by the project participants themselves (except for the video component). Plus, they also

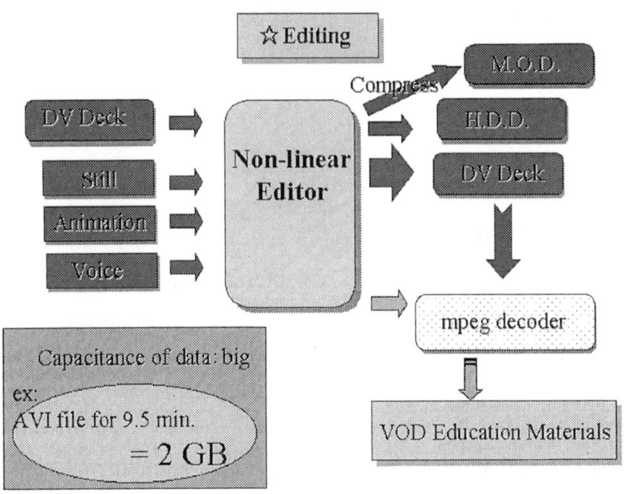

Figure 2. Schematic diagram of editing process

emphasized the importance of high-quality pictures.

The first draft was made into a slide-show using the 'Power Point' application. The equations were generated using the '*Tex*' and 'Mathtype' applications, and the figures using '*Mathematica* Notebook', 'Autoshape', 'Photoshop', 'Draw', *etc.* The editing was performed using a non-linear editor. Figure 2 is a schematic diagram of the editing process. The stills, animation, narration, and movies taken by the professional cameramen were transferred to the non-linear editor through a digital video (DV) deck. Various media, such as magneto-optical discs, hard discs, and digital video tapes were used to store the edited digital data. In general, the video and audio data files were very large. (For example, an AVI file for a 9.5 min. segment was about 2 GB, as such, became difficult to operate.) Therefore, the microsoft AVI files were transformed into MPEG-style files using an MPEG decoder, thereby completing the VOD Education Materials. A Home Page and Real Player were also useful in the production process.

3. Education Materials

3.1 *Mechanics*

Table I summarizes the contents of the 'Introductory Mechanics with Mathematica'. An example of the home page is shown in Fig. 3. The VU physics course includes introductory mechanics and electromagnetism. For the mechanics section, a digital textbook was prepared using *Mathematica.* The contents were transcribed from the book recently published by the current authors, "Mechanics Exercises with *Mathematica*" (1999, Baifukan Co. Ltd., Tokyo)[5]. A significant emphasis was placed on student exercises in physics using a computer.

3.2 *Electromagnetism*

The Electromagnetism contents are as follows:

I. Static electric phenomena

Chapter 1 Static electric fields   Chapter 2 Conductors

Chapter 3 Dielectric materials   Chapter 4 Steady Currents

II. Static magnetic phenomena

Chapter 5 Static magnetic fields due to currents, Currents and magnets

Chapter 6 Superconductors   Chapter 7 Magnetic materials

III. Maxwell's equations

Chapter 8 Electromagnetic induction

Chapter 9 Displacement currents, Maxwell's equations

An example of the home page is shown in Fig. 4. For electromagnetism, special attention was paid to providing a unified theory for application to practical problems involving various engineering situations.

Figure 3. An example of home page in Mechanics

Figure 4. An example of home page in Electronics

Table 1-a Contents of Mechanics Course: Elementary Mechanics using Mathematica

total time 11:48:08

Vol. 1	**Mathematica & Mechanics**		1:16:58
0-1	Contents		0:13:23
0-2	Introduction		0:13:23
	Chapter 1		
1-.1	Basic Operation of Mathematica		0:17:57
1-2.	Mathematica functions		0:06:07
1-3.	Physical Constants		0:26:08
Vol. 2	**Kinematics & Vectors**		1:52:37
2-1.	Fundamental Items	*total*	0:24:22
		0. Kinematics & Vectors	0:02:08
		1. Scalar & Vector	0:01:38
		2. Coordinate Systems & Position Vector	0:03:23
		3. Velocity Vector	0:03:48
		4. Acceleration Vector	0:05:47
		5. Products of Vector	0:07:38
2-2.	Vectors	*total*	1:28:15
		1. Law of addition & subtraction of Vector	0:48:04
		2. 3D Graphical Representation	0:40:11
Vol. 3	**Simple Motion**		1:53:43
3-1.	Fundamental Items	*total*	0:36:57
		0. Simple Motion	0:01:26
		1. Laws of Motion	0:09:38
		2. Dimensions & Units	0:03:38
		3. Motion of Falling Object at earth's surface	0:03:42
		4. Motion of Object & Effects of Air Resistance	0:06:54
		5. Simple Harmonic Oscillation	0:05:54
		6. Work & Low of Conservation Energy	0:05:45
3-2.	Free Fall		0:14:50
3-3.	Motion of Projectile		0:13:34
3-4.	Motion of Object & Effects of Kinetc Resistance	*total*	0:38:51
		1. Stopping an automobile	0:08:35
		2. Motion of Object & Viscosity Force -- Falling Raindrop	0:07:36
		3. Motion of Object & Inertial Force -- Baseball	0:22:40
3-5.	Simple Harmonic Oscillation		0:09:31

Table 1-b Contents of Mechanics Course (continued)

Vol. 4	**Motion of Planets (Satellites)**		1:32:57
4-1.	Fundamental Items	*total*	0:32:13
		0. Motion of Planets (Satellites)	0:04:47
		1. Central Force and Conservation of Angular Momentum	0:04:56
		2. Motion of Planets (Satellites)	0:11:45
		3. Mechanical Energy in Planetary (Satellite) Motion	0:02:55
		4. Orbits of Artificial Satellites	0:07:50
4-2.	Keplar's Law	*total*	0:16:42
		1. Solution by Function "NDSolve[]"	0:11:21
		2. Solution by Runge-Kutta Method	0:05:21
4-3.	Gravitational Potential Energy		0:14:55
4-4.	Three-Body Problem		0:29:07
Vol. 5	**Motion of System of Particle and Rigid Body**		1:32:57
5-1.	Fundamental Items	*total*	0:24:59
		0. Motion of System of Particles and Rigid Body	0:03:22
		1. Momentum of System of Particles	0:05:22
		2. Kinetic Energy	0:02:41
		3. Potential Energy	0:03:02
		4. Angular Momentum	0:02:38
		5. Rigid Body about a Fixed Axis	0:04:22
		6. Coplanar Motion of Rigid Body	0:03:32
5-2.	Collision	*total*	0:13:09
		1. Laboratory Coodinate System	0:04:50
		2. Representation of Motion of Center-of-Mass	0:01:45
		3. Center-of-Mass Coodinate System	0:06:34
5-3.	Moment of Inertia		0:11:46
5-4.	Rigid Body about a Fixed Axis	*total*	0:16:28
		1. Pulley	0:05:23
		2. Physical Pendulum	0:11:05
5-5.	Coplanar Motion of Rigid Body	*total*	0:26:35
		1. Coplanar Motion	0:03:37
		2. Billiards	0:16:22
		3. Slope	0:06:36

Table 1-c Contents of Mechanics Course (continued)

Vol. 6	**Oscillatory Motion (I)**			1:0
Vol. 7	**Oscillatory Motion (II)**			1:3
6-1.	Fundamental Item		*total*	0:3
		0. Oscillatory Motion		0:0
		1. Nonlinearity of Simple Pendulum		0:0
		2. Damped Oscillation		0:0
		3. Forced Oscillation		0:0
		4. Self-Excited Oscillation		0:0
		5. Phase Space		0:0
		6. Nonlinear Forced Oscillation		0:0
6-2.	Nonlinearity of Simple Pendulm		*total*	0:1
		1. Solution by Euler Method		0:0
		2. Solution by Runge-Kutta Method		0:0
6-3.	Damped Oscillation			0:1
6-4.	Forced Oscillation			0:1
6-5.	Nonlinear Forced Oscillation			0:1
6-6.	2dimensional Simple Harmonic Oscillation			0:1
6-7.	Coupled Oscillation			0:0
6-8.	Self-Excited Oscillation			0:1
6-9.	Phase Space		*total*	0:2
		1. Damped Oscillation (I)		
		2. Damped Oscillation (II)		
		3. Simple Harmonic Oscillation		
		4. Simple Pendulum		
		5. Self-Excited Oscillation		
Vol. 8	**Observable Coordinate System**			1:03:16
7-1.	Fundamental Item		*total*	0:1
		0. Observable Coordinate System		0:0
		1. Accelerated Translational Coordinate System & Galilean Transformation		0:0
		2. Rotatory Coordinate System		0:0
		3. Rotation of earth -- Coordinate System on the earth		0:0
7-2.	Corioli's Force, Cetrifugal Force		*total*	0:3
		0. Corioli's Force, Cetrifugal Force		0:1
		1. Rotation of Coordinate (I)		0:0
		2. Rotation of Coordinate (II)		0:0
7-3.	Foucault's Pendulum			0:0

4. Summary and Vision

The VU project was started in 1999 with the primary objective of examining the organization of "Virtual Learning Communities" and studying how such communities could be effectively created within multimedia environments, such as the SCS and Internet.

The development of distance education allows all types of students to accomplish their learning needs at their convenience. Yet due to the diversity of the student audience, it is sometimes a challenge to offer the appropriate learning materials. To address this difficulty, the creation of "Virtual Learning communities" is an attractive option, as the teaching and learning can be targeted for specific groups of learners.

The results of this research have practical applications and will be used, for example, in the organization of the Virtual University.

Acknowledgment

The authors would like to thank Dr. Y. Ohnishi at the KIT Information Service Center for his technical support. Also, we would like to thank H. Kii, K. Kawabata, and R. Yoshimoto for their technical assistance.

References

1. Y. Chikaura, S. Ohta, Y. Suzuki, Y. Tanaka, Physics Exercise by Computer --- for students in Science and Engineering Elementary Course, *Baifukan* (*Co. Ltd.*), Tokyo 1991. (in Japanese)
2. Y. Tanaka, Y. Chikaura, S. Ohta, Y. Suzuki, Mechanics with Computer Graphics --- for students in Science and Engineering Elementary Course, *Tokyo-Kyougakusya* (*Co. Ltd.*), Tokyo 1992. (in Japanese)
3. Y. Suzuki, S. Furukawa, S. Ohta, Y. Tanaka, Y. Chikaura, Atomic Physics with Computer Simulation --- for students in Science and Engineering Elementary Course, *Tokyo-Kyougakusya* (*Co. Ltd.*), Tokyo 1993. (in Japanese)
4. S. Furukawa, Y. Suzuki, Y. Chikaura, Y. Tanaka, Physical Mathematics with Computer Graphics --- for students in Science and Engineering Elementary Course, *Tokyo-Kyougakusya* (*Co. Ltd.*), Tokyo 1997. (in Japanese)
5. Y. Suzuki, Y. Chikaura, Mechanics Exercise with *Mathematica*, *Baifukan* (*Co. Ltd.*), Tokyo 1999. (in Japanese)
6. NEP (Network for Education of Physics) Reports on NEP Project (I), (II). http://www.nep.chubu.ac.jp/

Our home page address

Elecrtromagnetism : http://lecture1.vu.kyutech.ac.jp/otabe/

Mechanics : http://web.tobata.vu.kyutech.ac.jp/butsurigaku1.htm

SIMULATION PROGRAMS FOR PHYSICS EDUCATION USING VIRTUAL REALITY TECHNIQUE

JONG-HEON KIM[A], SANG-TAE PARK[A], HEEBOK LEE[A*], KEUN-CHEOL YUK[A] AND HEEMAN LEE[B]

[a]*Department of Physics Education, Kongju National University, Kongju 314-701, Korea*

[b]*Department of Computer Science, Seowon University, Chongju 361-742, Korea*

Virtual reality technology is now widely recognized as a major break through in the technological advance of science education. The use of computers in the physics classroom can increase student engagement, add realism to instruction, promote mastery and understanding of basic physics principles, augment laboratory experience, and encourage inferential learning. Accordingly, the current paper presents a recently developed computer-based virtual reality simulation that enables students to learn key physics concepts, such as wave propagation, ray optics, relative velocity, and electric machines, etc. in high school-level physics.

1. Introduction

Recently, Virtual Reality(VR) has evolved into a variety of configurations, for example, head-mounted displays(HMD), PCs, workstations, immersive rooms, large screen systems, and virtual tables. VR technology continues to develop very quickly, as such, there is a growing interest in the potential applications of VR techniques(e.g., Wayne, T. 1997).

VR technology can offer strong benefits to science education through facilitating constructivist learning activities and providing alternative forms of learning that can support different types of learners, such as visually-oriented learners(e.g., C. Youngblut, 1998). In traditional instructional environments, students are expected to learn by assimilation, for example, by listening to an instructor about a subject. Yet, current educational thinking suggests that students are better able to master, retain, and generalize new knowledge when they are involved in a learning by-doing situation. This pedagogical philosophy is called constructivism and its supporters vary, ranging from those who see it as a useful complement to teaching-by-telling to those who argue that the whole curriculum should be reinvented by students through gently guided discovery learning.

It has been reported that the use of computers in the physics classroom can increase student engagement, add realism to instruction, promote mastery and

*Corresponding autor. E-mail: heebok@kongju.ac.kr

understanding of basic physics principles, augment laboratory experience, and encourage inferential learning. Recently, VR simulations in physics education have become more attractive related to the use of computer-based technologies in class.

Accordingly, the current paper presents a recently developed computer-based virtual reality simulation that enables students to learn key physics concepts, such as wave propagation, ray optics, relative velocity, and electric machines, etc. in high school-level physics. With the capability of navigation from any Viscape-enabled Web browser (i.e., Netscape and Internet Explorer), this virtual reality physics simulation(VRPS) was developed for use in high school and university-level physics teaching, as well as general access to physics learning. The simulation involves assembling an instructional design, 3D modeling, an experimental set up constructed using 3D objects in a virtual world, keys to control, and 3D animation program for visualization of the 3D model output.

2. VR Applications in Education

As a computer-based technology, VR is a new communication medium and usually composed of a collection of technological hardware, including a computer, HMD, eye phones, motion-sensing data gloves, etc. Virtual reality is defined as a highly interactive, computer-based, multimedia environment in which the user becomes a participant along with the computer in a virtual real world. The key feature of VR is real-time interactivity, in that the computer is able to detect user inputs and instantaneously feed back to modify the virtual world(e.g., Winn, W., 1993).

Within the higher education community there has been an increase in the use of information technology, such as multi-media, with considerable success. Multi-media based systems provide students with a very rich source of educational material in a form that makes learning exciting. VR has an extremely wide application across a whole range of disciplines and the enabling technology has reached a sufficient level of maturity for it to be seriously applied to education, training, and research in higher education. Until fairly recently, the development of VR systems was inadequate to be seriously considered for anything other than a research tool. Plus the costs associated with a VR system were prohibitive for educational establishments (this is still true for fully immersive VR systems). However, recent technological developments in computer hardware and software now make it feasible to look at VR as an important teaching aid(e.g., Byrne, C. 1993).

It is certain that the flexibility provided by a VR system will be a major attraction for the educational community. Even though the initial cost is still quite high compared to conventional multi-media systems, this can be quickly offset when considering the different applications it can support. As such, it is not unimaginable that one day school laboratories will be replaced or augmented by a number of VR systems where experiments can be performed. The major difference being that the VR-based laboratory will be able to support chemistry, physics, and engineering sessions rather than just a single discipline.

The VR world can also be used to circumvent the physical, safety, and cost constraints that limit schools in the types of environments they can provide for learning-by-doing. For example, in the area of nuclear engineering, students would be able to further their understanding of a nuclear reactor by moving into an reactor using an HMD and 3D gloves. This type of activity can be performed in a virtual world. As such, VR learning environments can also support the notion of situated learning where students learn within the actual context of where that learning is to be applied.

VRPS is an educational system that uses a VR interface to bring together a 3D model of a real situation and a virtually real visualization of the model's output, such as images and moving objects. VRPS enhances students' understanding by providing a degree of reality unattainable in a traditional two-dimensional interface, and supports the creation of sensorially rich interactive learning environments. Furthermore, physical phenomena, which are neither easy to perceive nor measure in usual experiments, can be scientifically visualized in a virtual world and viewed in many different ways by traveling inside a VR laboratory.

3. Examples of Virtual Reality Physics Simulations

The current study used 3D Webmaster software for authoring VRPS programs. 3D Webmaster is a powerful Windows application that can create interactive 3D VR programs that work on Internet web browsers. Viscape is the browser platform that enables the user to display, move around, and interact with the worlds created using 3D Webmaster. Viscape also allows the user to adjust the viewpoint, display options, and device configurations and add final touches to the created world. These created worlds can be displayed on their own as fully interactive environments or embedded in 2D HTML pages on a PC.

Worlds created using 3D Webmaster are fast, realistic, and highly interactive. This authoring software enables the user to build virtual worlds using a variety of clipart objects, textures, and sounds supplied with 3D Webmaster, or create objects using 3D Webmaster's creation tools. URLs can be linked to objects to facilitate a direct link from one virtual world to another 2D or 3D page on the WWW.

The Superscape Control Language (SCL), a script language for desktop computers, helps to create realistic and useful worlds by assigning behavior to objects in the virtual world, performing complex actions, and modifying the virtual world based on the user's actions in the HTML documents. 3D

Webmaster's point-and-click interface can control all movement, interaction, and object manipulation in a virtual world based on a mouse and keyboard, or a peripheral proportional device, such as a Spacemouse or joystick. Only a few examples of VRPS are shown here due to lack of space.

3.1 *Velocity and Acceleration*

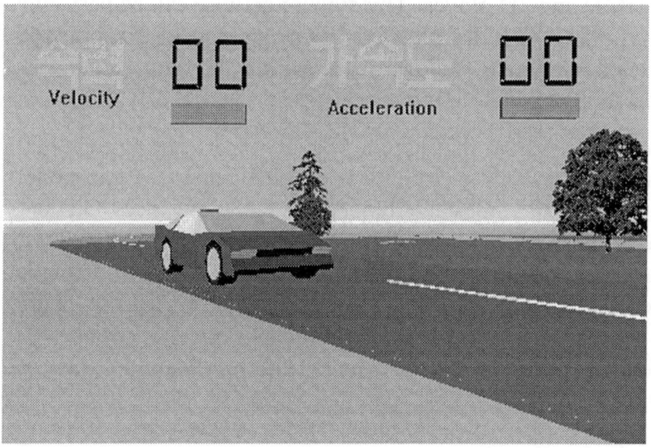

Fig. 1. VRPS for velocity and acceleration

Fig. 1 shows a VRPS display screen related to velocity and acceleration where the student sets the initial velocity and acceleration, then the car proceeds into a 3D spacious driving field. The movement of the car can be observed at various viewpoints selected by dragging the mouse or simply selecting the viewpoint from the car itself. As such, the velocity and acceleration can be perceived with changes in the driving conditions and from different viewpoints. It is also possible to measure the distance versus time and draw various graphs to study the linear motion. Consequently, the physics concepts of velocity and acceleration can be understood through learning by doing and with plenty of fun.

3.2 *Magnetic Force in Current-carrying Wire*

Fig. 2 shows a VRPS display screen for the magnetic force in a current-carrying wire where the student can switch the current on or off by the clicking mouse button and change the polarity of the magnetic field by flipping the horseshoe-type magnet. The student can also fly around the apparatus and check the connections. The voltage of the circuit can be control by adjusting the power supply.

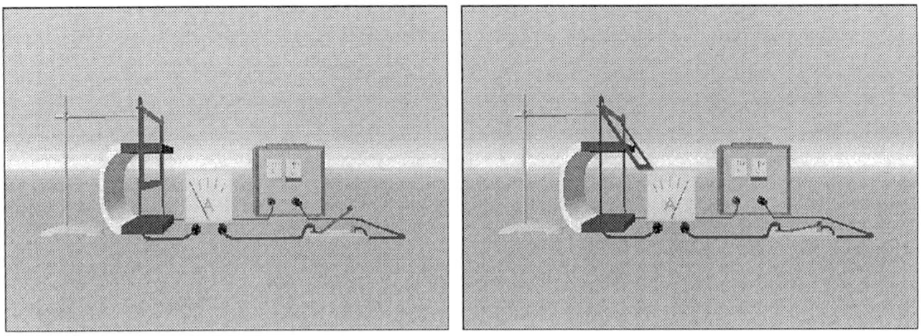

Fig. 2. VRPS for magnetic force in current-carrying wire

3.3 *AC Generator*

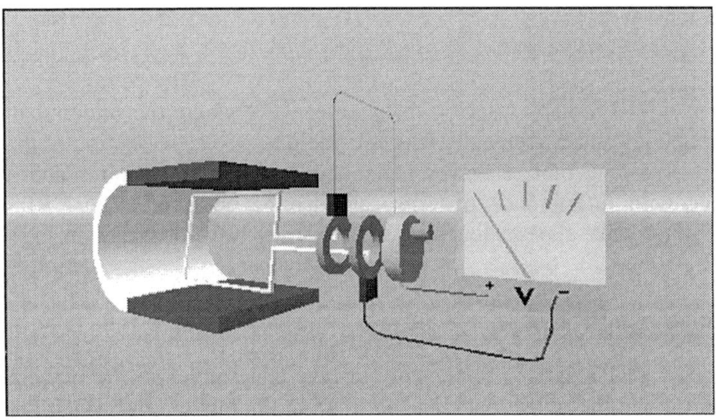

Fig. 3. VRPS for AC generator

Fig. 3 shows a VRPS display screen for an AC generator where the student can control the magnitude of the magnetic field and angular velocity of the rotor. The direction of the current and frequency of the induced voltage are the subjects of discussion in this class. Here, it is fun to fly over or travel through the brush ring. The VRPS experiment is more effective and efficient for teaching than performing the real experiment. The induced voltage data supplied by the VRPS AC generator can be analyzed by the students.

3. 4 *Waves*

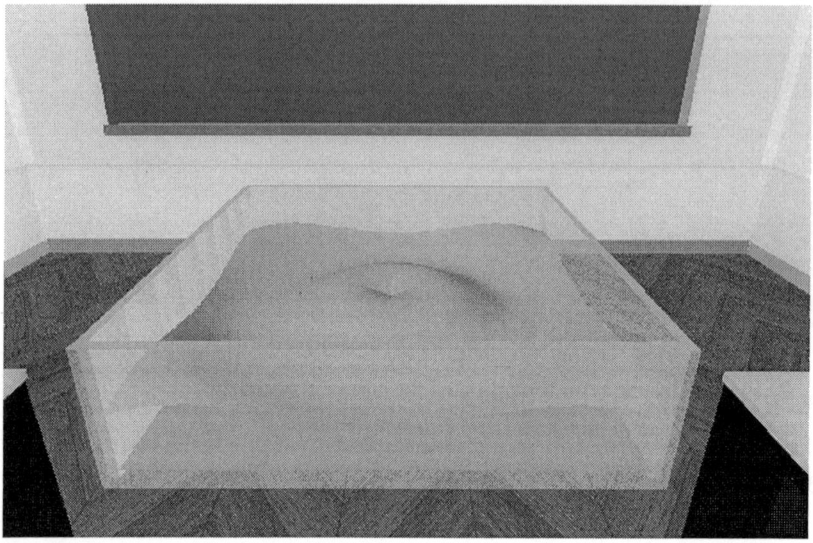

Fig. 4. VRPS for waves

Fig. 4 shows a VRPS display screen for a water wave where the student can control the frequency and wavelength. It is enjoyable to watch the motion of the waves. The overlapping of two waves generated at different locations and a standing wave with interference patterns can also be observed. The VRPS simulation of water wave appears like slow motion water wave propagation.

3.5 *Concave and Convex Mirrors*

Fig. 5 shows a VRPS display screen for concave and convex mirrors where the student can move the object back and forth from the mirror, and watch the image's magnitude and location depending on the location of the object. The distance of the object and image can be measured as a hands-on experiment.

3.6 *Exhibition*

Fig. 6 shows the display screen of an exhibition museum for displaying VRPS laboratories where students can visit the VR museum and fly over the museum yard by taking a helicopter. The background music enforces the realism along with a narrated tour of the museum. The exhibition museum consists of various VRPS laboratories.

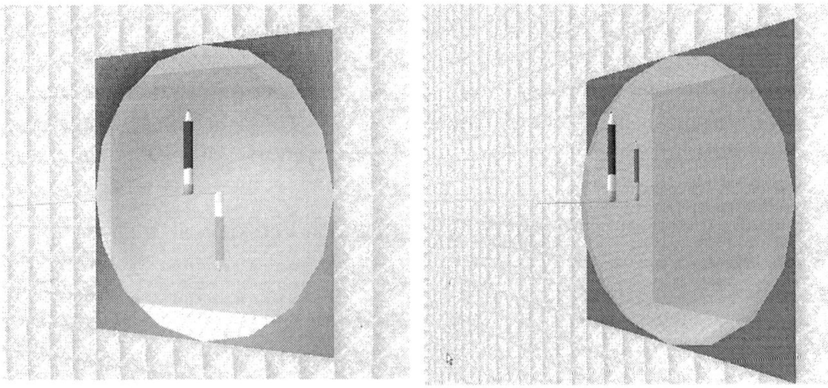

Fig. 5. VRPS for concave(left) and convex mirrors(right)

Fig. 6. Exhibition museum for displaying VRPS laboratories

4. Conclusion

The current paper presented VRPS programs that can enhance students' understanding by providing a degree of reality within rich interactive learning environments. Using VRPS, physical phenomena can be scientifically visualized in a virtual world and viewed in many different ways by traveling over and through the VR laboratories. The current authors developed several VRPS programs for key physics concepts, such as wave propagation, ray optics, relative velocity, and electric machines, etc. in high school-level physics, thereby demonstrating the usefulness and effectiveness of VRPS for physics education.

This work was supported by a Korea Research Foundation Grant (KRF-99-005-D00076).

References

1. Wayne, T. (1997). *Student Responses to Their Immersion in a Virtual Environment.* University of Washington, Human Interface Technology Laboratory of the Washington Technology Center, Seattle, WA. Technical Publication R-97-11.

2. Byrne, C. (1993). *Virtual Reality and Education.* University of Washington, Human Interface Technology Laboratory of the Washington Technology Center, Seattle, WA. Technical Publication R-93-6.

3. Winn, W. (1993). *A Conceptual Basis for Educational Applications of Virtual Reality.* University of Washington, Human Interface Technology Laboratory of the Washington Technology Center, Seattle, WA. Technical Publication R-93-9.

4. C. Youngblut. (1998). *Educational Use of Virtual Reality Technology.* Institute for Defense Analyses. IDA Document D-2128. Log:H98-000105.

EXPERIMENT FOR MEASURING PLANCK'S CONSTANT USING EIGHT COLORED LEDS

KAZUAKI NEMOTO
Hitachi First Senior High School, 3-15-1 Wakaba, Hitachi, Ibaraki, 317-0063, Japan

The current paper describes several experiments using multiple colored LEDs for teaching introductory Quantum physics. These experiments are all simpler than the *photoelectric effect.* LEDs are engineered to drop an electron based on a fixed potential difference and emit a photon of a definite color. As such, a range of LEDs with different colors can be used to illustrate the relationship between energy and frequency for photons. The measurement results demonstrate that the energy E is proportional to the frequency f, $E = hf$ with h being Planck's constant. An experiment using three colored LEDs is effective for beginners, where the measured value is approximately 6×10^{-34} Js. Whereas, an experiment using eight colored LEDs is preferable for a more detailed investigation, as a graph is produced with two straight lines from the origin and the slope represents Planck's constant. The author is interested in exchanging opinions with other researchers.

1. Introduction

An LED (*Light Emitting Diode*) is usually used as an indicator or for fiber-optic communications. However, it can also be used for scientific measurement.

This concept was originally presented at the Japan-UK conference on physics and chemical education in April 1998, using five colored LEDs, including red, orange, yellow, green, and blue. And the current author believes that this is a suitable experiment for introductory Quantum physics, because it is simpler than the *photoelectric effect.* The original paper[1)] on this experiment was written by I. Lawrence in 1996.

A similar experiment has since been included in a new textbook, *Advancing Physics AS*[2)], published in the UK in 2000. In this case, three colored LEDs are used, including red, green, and blue, which the current author believes is more suitable for beginners.

When testing the experiment using five colored LEDs, the current author became aware of two tendencies in the measurement values. Thus, an experimental device (Figure 1) with eight colored LEDs was constructed for a detailed investigation. The additional colors included were a reddish orange, yellow green, and bluish green.

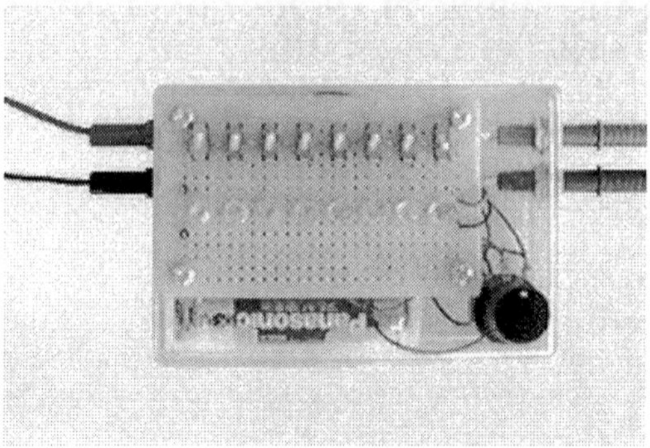

Figure 1 Experimental device with eight colored LEDs. The left-sided terminals are connected to an ammeter, while the right-sided ones are connected to a voltmeter. The device also includes built-in dry cells.

2. Theory

LEDs are engineered to drop electrons based on a fixed potential difference (p.d.) and emit photons of a definite color. The minimum fixed p.d., which is the threshold value under forward-bias conditions, is called the striking p.d. or turn-on p.d., and the product of the electron charge e caused by the striking p.d., defined as V_S, is equivalent to the energy E emitted as a result of recombination.

$$E = eV_S \qquad (1)$$

As such, a range of LEDs with different colors illustrates the relationship between energy and frequency for photons. The result of the measurement is that the energy E is proportional to the frequency f,

$$E = hf \qquad (2)$$

with h being Planck's constant.

3. Creation of Device

A circuit diagram of the device is shown in Figure 2. The type numbers for the LEDs (Manufacturer) were:

red ; TLRH180P, reddish orange ; TLSH180P,

orange ; TLOH180P, yellow ; TLYH180P,

yellow green ; TLGE183P (Toshiba), green ; E1L53-3G,

bluish green ; E1L51-KC, blue ; E1L53-3B (Toyoda Gosei)

Each LED was mounted in clear plastic, rather than colored plastic.

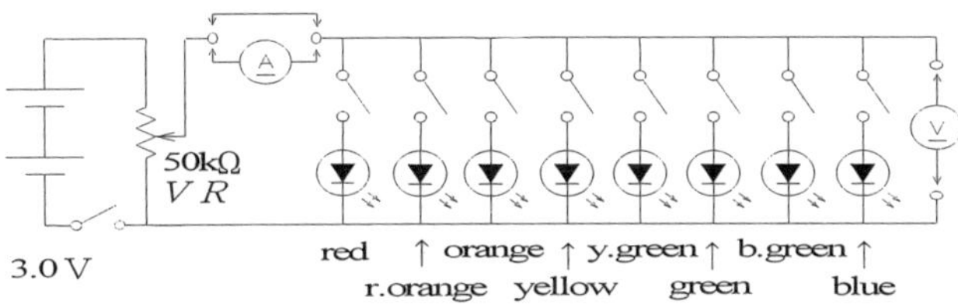

Figure 2 Circuit diagram of device.

4. Measurements and Results

The measurements consisted of the minimum p.d. required to turn an LED on†, and resulting light wavelength. The wavelength was measured in relation to the peak wavelength in the manufacturer's specifications. The results are shown in Table 1.

5. Discussion

5.1 *Experiment Using Three Colored LEDs*

Figure 3 shows that, to a good approximation, the energy was proportional to the frequency. The slope of the graph was 5.7×10^{-34} Js. This experiment is good for beginners, although the value is somewhat small for Planck's constant.

5.2 *Experiment Using Five Colored LEDs*

For comparison with the UK experiment, a five-colored LED device was fabricated. The current results for the five colors are shown in Table 1, while the results from the original paper[1)] are shown in Table 2.

Table 1 Tabulated results from experiment.

color	λ	f	V_S	eV_S
Red	644 nm	0.466 PHz	1.41V	0.225aJ
Reddish orange	623 nm	0.482 PHz	1.47V	0.235aJ
Orange	612 nm	0.490 PHz	1.50V	0.240aJ
Yellow	590 nm	0.508 PHz	1.55V	0.248aJ
Yellow green	574 nm	0.523 PHz	1.63V	0.261aJ
Green	525 nm	0.571 PHz	2.23V	0.357aJ
Bluish green	495 nm	0.606 PHz	2.27V	0.363aJ
Blue	470 nm	0.638 PHz	2.39V	0.382aJ

λ: wavelength, f: frequency, V_S: striking p.d., eV_S: energy

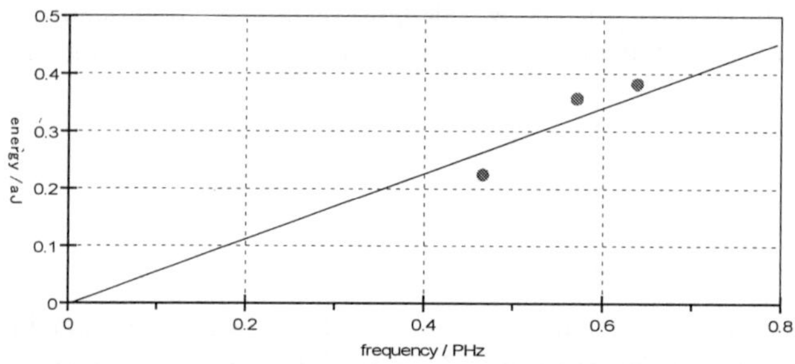

Figure 3 Graph of results from Table 1. The colors included are red, green, and blue.

Table 2 Tabulated results from the UK experiment.

color	λ	f	V_S	eV_S
Red	650 nm	0.462 PHz	1.47V	0.235aJ
orange	620 nm	0.483 PHz	1.48V	0.237aJ
yellow	585 nm	0.512 PHz	1.63V	0.261aJ
green	563 nm	0.533 PHz	1.69V	0.270aJ
blue	470 nm	0.638 PHz	2.38V	0.381aJ

The current values ranged between 2 ~ 10 μA. Generally, LEDs are used under 20 mA conditions.

The plot in Figure 4 is based on the results in Table 2, and although it is similar to a graph demonstrating the *photoelectric effect*, it does not exhibit a directly proportional relationship, as predicted by Eq.(2).

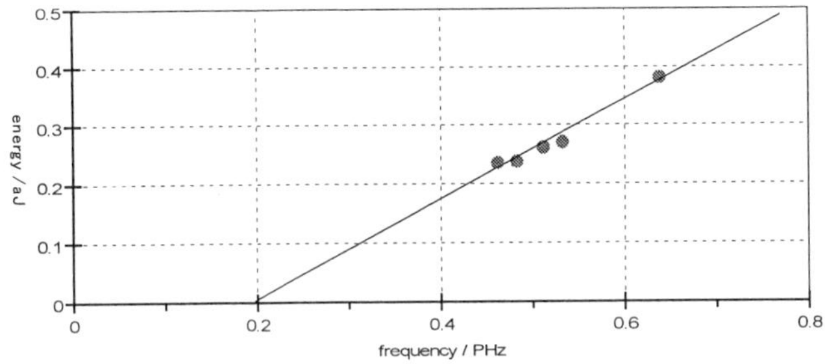

Figure 4 Graph of results from Table 2. The colors included are red, orange, yellow, green‡, and blue.

The plots in Figure 5 are the result of Table 1. The distribution of the five points resembles that in Figure 4, except for the green LED‡.

Figure 5 shows the dual tendencies of the measured values. Due to the directly proportional relationship between the energy and the frequency, each tendency can be represented by a line passing through the origin.

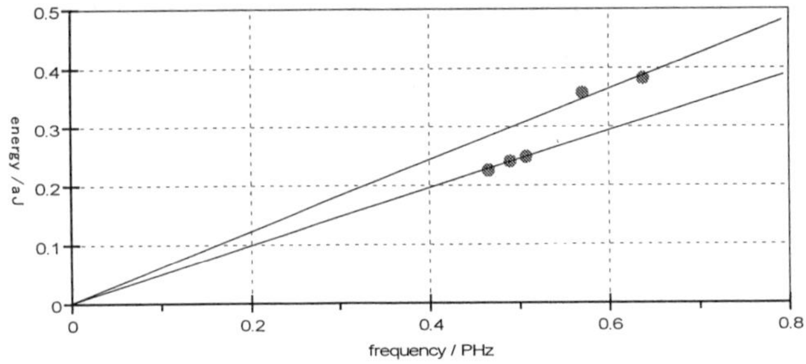

Figure 5 Graph of results from Table 1. The colors included are red, orange, yellow, green‡, and blue.

5.3 *Experiment Using Eight Colored LEDs*

For a more detailed investigation, an eight-colored LED device was fabricated. Figure 6 shows a more defined dual tendency behavior.

The slope from red to yellow green was 4.9×10^{-34} Js, while the slope from green to blue was 6.0×10^{-34} Js. It is assumed that the LED structures can be classified into two properties: a low frequency group, from red to yellow green, and a high frequency group, from green to blue.

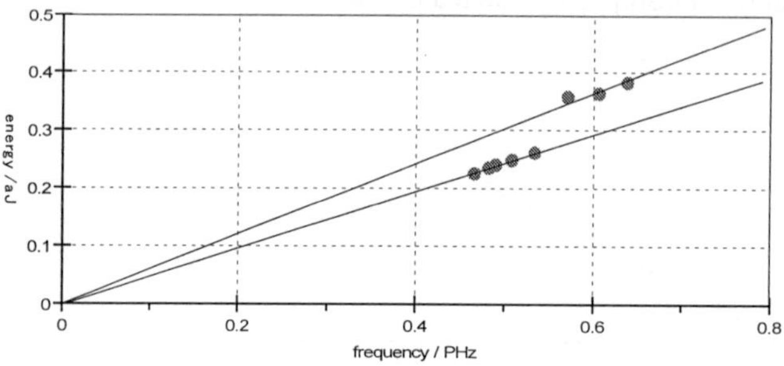

Figure 6 Graph of results from Table 1. The colors included are red, reddish orange, orange, yellow, yellow green, green, bluish green, and blue.

The wavelength for green was 525 nm in Table 1, yet 563 nm in Table 2. As such, the color in the original paper[1)] was not green, but rather yellow green.

Consequently, there are two problems that remain to be solved. The first is that the graph has two slopes, and the second is that each value on the slope is smaller than Planck's constant. This means that the input energy as the supplied p.d. is less than the output energy as light.

A study of the structure of LEDs is necessary to solve these problems. It may also be necessary to measure the wavelengths directly.

6. Conclusion

The current paper described several experiments for measuring Planck's constant using multiple colored LEDs.

In the experiment using three colored LEDs, even though the graph included some scattering, the measurements still showed that the energy was approximately proportional to the frequency. Although the slope was smaller than Planck's constant, this experiment is recommended for introductory Quantum physics, because it is simpler than the *photoelectric effect*.

In the case of five colored LEDs, a dual tendency was observed and two lines passing through the origin were drawn giving rise to two slopes. Therefore, an eight-colored LED device was developed and fabricated to investigate the problem in detail.

In the case of eight colored LEDs, the same dual behavior was also observed and two slopes obtained. The slope of the low frequency group was 4.9×10^{-34} Js, while that of the high frequency group was 6.0×10^{-34} Js. These two values are both smaller than Planck's constant, indicating that the input energy as the supplied p.d. is less than the output energy as the light.

Therefore, two problems remain to be solved through a detailed investigation, involving a structural study of LEDs. Accordingly, the main purpose of this paper is to present experiments using multiple colored LEDs for teaching introductory Quantum physics. It is hoped that the experiments will be widely used. The current author intends to expand these experiments, and is interested in exchanging results with other researchers.

Acknowledgements

I would like to acknowledge many valuable discussions with members of the APEJ and the Rika-Kyoiku Mailing List. I would also like to thank Y. Fukazawa for his helpful remarks concerning the draft copy of my paper.

References

1. I. Lawrence, Quantum physics in school, *Physics Education*, vol. 31-5, pp.278-286, 1996

2. J. Ogborn and M. Whitehouse, *Advancing Physics AS*, IOP, UK, pp.161, 2000

REMOTE CONTROL LABORATORY FOR NUCLEAR EXPERIMENTS VIA INTERNET

HEEBOK LEE[a*], JONG-HEON KIM[a], SANG-TAE PARK[a], KEUN-CHEOL YUK[a] AND HEEMAN LEE[b]

[a]*Department of Physics Education, Kongju National University, Kongju 314-701, Korea*

[b]*Department of Computer Science, Seowon University, Chongju 361-742, Korea*

Experiments are an essential part of science education for understanding natural phenomena and the related principles. Therefore, remote control experiments via the Internet are a key solution for distance learners in science education. Remote experiments can also be used for time-consuming experiments that take several days, collaborative experiments between distance learners, experiments involving expensive laboratory equipment that is unavailable to most students, and experimental procedures that are dangerous. Accordingly, the current study developed a general method for creating a remote control laboratory system using the Internet and interface techniques, thereby enabling physics students to control real instruments through the Internet and conduct remote physics experiments. A remote control nuclear experiment system is constructed as a sample application.

1. Introduction

Distance education has become an integral part of today's world where computers and networks can be used to improve the quality of education from regular school systems to personal, community, and national development through lifelong learning. Science and technology education has also become a more important subject for continuing education in the present high technology-based society.

As such, the use of multimedia through computers and the Internet has become a powerful tool for providing knowledge and understanding, supporting collaborative learning between distance learners, enabling effective evaluation of learning with instant feed-back, and supplying interactive learning environments using virtual reality, etc(e.g., S. C. Brofferio, 1998).

Since laboratory experiments are an essential part of science education providing vivid interaction between learners and nature (e.g., Alessandro Ferrero, Vincenzo Piuri, 1998), laboratory activities are required even for distance learners. However, it is impossible to replace all laboratory activities with alternative methods using multimedia techniques, such as computer

*Corresponding author: heebok@kongju.ac.kr

simulation programs, video, animation, virtual reality, etc. One alternative is to create experimental kits that are sent to learners and then returned at the end of the course. Yet, in this case, high costs limit the experiments to lower level lecture courses. Furthermore, physics experiments are becoming increasingly computerized and the price of the experimental equipment more and more expensive.

Accordingly, the current paper proposes a novel experimental technique for physics education that can be used for distance learning and in regular schools based on introducing a remote control experiment system through the Internet. Therefore, the proposed remote control laboratory experimentation represents an extension to the current ways in which the Internet is utilized for science education. The remote control laboratory for science education is implemented in an integrated environment to enable the user to control real devices from a remote site and conduct actual experiments in the remote control laboratory through a computer network. The remote control laboratory consists of a cluster of general-purpose and specialized instruments interfaced to a set of personal computer systems connected to the Internet. With the ability to configure instruments and perform data analysis remotely via the Internet, the laboratory also facilitates the sharing of expensive instruments and equipment.

State of the art technology can now provide remote access to laboratory equipment and instruments through the Internet. This is especially useful in physics education; (1) when the laboratory class time is too short to perform inquiry experiments, analyze data, discuss, etc, (2) when part-time students or distance learners need to conduct laboratory experiments remotely, (3) if the laboratory equipment is too expensive and unavailable in local schools, and (4) if the experimental procedure is too dangerous.

Therefore, the current paper presents a general method for developing a remote control laboratory system using the Internet and interface techniques, thereby enabling physics students to control real instruments and conduct remote physics experiments. Remote control nuclear experiments and an electric experiment were constructed as applications of the proposed system.

2. System Architecture

When students log on to the system through the Internet, they are able to control both the computer and the connected equipment. A video camera can also be used to provide a live broadcast of what is happening in the laboratory. Remote laboratory experimentation is created by utilizing the Web server, HTTP, HTML, CGI, Web camera, and experimental equipment. The data acquisition system, HP34970A, consisting of many multiplexer switches, DA converters, digital multimeter, etc., is used to interface with other physics experimental equipment. A multi-channel analyzer(MCA) with a nuclear detection system is also used for nuclear experiments. The clients are IBM-compatible personal computers running MS-windows 98. CGI is the major tool for communication between the HTTP server and the experimental equipment. A web page is provided to enable the user to send instrument parameters to the CGI program. A Windows NT web server (IIS) and Windows 98 Personal Web server are both used to control the interface. The standard CGI and Visual Basic programs are used to support the standard input/output streams. The HP standard instrument control libraries(SICL) are used to control the GPIB interface card and HP 34970A. The GammaVision software by ORTEC is also used to control the nuclear detection system.

The server connected to the laboratory equipment manages the client requests and sends the parameters to the CGI program, which controls the real instruments in the remote control laboratory. Then the CGI program returns the information collected from the server-side apparatus to the client computer. The client keeps the transmitted data in their own directory and uses the data for analysis and writing a report using Microsoft Excel or other programs. Based on this general method, other remote experiment systems can be developed, as the only requirements are the development of a program to control the different instruments and the design of the procedure and new Web contents for the new experiment. A video camera and Webcam software can be used to produce live broadcasts to monitor the remote laboratory equipment. This is very important, as the user then feels directly involved in controlling the real instruments, rather than manipulating a simulation program. Figure 1 shows the hardware structure of the proposed remote control laboratory system.

The list of hardware used in the system is as follows.

- IBM compatible PC computer with data acquisition cards, GPIB interface card, and Ethernet card.
- Programmable Instruments, such as data acquisition system, digital multimeter, programmable power-supply, physics experimental apparatus, and GPIB card.
- IBM compatible PC computer with video camera and Internet-based video server.
- ORTEC nuclear detection system (scintillation counter, high voltage power supply, multichannel analyzer(MCA), electronic signal processor modules, etc.)

The software structure for the proposed remote control laboratory, for both the server and the client, is shown in Figure 2. In the proposed realization, the server is implemented on a Win NT server using the TCP/IP protocol, and an IIS HTTP server. The clients are IBM-compatible personal computers running

MS-windows 98; the WWW browser is Internet Explorer 5.0 or higher due to the use of ActiveX and a video stream server program.

A Web page is designed to operate the remote laboratory according to a time schedule for a series of experiments. A laboratory manager then sets up the remote control experiments according to the timetable. As such, when performing the experiments, it is irrelevant whether the student is in a nearby room or on the other side of the world.

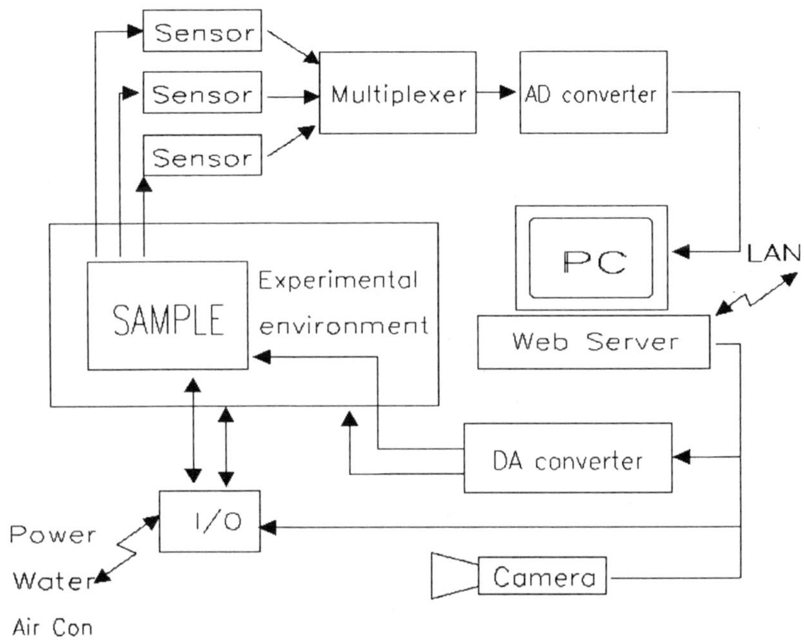

Figure 1. Hardware structure of remote control laboratory system

3. Experimentation Set-up in Remote Control Laboratory

To set up a remote experiment it is important to understand how to control the remote instruments through the Internet. CGI and TCP/IP are the two major tools used for communication between the users, the local instruments, and the HTTP server. An interface software can be developed for controlling remote instruments.

A Web page is provided to enable the user to send instrument parameters. The HTTP server accepts the parameters, activates the CGI program, and transmits the instrument parameters to the CGI program(e.g., B. Cabell V, J. J. Rencis and H. T. Grandin, 1997). Next, the CGI program establishes GPIB and

RS232C connections to the Lab PC, which then activates the local instrument according to the parameters transmitted from the server. Finally, the local instruments hand over the experimental results to the server and the CGI program sends the experimental data and parameters to the client. In principle, any instrument can be controlled remotely through interface software if the local control program supports the specifications of the interface software.

As a sample application, a remote control experiment system was constructed for nuclear experiments.

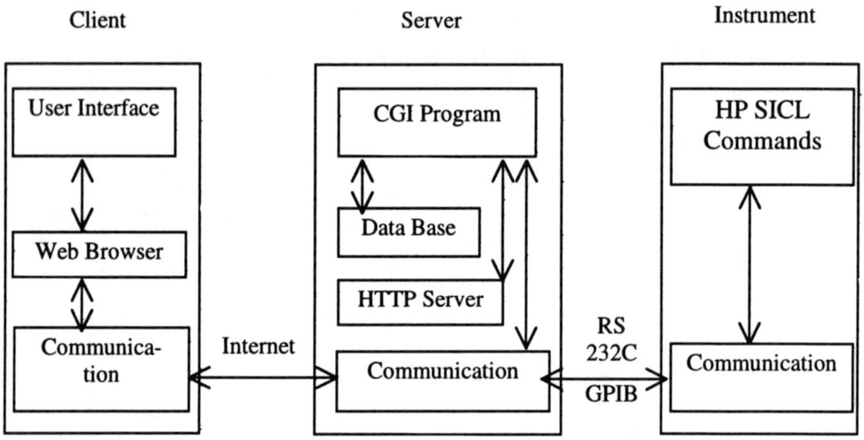

Figure 2. Three-tier client/server software environment

4. Nuclear Experiment System

Nuclear experiments are rather worrying due to the use of radioactive materials. On the other hand, such experiments are very interesting because radioactivity is a curious subject related to nuclear physics and modern technology, such as nuclear energy, medical radiotherapy, atomic bombs, etc(e.g., EG&G ORTEC, AN34). Also, nuclear experiment systems are very expensive and not available at every educational institute. Therefore, a remote control nuclear experiment system was developed to overcome such constraints.

Examples of nuclear experiments include Geiger counting, alpha- beta- gamma-ray spectroscopy, Compton scattering, coincidence techniques, nuclear lifetime, radiation biology, environmental studies, etc.

After implementing the remote control instrument, the remote laboratory is set-up based on the experiment contents and a Web page created for user instruction. The user then follows the instructions to conduct the remote experiment. While

allow other users to initiate another experiment. Plus, users can only perform those experiments included in the time schedule and setup by the lab manager.

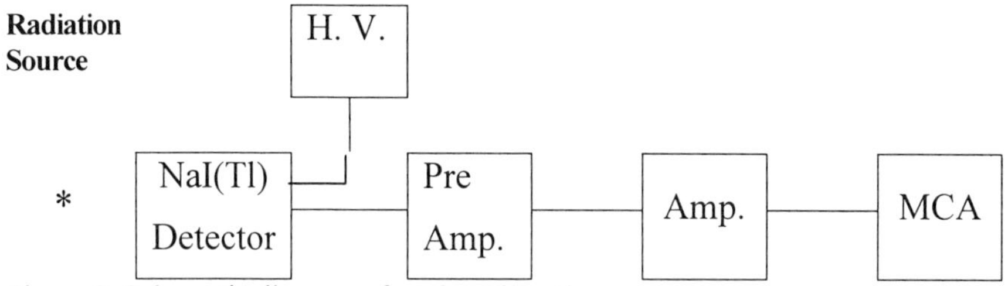

Figure 3. Schematic diagram of nuclear detection system

The user activates and conducts the remote experiment by performing the following steps:

Connect to Web page of remote control laboratory.

Identify user authorization in Web server and obtain valid session ID for remote laboratory.

Select available experiments in remote laboratory.

Show corresponding Web pages including tutorial and basic instructions.

Follow experiment steps as described in Web page.

Conduct remote experiment and view real-time image(Webcam) on browser,

Terminate session by releasing server connection and session ID.

Analyse collected data.

Report results and discussion to instructor.

Check next experiment.

5. Conclusion

The current paper proposed a prototype for Web-based remote control laboratory experimentation. A CGI environment is adopted for the local instrument control, whereas the remote instrument control is achieved through the Internet based on a three-tier client-server architecture between the WWW server, CGI program, and client. Commands and parameters are submitted to the CGI program via an interface Web page to set the experimental parameters. Remote users can control the real instruments and observe the real time results of the experiment. In the future, such experiment systems are expected to become more flexible and user-friendly. Remote control laboratory experiments can be very useful in cases such as time-consuming experiments, science education for distance learners, and very expensive or dangerous experiments. A remote control nuclear experiment system was constructed as a sample application.

This work was supported by a Korea Research Foundation Grant (KRF-99-005-D00076).

References

1. S. C. Brofferio, (1998). *A University Distance Lesson System: Experiments, Services, and Future Developments.* IEEE Transactions on Education, Vol. 41, No. 1, pp. 17-24.

2. Alessandro Ferrero, Vincenzo Piuri, (1998). *A Simulation Tool for Virtual Laboratory Experiments in A WWW Environment*, IEEE Instrumentation and Measurement Technology Conference, St. Paul, Minnesota, USA, May 18-21.

3. B. Cabell V, J. J. Rencis and H. T. Grandin, (1997). *Using Java to Develop Interactive Learning Material for the World-Wide Web*, International Journal of Engineering Education . Vol. 13, No. 6, pp. 397-406.

4. EG&G ORTEC, *Experiments in Nuclear Science.* AN34

MEASUREMENT OF WAVELENGTH USING CCD CAMERA

HEEBOK LEE[a*], SANG-TAE PARK[a], JONG-HEON KIM[a], KEUN-CHEOL YUK[a] AND HEEMAN LEE[b]

[a]*Department of Physics Education, Kongju National University, Kongju 314-701, Korea*

[b]*Department of Computer Science, Seowon University, Chongju 361-742, Korea*

The current study developed a useful computer program for teaching optics in physics. The main advantage of the program is that students can gain an idea of the concept of wavelength from the RGB color data from photos that are easily obtainable from many CCD devices without the use of expensive spectroscopy. In addition, the temperature distribution of a light source can also be determined, assuming that it emits black body radiation.

1. Introduction

In today's information society many students have access to a variety of electronic devices in their homes, such as TVs, DVDs, camcorders, digital cameras, computers, cellular phones, etc. As such, the idea of using these commonly available electronic devices as teaching aids should be highly encouraged. Many students enjoy using computers to listen to music, play electronic games, and surf the World Wide Web(WWW). Computers now play a major role in education and have become an essential teaching aid in class activities.

Digital still or video pictures can be easily obtained from a WWW browser, VCR, digital camera, and scanner, etc. All these devices use RGB color coordinates to represent colors. Color is the human perception of light corresponding to specific wavelengths. Therefore, the program developed in the current study gives students hints about the concepts of color and wavelength. In addition, the proposed program can also be used to determine the temperature of a light source from RGB data in a picture of the light source, assuming that the light source emits black body radiation.

The main advantage of the proposed computer program is the adaptation of photos, which can be easily obtained from many commonly available CCD-based electronic devices, as education tools instead of using expensive

*Corresponding author: E-mail: heebok@kongju.ac.kr

spectroscopy. The following section describes the principles of how the program works.

2. Basic Color Theory

What we perceive as light is a narrow frequency band within the electromagnetic spectrum, the other frequency bands being radio waves, microwaves, infrared waves, and X-rays. Each frequency value within the visible band corresponds to a particular color. The low-frequency end produces the color red(4.3×10^{14}Hz), while the highest frequency produces the color violet(7.5×10^{14}Hz). Since wavelength units are somewhat more convenient to deal with than frequency, the remainder of this paper uses wavelengths. Color perception is the interaction of light with the human visual nervous system. A light source, either the sun or a light bulb, emits over a wide range of wavelengths, including the visible frequencies between 400 and 700 nanometers(nm), and the color white is perceived when all the visible wavelength are presents.

The retina contains photoreceptors that receive incoming light of different wavelengths. These photoreceptors absorb photons and then transmit chemical signals to the brain. There are two types of photoreceptors: rods and cones. Rods are night-vision receptors and have no color sensitivity, while cones have color sensitivity, yet require a higher level of light intensity than rods. Cones have three types of photopigments, blue, green and red, which have a maximum sensitivity at 430nm, 530nm, and 575nm, respectively. Light with a single wavelength can activate all three types of cone, e.g., at a wavelength of 480nm, there will be a strong blue activation, plus some red and green component, as shown in Fig. 1.

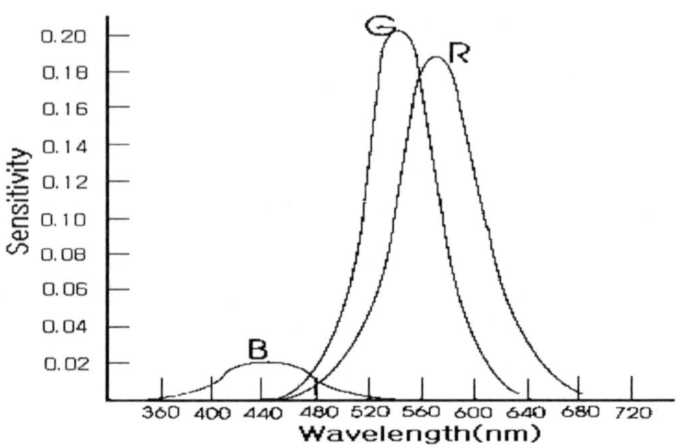

Fig. 1. Sensitivity of cones in human retina

When white light is incident on a surface, some wavelengths are absorbed and others are reflected. The three types of photopigments in the retina then respond to the reflected wavelengths. If low frequencies are predominant in the reflected light, the object is perceived as red. In this case, the perceived light has a dominant frequency at the red wavelength and humans perceive the dominant frequency as the representative color. The dominant wavelength is called the color or hue of the surface.

Lights from two or more sources with different intensities are combined to produce a range of colors. Normally three colors, called the primary colors, are used to produce a range of colors called a color gamut. Although it is impossible to represent the entire visible spectrum from the combination of these three primary colors alone, they can provide a color gamut that covers most colors. Fig. 2 shows the amount of Red, Green, and Blue needed to produce the specific colors that are visible to humans. The curves were obtained by averaging the judgement of a large number of observers and are referred to as the RGB color matching functions. As such, colors within the vicinity of 500nm can be matched by subtracting red light from a combination of blue and green light.

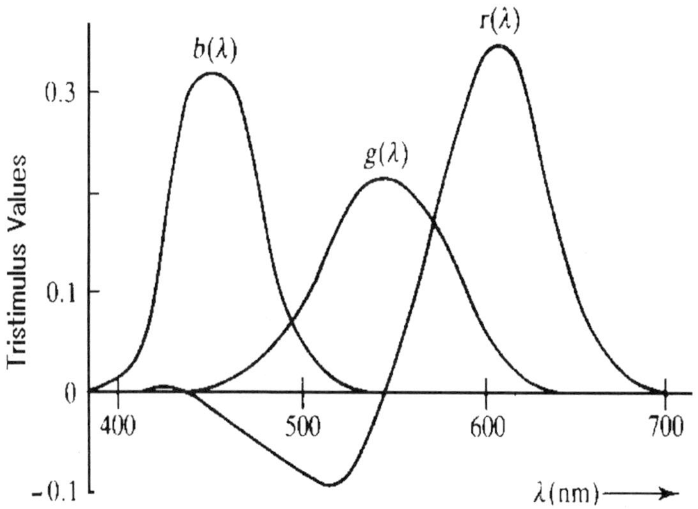

Fig. 2. RGB Color Matching Function

Most computer applications use the colors in the three-component system of RGB derived from the mechanisms of human color perception. While no finite set of color light can be combined to display all possible colors, three standard primaries were defined in 1931 by the International Commission on

Illumination, known as the CIE. The three CIE standard primaries are imaginary colors and defined mathematically with positive color-matching functions. Accordingly, this provides an international standard definition for all colors and eliminates negative value color matching. Fig.3 shows the CIE color matching functions.

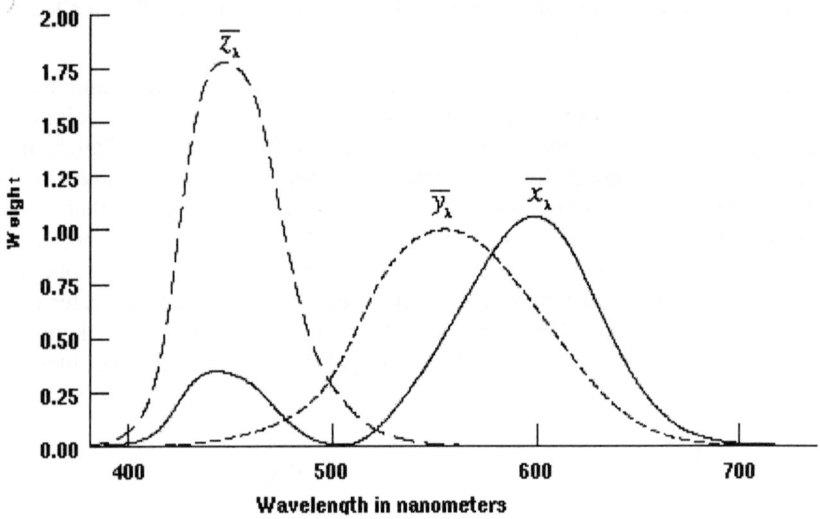

Fig. 3. CIE Color Matching Function

The CIE tristimulus values of *X, Y,* and *Z* can be obtained from the CIE color matching functions(Fig. 3) given a light spectrum with $P(\lambda)$ (CIE, 1971, and CIE, 1986). To calculate the CIE *X, Y,* and *Z* values for light with a particular spectrum, the products of the color matching function weights for each wavelength and intensity emitted at that wavelength are added for the whole the visual spectrum range. The constant *k* is 680 for a self-luminant light source.

$$X = k\sum_{380}^{780} \overline{x_\lambda} P(\lambda)$$

$$Y = k\sum_{380}^{780} \overline{y_\lambda} P(\lambda) \qquad (1)$$

$$Z = k\sum_{380}^{780} \overline{z_\lambda} P(\lambda)$$

When discussing color properties, it is convenient to normalize the amounts against luminance values as in $x+y+z=1$. Thus, any color can be represented by just x and y values. Plus, since x and y values are normalized against luminance, the x and y parameters are called the chromaticity values.

$$x = \frac{X}{X+Y+Z}$$
$$y = \frac{Y}{X+Y+Z} \qquad (2)$$
$$z = \frac{Z}{X+Y+Z}$$

When the normalized values x and y are plotted for the visible spectrum, a tongue-shaped curve is obtained, as shown in Fig. 4. The edge of the tongue-shaped curve corresponds to pure colors, which are labeled according to the wavelength from the red-end to the violet-end of the spectrum. The line joining the violet-end point to the red-end point is called the purple line, which is not part of the spectrum. The point W corresponds to the white-light position., which is used as a standard approximation for average daylight. The points R, G, and B are the three primary colors. The color gamut using these primary colors is the area inside the triangle. Since the triangle can not encompass all colors in the CIE diagram, no three primary colors can generate all visible colors.

3. Calculation of Wavelength and Temperature from RGB Data

3.1 *Calculation of Wavelength*

Many electronic devices, including computer monitors, video cameras, and digital cameras, use RGB color coordinates. Table.1 shows the primary color chromaticity of three broadcasting systems(Martindale and Paeth, 1991).

Table 1. Primary color chromaticity

Broadcasting system	Red		Green		Blue		White Point	
	x_r	y_r	x_g	y_g	x_b	y_b	x_w	y_w
NTSC	0.63	0.33	0.21	0.71	0.14	0.08	0.31	0.3162
PAL	0.54	0.33	0.29	0.6	0.15	0.06	0.313	0.3291

No finite set of three primary color lights can be combined to display all possible colors. However, the three primaries define a triangle on the CIE diagram and any color within the triangle can be formed by mixing the primaries. The RGB coordinates do not include any pure spectral colors. The relationship between the RGB and CIE color coordinates is as follows:

$$x_r R + x_g G + x_b B = x_w$$

$$y_r R + y_g G + y_b B = y_w \qquad (3)$$

$$z_r R + z_g G + z_b B = z_w$$

The RGB color values used in a computer are usually represented by integer values using bits or bytes. Therefore, the color value should be normalized before using the above relationship. Then, the normalized RGB can be converted into CIE color coordinates using Eq. (3). Once the CIE color coordinates are identified, the dominant wavelength can be obtained from the color coordinates. For the color point C1 in Fig. 4, a straight line can be drawn from W(white point) through C1 to intersect the spectral curve at point A1. As such, the color C1 can be represented as a combination of white light and spectral color A1 and the dominant wavelength for C1 is A1. Therefore, this method only requires the identification of the intersection point on the spectral curve. However, there are certain limitations for colors whose intersection points are on the purple line(i.e. a non-spectral color). Color C2 has intersection points on the purple line and the complementary spectral color is A2. In this case, color C2 is generated by subtracting the spectral dominant wavelength A2 from white light.

The dominant wavelength can only be obtained inside the triangle in Fig. 4, except for the shadowed area where no spectral color exists. Two discrimination functions are defined for identifying the area where the dominant wavelength can be calculated. The line y1 in Fig. 4 extends from W to the left-end of the purple line, while line y2 extends from W to the right-end of the purple line. For the NTSC system, y1=1.384x-0.11376 and y2=0.0406x+0.357 are used. For the color point C1 in the triangle area, the straight line from W through C1 that intersects the spectral curve is as follows:

$$y = ax + b \tag{4}$$

where $a = \dfrac{(Y_w - CIE_y)}{(X_w - CIE_x)}$, $b = Y_w - aX_w$

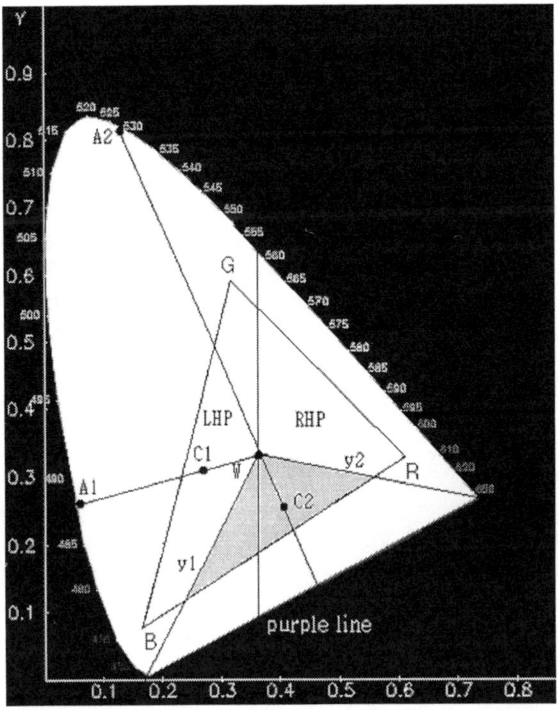

Fig. 4. Calculation of wavelength

To calculate the dominant wavelength from the RGB data, the corresponding CIE coordinates are converted to determine whether the point is outside the shadowed area or not. If the point is outside the shadow area, the line y can be calculated ,using Eq. (4) and the corresponding wavelength determined at the intersection point between the y line and tongue-shaped curve calculated from Eqs. (1) and (2) using the matching function, as shown in Fig. 3.

3.2 *Calculation of Temperature*

The temperature of a light source, such as a candle, alcohol lamp, or carbon arc lamp, can be determined from the RGB color. A black body radiator is a source whose spectral distribution can be described by the Plank function(Hunt, 1987). Many natural light sources can be approximated to an ideal black body radiator where the spectrum is entirely determined by its temperature. From Planck's radiation law, for an ideal black body at temperature T (Kelvin), the emittance at a given wavelength can be calculated using the following expression,

$$P_\lambda d\lambda = \frac{a\lambda^{-5}}{e^{\frac{b}{\lambda T}} - 1} d\lambda \tag{5}$$

where *a* and *b* are the constants.

The CIE *XYZ* coordinates of a black body with temperature T can be determined using Eqs. (1), (2), and (5). As such, the color and its temperature can be tabulated. Therefore, if the RGB data of a light source is converted into CIE coordinates, the corresponding temperature can be obtained from the table. Fig. 5 shows the loci of a black body radiator based on CIE color coordinates.

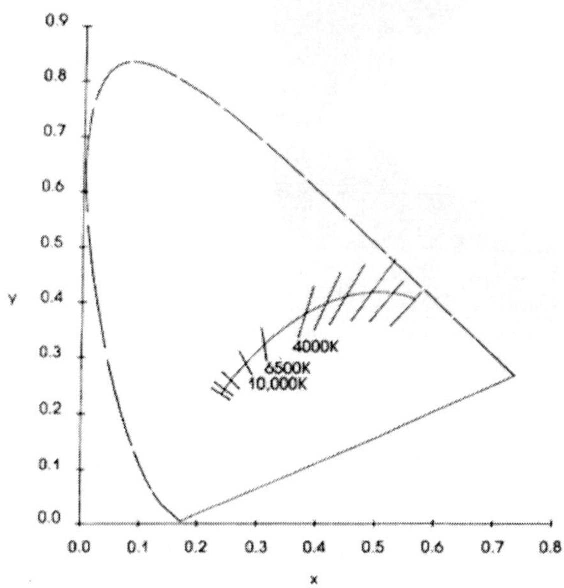

Fig. 5. Loci of Black Body Radiator

3.3 *Utilization of Proposed Program in Physics Classes*

Fig. 6 shows the display screen of the proposed program running in Windows 98. A photo of an alcohol lamp taken with an ordinary digital camera is displayed on the left. The RGB data is read from the point where the mouse pointer is located, then the calculated CIE coordinates are shown in the CIE chart at the center of the display screen. The calculated wavelength and temperature of the RGB data are also displayed.

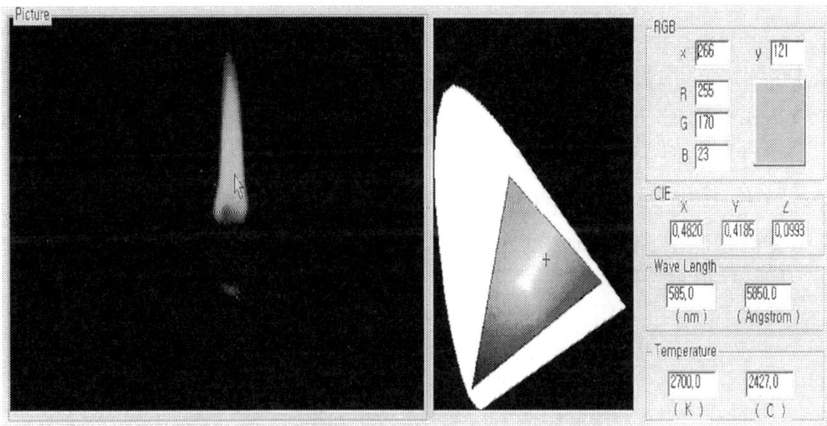

Fig. 6. Screen of Proposed Program

Students enjoyed using this program for their optics laboratory in physics classes and became familiar with the idea of determining wavelengths from various color photos taken with a CCD camera instead of a spectroscope. The resolution of the wavelength is about 10 angstroms, which is accurate enough for high school and college experiments. The proposed program is also useful for various other optics experiments, such as the analysis of interference fringes and dispersion of light by a prism or grating.

4. Conclusions

The current study developed a computer program that can convert the RGB data in photos into the corresponding wavelengths. The advantage of the program is that students can get an idea of the concept of wavelength from the RGB data in digital pictures that are easily obtainable from many CCD devices without the need of expensive spectroscopy. In addition, the temperature distribution of a light source can also be determined assuming that the light source is a black body radiator.

This work was supported by a Korea Research Foundation grant (KRF-99-005-D00076).

References

1. Commission Internationale de l'Éclairage (CIE). (1971). *Colorimetry*, No. 15, 1986.

2. Commission Internationale de l'Éclairage (CIE). (1986). *Standard on Colorimetric Observers*, CIE S002.

3. R.W.G. Hunt. (1987). *Measuring Colour*. Ellis Horwood Limited.

4. David Martindale and Alan W. Paeth. (1991). *Television Color Encoding and "Hot" Broadcast Colors*. James Arvo, editor, pages 147-158. Academic Press, Boston.

HOW CAN THE COMPUTER HELP STUDENTS IN THIS AGE OF LIFE LONG LEARNING?

JOAN SOLOMON
Centre for Science Education. The Open University

MARGARET MAZZOLINI
Centre for Astrophysics & Supercomputing, Swinburne University of Technology

Both the UK and Australia have long experience of distance learning and the problems which can emerge because of the students' lack of social contact. In this paper the authors first consider the theoretical basis for the topic to be found in the work of Mead, Schutz and Kuhn. Then they consider the early research in both countries which show just how important this lack could be and how comparatively easily it could be repaired to provide a more genuine network-based experience. The authors present an account of some new work using on-line education, and the importance of synchronicity, including comments from the lecturer. In the final section the authors make links with the classic work of Perry in more traditional education. The difficulty of ambiguous information produces an urgent need for more social support, which can be satisfied by new computer intervention.

The UK Open University, dedicated to life-long learning, received its first cohort of science students in 1970 Because it was a "distance" university, students would receive their learning materials (at that time almost exclusively books) through the post. However there were always tutor groups which met about once a month with the regional tutor. The location was not always close enough to the students and many could not or did not choose to attend these tutorials. The science students also received a small experimental kit to carry out some exercises at home, and all students had assignments to do. This was the single place which, apart from the rather poorly attended tutorial groups, could provide any social opportunities for the students and it was, and is, enjoyed enormously by them. The main point of this paper, and at least one of the answers to the question posed in the title, connects this pleasure in social contact, and indeed the need for it, to social learning theories.

1. The Social Construction of Knowledge

The general public, until well into the last century, believed that scientific cognition was a supremely solitary affair. If science were indeed like this it would be an ideal subject for distance education.

However a new and more social approach to thinking and learning dates back to the more mature works of G.H.Mead. By 1934 he was writing unequivocally that

'We must regard mind, then, as arising and developing within the social process, within the empirical matrix of social interactions.' (page 133.)

We may more easily accept that the infant's mind developes through exchanges with people. But Mead was emphatic that the full process of experience is only possible for a group of interacting adults who can discuss and reconstruct what has happened and hence develop an understanding of what was experienced by each one of them. Mead was a neighbour and close friend of the American educational philosopher John Dewey and the two exchanged so many ideas that it is sometimes difficult to disentangle one man's thought from the other's. That makes it all the more surprising that the social aspects of learning took so very long to permeate into school or university education.

The next steps in what has now come to be known as the sociology of knowledge were taken by Alfred Schutz. In his philosophy the social construction of thought is practised by all of us almost all the time as we ask others if they '*know what I mean?*' This expresses the need of each of us to be reassured that others understand us and think in a similar way. Once again these factor are clearly for science education, especially perhaps for distance education, but one which has only rarely been used in educational research (Solomon 1984). This analysis was very similar to that of Berger and Luckmann (1966) who described not only the foundations of knowledge in everyday life, but also the institutions of science - schools and universities - as having *'conceptual machinery'* by means which they maintained their consensual knowledge system.

In 1962 Thomas Kuhn published his Structures of Scientific Revolutions in which he described the paradigm shift which becomes a revolution in scientific theory. He saw this as a social phenomenon both because it was generated by a group of like-minded scientists, and because it succeeded in convincing others that they should see the world in a new way that was not more or less logical, but seemed to have greater explanatory power. Only a few years later John Ziman's book Public Knowledge (1968) completed this stage in our understanding of the social nature of science by explaining that its goal was to produce a consensus of universally accepted knowledge. This made the internal relations between scientists, and the ways in which they communicated with each other, of the greatest importance.

2. Science Education as Socialisation

There was no great problem in accommodating this general perspective to school science education. Clearly it took place in a social space within a classroom. By argument, analogy and experiment the teachers try to 'socialise' their students into the scientific way of thinking. However when the students leave the science classroom they very often revert to the life-world way of thinking which is continually reinforced by everyday talk. This is how it was put in a valuable book on understanding talk

> *'As social beings we need to legitimate the world picture we are continuously constructing and maintaining. So we hold out to others - in talk - our observations, discoveries, reflections opinions, attitudes and values, and the responses we receive profoundly affect both the world picture we are creating and our view of ourselves'* Martin et al 1976 p15.

In conventional universities science education where there is pressure to absorb a large quantity of new content which is not intuitive, students often feel the need to talk it over with others. What could they do in distance education?

Social learning must be, of course, an even rarer event in distance education. during the one week of residential summer school at the Open University, the students are to be heard socialising and sharing their learning problems with each other with great enthusiasm late into the night. As one very experienced science lecturer remarked *'there are dramatic changes in how they talk and take part in the work'* over the course of just this one week. It does seem that this study week is valued for more than just tutoring on the topics of the course.

3. Research Findings on Telematic Communication in Distance Education

One of the earliest reports on the use of audio teleconferencing involved the continuing education of nurses (Hart 1990). A previous study had shown a very high level of dissatisfaction (83%) with inset provision amongst nurses who worked outside metropolitan areas in New South Wales. Both new graduates and experienced nurses in rural situations were invited to join in the teleconferencing programme and to suggest topics that would be useful to them. There were a considerable number of initial and continuing problems with the teleconferencing set-up and sound quality, nevertheless the results were interesting.

99% of experienced nurses and 97% of new nursing graduates in this study found that the teleconferences had provided a good opportunity to have contact with other nurses and to share experiences. Almost all the suggestions made for improving future teleconferences involved strategies *for increasing nurses' participation.* The new nurses commented that it helped them to hear other nurses' comments and that it reduced their sense of isolation.

A little later Thompson (1996) drew on her experience of teleconferencing to make a sharp distinction between teletutoring where the tutor has control, and true teleconferencing where the students join with their peers in real discussion which has, she wrote, *'the potential to contribute to the growth of understanding.'* Both she and Edwards (1996) commented on how frequently the students use these opportunities for peer interaction to reassure themselves that others are suffering from the same problems with assignments as they do. Some of their talk, as Edwards interpretes it, is for establishing a sense of belonging to a student community, and being 'a proper student'.

From these studies we may draw some tentative generalisations about the use of communication technologies in distance learning. It may lead to an interchange of ideas between distant students, which Trentin (1996) called *'network-based education'*. Kirkwood (1998) added

> *Quality in an educational system is not enhanced by the application of technologies as such: it is how they are used that matters.* page 239

Advances in information technology has opened up other possibilities for OU students. Here again there are signs that this is being taken up along two different lines - one is for tutoring and the other for the social construction of learning. First and second year science students may drop into the OU Internet site called 'coffee shop' where they can ask for help and exchange methods for solving problems or finding data to help them carry out assignments. At a post-graduate level the author of an OU module called 'Science and the Public' has provided video-conferencing facilities for the students. The reported results of this were rather disappointing at first with only a very small core of enthusiasts taking part. It was only when the sessions were deliberately orchestrated to be more provocative that more and richer opinions began to be exchanged. This early need for a 'facilitator' of Internet discussion has gradually become less acute as the use of E-conferencing becomes more familiar.

4. An Example of Telematic Communication in Online Education

Online education – distance education featuring the use of the Internet – is often referred to as the next or fourth generation of distance education (Taylor 1995). One example of an online education program in which telematic communication occupies centre stage is Swinburne Astronomy Online (SAO) (Germany & Mazzolini, 2001). SAO is an online suite of programs offered by Swinburne University of Technology in Australia, leading to a Master of Science in Astronomy (by coursework). This program currently teaches astronomy to approximately 200 fee-paying students who are located in 34 countries and whose birthdates span five decades. SAO attracts groups such as amateur astronomers, science educators, people engaged in astronomy outreach and science journalists, all of whom generally want to learn about the conceptual foundations and key issues in contemporary astronomy.

SAO is a "hybrid online program", in that course *content* is not provided online. Instead, content is supplied as high-bandwidth multimedia material on CD-ROMs, and students go online to research essays and projects and, in particular, to take part in newsgroup discussions. By design but also by necessity - given that SAO students are located in almost all time zones - course communications are 'asynchronous'. In this context, asynchronous indicates that students carry out 'conversations' about the course material with each other and with the instructors by posting to newsgroups at times to suit themselves.

Asynchronous newsgroup discussions may sound like a clumsy way to interact, but they can be surprisingly flexible and engaging, providing socialization so frequently missing in distance education. In SAO, conversation threads build up on the newsgroups as geographically separated students come online in their particular time zones. Relationships are established between students who have never met or visited each other's countries but share a passion for astronomy.

The newsgroup component of the program was designed from within a constructivist philosophy to encourage active learning online. Students are expected to post their own questions or extension comments about current course content and attempt to answer each other's questions. Instructors act as 'guides on the side', aiding the discussions but not dominating them. At the end of the teaching semester, students choose which of their postings they wish to nominate for assessment. Counting newsgroup contributions towards assessment provides a 'carrot' to recognize the time and effort students put in to researching and composing their newsgroup contributions, and a 'stick' to encourage newsgroup contributions from all, not just from the most confident (Mazzolini 2000).

SAO experience suggests that synchronous communications can also support discussions that encourage deep learning. The depth of the discussions can be very impressive, even inspiring, for instructors more used to on-campus students displaying the "but will it be on the exam?" syndrome. John, an SAO student who is also an engineering lecturer in Europe, commented on interactions in his first SAO course,HET602:"The HET602 group are a terrific bunch, even though class is virtual. And the discussions and questions are imaginative and demand thinking caps. It's more like the classic Greek form of tuition, with the course tutors and students sitting round a forum discussing the subject." Brian, another SAO student who is also a university lecturer (in Australia) reflected: "I think that giving the students the responsibility for answering questions amongst themselves in this way is a brilliant idea, which has the potential to contribute to life long learning ..."

Posting to asynchronous newsgroups may aid socialisation and support learning, but it can prove to be daunting for some, at least initially (Salmon 2000). Learning through online discussions will never suit the learning styles of everyone equally well, but instructors can do a lot to overcome initial nervousness by encouraging a positive, supportive atmosphere. Helen, an Australian who studied with SAO in her 'spare time' as mother of a new baby, made these comments about the newsgroup discussions: "I enjoy reading the latest entries each night, but admit to having some nervousness about submitting queries and replies to others queries. However, I feel it is good because it is pushing me to be more confident and actually respond rather than 'sit in the back of the class' and say nothing, which is what I usually find myself doing in lecture situations. It has also made me spend time doing further reading, in books and on the internet, to come up with non-trivial queries, so that is also a good thing. So, overall, I find the newsgroups challenging (but then why do something new if it doesn't challenge you?) but useful."

5. Tentative Conclusions and More Challenges

We have seen that a considerable proportion of students, both in conventional universities and in distance learning, find that there is a form of cognitive socialising or networking which makes their learning easier, as theories of the social construction of knowledge would have predicted that they might.

It is clear that there is a need for further research in this field, which should include the learning styles of the students most in need of help, and the contexts where students feel most bewildered and most in need of social contact. When learning science in a social context (STS) personal views on civic and social justice and the anxiety produced by hazard and risk, make expressing an individual opinion even more difficult but also very important. This suggests that discussion may be invaluable for clarifying the students' own moral viewpoints.

However there is reason for believing that this is not the only context where ambiguous situations raise uncertainties. A further kind of cognitive uncertainty is endemic to all properly challenging university learning. According to William Perry (1968) facing up to epistemological dilemmas is essential for reaching a higher level outcome. School and lower college level students are taught well-accepted scientific theory, but at some time during a university degree course we may expect them to come up against a situation for which there are two or more rival explanations or views. Perry's investigations were based on a longitudinal interview study of humanities students, but there is every reason to suppose that the same would happen to science students.

Briefly, Perry's theory sees students going through three phases. In the first the students suppose that one explanation is definitely right and the other wrong but, for some reason, the authorities have not yet been able to decide on the right one. In the second phase students begin to accept that the uncertainty itself is legitimate, because either theoretical position is possible. In the third phase the students begin to take up a personal position and to see how commitment to one of these theories could be used in their personal studies or career. Sometimes this last stage may also include moral positions or even issues of responsibility, where the issue may allow for this (eg. animal conservation, medicine, environmental work, etc.). Perry claimed that the need to move from one stage to another put students under considerable pressure. After years of learning scientific theories which seemed to carry a guarantee of orthodoxy, the movement from stage one to stage two might well be resisted. Indeed some of Perry's students never did make the transition, and found the confrontation too intellectually stressful for comfort.When physics students meet wave/particle duality, and when biology students face up to the genetic determinism of the new biology, this kind of uncertainty may be, and possibly should be, unavoidable. Once again we can see a real need for a social context of learning which a new use of computer technology seems to be able to satisfy.

References

Berger, P., and Luckmann, T (1966) The Social Construction of Reality. Harmondsworth. Penguin.

Edwards, D.(1996) *Troubled times? Personal identity, distance education and open learning.* Open learning (Feb), 3-11.

Germany, L., and Mazzolini, M. (2001) *Swinburne Astronomy Online – A World-wide Classroom*, Tenth United Nations/European Space Agency Workshop on Basic Space Science: Exploring the Universe – Sky Surveys, Space Exploration, and Space Technologies, University of Mauritius, Reduit, Mauritius, 25 – 29 June.

Hart, G. (1990) *Peer learning and support via audio-teleconferencing in continuing education for nurses.* Distance Education. 11(2), 308-319.

Kirkwood, A. (1998) *New media mania: Can information and communication technologies enhance the quality of open and distance learning?* Distance Education. 19 (2), 229-241.

Kuhn, T. (1962) The Structure of Scientific Revolutions. Chicago. University of Chicago Press.

Martin, N., Williams, P., Wilding, J., Hemmings, S., and Medway, P. (1976) Understanding children talking. Harmondsworth. Penguin.

Mazzolini, M. (2000) *Assessment Techniques in an Online Astronomy Course* Publications of the Astronomical Society of Australia. 17, 141 – 144.

Mead, G.H. (1934) Mind, Self and Society. University of Chicago Press.

Perry, W. (1968) Forms of Intellectual and Ethical Development in the College Years. New York. Holt, Reinhart and Winston.

Salmon, G. (2000) E-Moderating: The Key to Teaching and Learning Online. London. Kogan Page.

Schutz, A. and Luckmann, T. (1973) Structures of the Life-world. London Heinemann.

Solomon, J. (1984) *Prompts, cues and discrimination: the utilisation of two separate knowledge systems.* European Journal of Science Education. 6(3), 277-284.

Taylor, J.C. (1995) *Distance education technologies: The fourth generation* Australian Journal of Educational Technology. 11(2), 1 – 7.

Thompson, D. (1996) *Audioteleconferencing: myths and realities.* Open Learning (June), 20-27.

Trentin, G. (1996) *Internet: does it really bring added value to education?* Educational Technology Review. No 6. pages 10-13.

Ziman, J. (1968) Public Knowledge. Cambridge. Cambridge University Press.

DEVELOPMENT OF MULTIMEDIA CD FOR HIGH SCHOOL PHYSICS BASED ON SF MOVIES

W. CHOI* & J. SONG°

Taegu University, Kyungsan, Kyungbuk, 712-714, Korea.

°Seoul National University(SNU), Shilim-dong,Kwanak-gu, Seoul, 151-742, Korea

Science fiction(SF) movies have a variety of advantages that are usually lacking in school science. Previous studies have already shown that both students and teachers have a positive attitude toward the use of SF movies in school science lessons. However, despite the advantages, using SF movies is an extra burden on science teachers, because it is hard for individual teachers to analyse and edit SF movies and then develop the necessary teaching materials. Therefore, to reduce this burden, the current authors developed a multimedia CD, named 'SF & Physics', which can easily be used by ordinary teachers in their science lessons. The CD focuses on the mechanics part of high school physics and contains lots of interesting movie clips taken from a variety of SF movies. In the process of developing the CD, comments and advice were collected from physics teachers and science educators. The CD was then shown in two high school classes and the results closely observed. It was found that both teachers and students were satisfied with the use of the CD and insightful recommendations were collected for future classroom implementation.

1. Introduction

One of the most difficult tasks for physics teachers is how to motivate students to study a subject that is normally perceived as difficult and boring. Students also tend to think of scientists, especially physicists, as rather strange characters with a very high intelligence yet unsociable personality (e.g. (1)-(2)). Therefore, students often believe that they are unsuited for studying such an intellectual subject. In contrast, science fiction(SF) has always enjoyed a great deal of popularity from the public, including students. Currently, SF, and particularly SF movies, is part of mainstream culture and enjoyed by all part of society.

Why are SF movies so popular, yet not physics ? Ironically, most SF movies contain a lot of physics. Therefore, the clear challenge is how to utilize the advantages of SF movies to teach physics and provide a new source of motivation. Over the last three decades, a small number of science educators have developed special courses utilizing SF and SF movies for more effective science teaching (e.g. (3), (10)). However, there have been few concrete examples of approaches using SF movies for physics teaching.

In a previous study, the use of SF movies in physics classrooms was demonstrated based on concrete examples specifically derived from two recent SF movies, 'Deep Impact' and 'Armageddon' (11). However, despite its advantages, the use of SF movies is an extra burden on science teachers as it is difficult for individual teachers to analyse and edit SF movies, and then develop the necessary teaching materials.

Accordingly, to reduce this burden, the current authors developed a multimedia CD, named 'SF & Physics', which can easily be used by ordinary teachers in their normal science lessons. The CD is the materialization of 'Using IT', one of six methods suggested in a previous study by the current authors and believed to be the most effective one.

Therefore, the present paper explains the procedures involved in developing the CD and some features of the CD. In addition, the results of implementing the CD in a classroom environment are presented, along with the reponses from the students and teachers.

2. Development of 'SF & Physics' CD

SF movies have various advantages that school science education does not have, plus students generally show amazing responses to their use in science classes. However, as mentioned earlier, it is extremely difficult for individual science teachers to analyse and edit SF movies and identify the right scientific knowledge relevant to particular SF movies. Accordingly, the development of multimedia teaching materials that teachers can use along with school textbooks and that students can also use to study by themselves is vitally important.

Consequently, the current authors developed a multimedia CD named 'SF & Physics' for teaching physics and investigated the effectiveness and problems of using SF movies in a science class.

The 'SF & Physics' CD is aimed at the mechanics part of high school physics. A variety of popular recent SF movies were selected and analyzed. During the development process, responses were also collected from science teachers and experts on IT-based physics teaching. Finally, the CD was tested in a science class and the responses evaluated. The procedure involved in creating 'SF & Physics' can be summarized as follows:

Stage I – Planning : reviewing relevant data and selecting the right authoring tool

Stage II - Organizing Contents : analyzing the contents of SF movies and constructing a scenario and storyboard on the basis of the movie content analysis

Stage III - Developing CD : developing a multimedia CD using MS PowerPoint 2000 on the basis of the storyboard. Opinions and advice were gathered from physics educators and through an on-line homepage(http://user.chollian.net/~nettrek).

Stage IV - Analyzing Teachers' Responses : showing the CD to physics classes, and surveying the responses of the teachers

Stage V - Revising and Pilot Trials : revising the beta version of the CD, developing student activity sheets, and carrying out two pilot trials in high schools

Stage VI - Re-revising CD : revising the CD again on the basis of the results of the pilot trials

The following lists some of the basic considerations applied during the development process:

- help teachers teach physics using contents and scenes from SF movies
- allow convenient use in ordinary science classes
- illustrate various methods of using SF movies
- include useful contents that can both motivate and help students understand scientific principles
- use MS PowerPoint, which is widely used by school teachers in Korea
- produce moving picture files(AVI, MPEG) by capturing scenes from SF movies
- include a variety of learning activities to help students actively engage in lessons
- be economic, in terms of time and money, as regards use by teachers
- develop a CD with follow-up materials provided via the Internet
- include various moving clips for promoting student interest, while excluding non-educational aspects

The most important points considered when developing the CD were how to motivate students to study science and how to help ordinary school teachers use the CD with no extra burden. Accordingly, the following strategies were applied:

- use of exciting and interesting clips and images to make students motivated

- avoid boring scenes by introducing animation effects and including a lot of moving clips
- exclude unnecessary icons or keys to make the slides simple and straightforward
- make the CD as flexible as possible to meet a variety of teaching conditions
- avoid textbook-like layouts by creating colorful and interesting designs
- use the right size of text and layout, considering classroom conditions (e.g. 43″projection TV)
- include as many moving clips as possible
- introduce background music when a new sub-unit is introduced
- avoid distraction due to unnecessary background music
- make active connections between the move scenes and relevant scientific concepts
- make students take part in the learning activities

Figure 1 and Table 1 show the general outline and organization of 'SF & Physics', respectively.

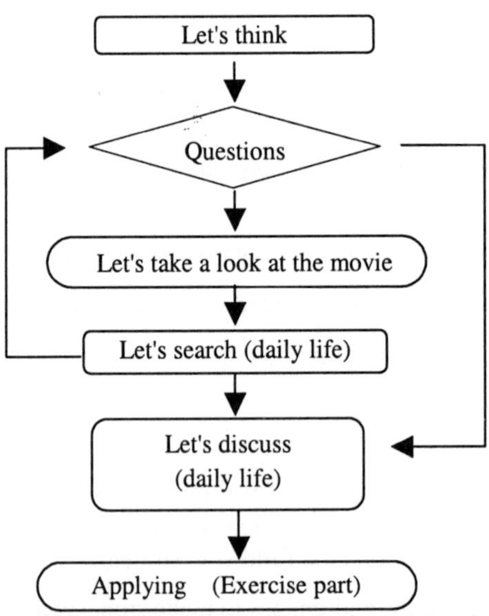

Figure 1. General outline of 'SF & Physics' CD

Table 1. General outline of 'SF & Physics' CD

Sub-unit	Learning point	Movie source	Movie-clip contents	Study contents
Description of Motion	distance and displacement	Superman 1	motion of Superman and his friends	scalar, vector, vector addition and subtraction, vector analysis
	Speed	Star Wars 4	Captain Solo explains spaceship's speed	speed, average speed, instantaneous speed, lineal uniform motion.
	Velocity	Star Wars 1	motion of race car in pod race	velocity, average velocity, instantaneous velocity
	relative velocity	Star Wars 6	pursuit scene between Empire army and Luke's Speeder Bike	lineal relative velocity, plane relative velocity
	Acceleration	Terminator 2	overturned car	acceleration, average acceleration, instantaneous acceleration, acceleration formula, acceleration graph
Laws of Motion	Force	Superman 1	The child (Superman) lifts a car	force, force unit, force addition and subtraction, force equilibrium
	Friction	Robo Cop 3	overturned patrol car	force of friction, force of static friction, force of kinetic friction
	Elasticity	Flubber	performance test of flubber on basket court	elasticity, Hook's law, force of elasticity, spring and force of elasticity
	law of motion(1)	Inspector Garget	The car stopped suddenly	inertia, law of inertia
	law of motion(2)	Back To the Future 3	increase speed by pushing car(time machine) with train	Newton's second law, motion formula, mass and weight

	law of motion(3)	Cowboy Bebob	The person who gets the gun moves backward due to the recoil after firing	Newton's third law
	Gravitational field(1)	Inspector Gadget	Free fall of Garget from helicopter	free fall, projectile motion
	Gravitational field(2)	Inspector Gadget	parabolic motion by spring	parabolic motion
Circular Motion and the Law of Universal Gravitation	uniform circular motion	Star Wars 6	dangled Speeder Bike	uniform circular motion, force of inertia, central force
	simple harmonic motion	Star Wars 6	dangled log	displacement, velocity, acceleration, force, period
	the law of universal gravitation	Deep Impact	Explanation by president about comet	Kepler's Laws of Planetary Motion, The law of universal gravitation, gravity, speed and period of artificial satellite
Momentum and its Conservation	momentum and impulse	Back To the Future 3	falling on his back after being hit by a bullet	momentum, impulse
	conservation of momentum	Matrix	Two persons meet together in the air	conservation of momentum, momentum addition and subtraction, collision and coefficient of restitution

3. Teachers' Responses to Beta Version of CD

The beta version of the CD was played in front of two groups of postgraduate students majoring in science or physics education at two metropolitan citiy universities in Korea: i.e, 17 graduate students majoring in physics education at Pusan National University and 6 graduate students majoring in science education at Seoul National University. Most of the audience already had some practical experience as secondary teachers, and a total of 14 answered all the items in the survey. Table 2 shows the results of the survey.

In addition to the nine questions included in the survey, the respondents also expressed some individual opinions as follows: 'the interaction between the teacher and the students needs to be active and systematic'; 'more everyday life topics should be presented to allow teachers to explain the movie scenes and physical phenomena'; 'using this CD will be useful for identifying and changing students' misconceptions'. Finally, the CD was also reviewed by several specialists in 'teaching physics using IT'.

Based on the responses collected from the survey and expert review, the following points were considered when revising the CD:

- help students read the text by reducing the amount of text and enlarging the font
- make the CD more convenient by making the layout simpler and more consistent
- encourage more student-centered learning by introducing activities like 'Identifying Misconceptions' and 'Discussion'
- make students concentrate on physics, rather than on the movie scenes or stories

4. Pilot Trials and Student Responses to CD

After revising the CD based on the teacher responses to the beta version, the new version was shown in actual science classes. One researcher and one assistant visited and observed the lessons at two high schools, one in a rural district and the other in a local city. The numbers of students attending the classes were 14 students at Sabul High School in Sangju and 40 students at Jinhae High School in Jinhae City. Detailed observations were made with the help of observational record sheets and video recording.

In the case of Sabul High School, the CD was given to the teacher in advance to allow for lesson prepartion. The classroom was equipped with a 43″projection TV and computer. However, the mouse was too old to use, and since there was no codec, this had to be set up during the class. On the whole, the students showed a good concentration. In the case of Jinhae High School, the physics lab was equipped with a 29″TV and 24″TV on both sides of the class that were connected to each other. Again the CD was provded in advance to allow for lesson preparation.

After the class, the students were asked to fill out a survey, which included basically the same questions that were used for the teacher survey. In general, the students complained that the content of the CD was slightly different from 'Dynamics in Common Science' and thus needed to be more specifically revised for Common Science. Since some of the students had no knowledge of the movies included in the CD, they requested that teachers add explanations about the movies. The other responses were very similar to those from the teachers. The results of the survey are shown in Table 3.

Table 2. Analysis of responses from science teachers to CD

Question	Possible Answers	Frequency
1. Do you think SF movies are good material for teaching science?	- They help a lot.	2
	- They help a little.	11
	- They have no effect.	0
2. What result do you expect when teaching science with SF movies?	- Making students interested in science	*4.7
	- Improving learning efficiency	4.0
	- Helping students to have good grades	3.5
	- Applying science to daily life	3.4
	- Understanding scientific concepts	3.2
3. Is this multimedia CD appropriate for use in class?	- It is convenient to apply	1
	- Teachers will get help from this multimedia CD, yet there are some vulnerable points	8
	- It is better to revise the CD for students.	3
	- Other	1
4. This CD was developed using PowerPoint. Is this the best program for developing a multimedia CD?	- It is adequate for teachers to use, yet it has some weak points when expressing designed things.	7
	- It is appropriate for teachers to use and develop multimedia CD.	5
	- It is appropriate for teachers to develop multimedia CD.	1
5. Is the visual design effective?	- It is appropriate and good.	3
	- It needs to be revised a little.	8
	- It needs substantial revision.	2
	- Other	3
6. Do you think this multimedia CD is convenient to use?	- It is simple and convenient.	6
	- It needs to be revised a little.	8

7. How would you use this multimedia CD?	- For arousing interest	8
	- As supplementary material in the development stage	3
	- For confirmation learning or formative testing	3
	- For the the whole learning process	1
8. Which aspect needs to be revised?	- Structure: content and development process	4
	- More detailed information required on the subject	1
	- More variety of methods (e.g. discussion, Let's think)	9
	- Needs to be provided through the Internet	0
	- Other	1
9. When a multimedia CD is developed, what aspects should be considered?	- Content and sequence	12
	- Detailed analysis of lesson contents	1
	- Visual effects	1
	- Internet back up	1

* For this item, the average scores from a 5-point scale were calculated

Figure 2. Main page of 'SF & Physics' CD

Table 3. Analysis of responses from students that participated in pilot trails

Question	Possible Answers	Sabul (14)	Jinhae (40)
1. Do you think SF movies are good material for teaching science?	- They help a lot.	12	37
	- They help a little.	2	0
	- They have no effect.	0	1
2. What result do you expect when teaching science with SF movies?	- Making students interested in science	*4.9	4.8
	- Understanding scientific concepts	4.1	4.4
	- Applying to daily life	3.6	4.5
	- Improving learning efficiency	4.1	4.7
	- Helping students to have good grades	3.6	3.7
	- Positive change in learning attitude	4.5	4.7
3. Is the visual design effective?	- It is effective and good.	9	28
	- It needs to be revised a little.	5	11
	- It needs substantial revision.	0	0
	- Other	0	0
4. Do you think this multimedia CD is convenient to use?	- It is simple and convenient.	6	31
	- It needs to be revised a little.	8	9
5. How would you use this multimedia CD?	- For arousing interest	7	22
	- As supplementary material in the development stage	2	12
	- For confirmation learning or formative testing	3	1
	- For the whole learning process	2	3
6. What part needs to be revised?	- Structure: content and development process	1	17
	- present more detailed information on the subject	1	12
	- use more variety of methods (e.g. discussion, Let's think)	6	5
	- needs to be provided through the Internet	6	5

7. When developing a multimedia CD, what aspects should be considered?	- Content and sequence	1	17
	- Detailed analysis of lesson contents	4	12
	- Visual effects	1	2
	- Internet back up	8	4

* For this item, the average scores from a 5-point scale were calculated.

5. Observations from Pilot Trials

In addition to the student survey, the video tapes of the classes were analyzed to identify potential problems and revisional points for the CD. The following are the results:

① Equipment for multimedia materials, such as projection TVs and suitable computers, must be prepared before playing the CD.

② Preliminary inspection of the computer system is crucial. The CD uses 'mpeg4 codec' to improve the moving clips. Although there is a manual and software for setting the codec, the user must also have basic knowledge of the computer system and its proper settings to play the moving clips.

③ The user should know how to use PowerPoint. No matter how easy a program is to use, if the user does not know how to use it, it is completely useless.

④ Each school and teacher has their own teaching schedule and method of teaching physics. This means that the teaching procedures will not always be the same as the ones intended in the CD. Therefore, hyper-linked sub-units must be arranged to allow more flexibility.

⑤ Proper sound effects and animation effects are important.

⑥ The time used for playing moving clips must be controlled. If the moving clips are too long, students only concentrate on the movie and miss the aim of the lesson, plus the teacher's explanation time is also cut short. The appropriate playing time for a moving clip should only be one or two minutes. If the time exceeds 3 minutes, the student concentration on the clip also decreases rapidly.

⑦ Teachers must read the manual and prepare the lesson in advance. No matter how excellent the materials are, they are useless if teachers do not prepare the lesson in advance. Before the class, the teacher must play the CD at least once and think of how to optimize the effects of using the CD.

⑧ The number of letters and font must be appropriately arranged. If the font is too big, it cannot be fully explained on a single slide. Conversely, if the font is too small, students can not read the text well.

⑨ It is important to prepare relevant hand-outs. When students are bored and do not concentrate on the lesson due to a high number of slides, hand-outs can be very helpful. The provision of exercises or supplementary materials through hand-outs can also considerably improve the efficiency of the lessons.

6. Summary and Conclusions

Every science teacher is familiar with the fact that students tend to lack interest in science class, and the higher the grade, the less the interest. As such, the use of multimedia in science classes can be quite effective for students who are familiar with computer and multimedia games. Plus, the use of SF movies in science classes can provide both motivation and multimedia materials.

In a previous study by the current authors, six ways of using SF movies for teaching science were suggested and teacher's opinions about these methods investigated. Most science teachers had positive attitudes, however, they pointed out the extra burden of preparing such materials.

Accordingly, the current study developed a multimedia CD, named 'SF & Physics', which can easily be used by ordinary teachers in their science lessons. In the process of developing the CD, comments and advice were gathered from physics teachers and science educators.

The new multimedia CD aimed at the Mechanics of Physics Ⅱ in high schools consists of 4 units, and each unit has various sub-units and an exercise part. It was found that despite some partical difficulties, teachers and students both seemed to be satisfied with the CD and students showed a more positive response than teachers about the use of SF movies.

The general advantages of teaching science using SF movies can be summarized as follows: First, SF movies, rather than any other teaching materials, can make students more interested in science, concentrate on the lesson, and have a more positive attitude toward science; Second, SF movies can supply a variety of useful information on science, society, and technology. One good movie can be more effective to illustrate the issues of STS than a series of lectures or explanations ; Third, the proposed CD can encourage the use of the current AV and IT facilities in each school, and will eventually lead to improved teaching methods with the help of new educational technologies; Fourth, SF movies can be applied to every lesson stage, such as the developmental part or formative testing. With SF movies, teachers can diversify their teaching methods, beyond the boundary of traditional ways of teaching physics.

However, despite all these advantages, teachers are not advised to teach using movies without caution, as it takes a lot of time to prepare lessons, requires professional knowledge and experience, and the movies usually contain scientific knowledge well above the level of students' understanding.

Fortunately, these problems can be solved by developing and distributing well-prepared CD-ROM teaching materials, like the one illustrated in the current study. Since it is difficult for each individual teacher to develop good CD-ROM materials, it is important to develop CDs more systematically, for instance organized by Local Education Centers, and then distribute the CDs to science teachers in local schools. Finally, it should also be noted that is very important to obtain permission for use of clips from the movie producers before developing the CD, otherwise it will not be possible to distribute the final products.

References

1. Calame, G. P., The Science in Science Fiction: A Seminar course, *American Journal of Physics*, vol. 41, pp. 184-187, 1973.

2. Schmidt, S. A., Science fiction courses: An example and some alternatives, *American Journal of Physics*, vol. 41, pp. 1052-1056, 1973.

3. Zander, A. R., Science and fiction : An interdisciplinary approach, *American Journal of Physics*, vol. 43, pp. 9-12, 1975.

4. Marks, G. H., Teaching biology with science fiction, *The American Biology Teacher*, pp. 275-279, 1978.

5. Friedman, A. J., Contemporary American physics fiction, *American Journal of Physics*, vol. 47, pp. 392-395, 1979.

6. Freedman, R. A. & Little, W. A., Physics 13: Teaching modern physics through science fiction, *American Journal of Physics*, vol. 48, pp. 548-551, 1980.

7. Borgwald, J. M. & Schreiner, S., Classroom Analysis of Rotating space Vehicles in 2001: A Space Odyssey, *Physics Teacher*, vol. 31, pp. 406-409, 1993.

8. Dubeck, L. W., Moshier, S. E. & Boss, J. E., *Fantastic Voyages: Learning Science Through Science fiction Films*, AIP, 1994.

9. Song, J. & Choi, W. S., "Waiting for a 'DEEP IMPACT' of 'ARMAGEDDON': How to use SF movies in science classrooms", *Proceedings of '99 International Conference of Physics Teachers & Educators*, In X. Luo & K. Zhao(Eds.) Guilin, china: Guangxi Normal University, 304-307, 2000.

SIMPLE DEVICES FOR EXPERIMENTING THE PROPERTIES OF LIGHT

ANN CHIN HANN

Dept. of Science Education, Inchon National University of Education, Inchon, Korea

The purpose of the current study was to develop simple and low cost devices for experimenting with the properties of light in an ordinary classroom. Although children have a high interest in investigating the properties of light, such experiments are difficult due to inappropriate light sources, poor equipment, and the costs of providing a darkroom. Accordingly, two new simple devices are developed for conducting light experiments in an ordinary classroom. The proposed equipment enables experiments related to the laws of reflection and refraction, total internal reflection in water, critical angle, and other basic properties of light.

1. Introduction

Light is one of the most familiar natural phenomena that we experience in everyday life. Therefore, it is easy to find activities related to light in the textbooks for both primary and secondary schools. However, it is also difficult to carry out experiments with light due to lack of equipment and a darkroom.

Primary school children in Korea are supposed to learn about the properties of light in fourth grade. They study the straight propagation of light both in air and water, the law of reflection, the properties of refraction when light propagates from air to water and vice versa, the pin-hole effect of light etc.(Ministry of Education, 1999). However, although children have a high interest in investigating the properties of light, experiments with light are not easily carried out due to inappropriate light sources, poor equipment, and the costs involved in providing a darkroom.

Therefore, appropriate equipment needs to be developed to enable primary and secondary school students to carry out experiments related to the properties of light more easily.

2. Purpose of Current Study

Previous authors, such as Hewitt(1993), Highsmith(1975), Wilson(1989), and Herbert and Ruchlis(1968), have not included ways of experimenting with the basic properties of light that can be conducted in an ordinary classroom that is not a darkroom.

Accordingly, the purpose of the current study was to develop simple and low cost devices that can facilitate experiments on the basic properties of light, such as the straight propagation of light, reflection, refraction, critical angle, and total internal reflection, in an ordinary classroom that is not a darkroom.

3. Teaching of Light in Primary School

3.1 *Straight Propagation of Light*

1) propagation of light in air
2) propagation of light in water
3) change of shadows
4) pin-hole camera

3.2 *Reflection of Light*

1) reflection in a mirror
2) angle of incidence and angle of reflection
3) reflections with two mirrors
4) looking at someone through a mirror
5) making a periscope
6) images between two mirrors
7) making a kaleidoscope

3.3 *Refraction of Light*

1) refraction when light propagates from air to water
2) refraction when light propagates from water to air
3) refraction by convex lens
4) refraction by concave lens
5) looking at an object through a lens
6) observing light passing through a prism
7) using mirrors & lenses

4. Teaching of Light in Secondary School

4.1 *Properties of Light*

1) reflection of light
2) refraction of light

4.2 *Optical Instruments*

1) microscope
2) telescope
3) projector
4) camera
5) eyes

4.3 *Light waves*

1) speed of light 2) interference & diffraction

3) polarization 4) diffusion of light & color

5. Simple Equipment Developed in Current Study

5.1 *Types of Experiment That Can be Performed Using Proposed Equipment*

Experiments related to the laws of reflection and refraction are difficult to carry out in an ordinary classroom due to inappropriate light sources and poor equipment. Therefore, to overcome these difficulties, two simple devices are proposed, the first is for experimenting with the law of reflection, and the second is for experimenting with the law of refraction. The types of experiment that can be performed using the proposed equipment are as follows:

1) law of reflection 2) propagation of light in air and water

3) propagation of light from air to water and vice versa

4) law of refraction 5) critical angle

6) total internal reflection in water

5.2 *Materials and Diagrams of Proposed Equipment*

1) Materials needed to construct proposed equipment

One of the key points in experimenting with the properties of light is putting smoke and chemicals in the device so as to see the propagating light clearly. As such, the current study uses mosquito smoke for the former and phosphorus red for the latter. Yet, only mosquito smoke is required for the experiments on the law of reflection.

2) Diagram of proposed equipment (Figure 1)

5.3 *How to Use the Proposed Equipment*

A laser pointer is used as the light source due to its unique characteristics, i.e. it is highly directional, coherent, and very bright. For experiments on the law of reflection, the half-circle type equipment is used after inserting a small amount of mosquito smoke. As regards the circle-type equipment, water and a small amount of phosphorus red is used to fill one half of the device, as shown in the diagram, then a small amount of mosquito smoke is inserted in the remaining section. This set up can then be used to demonstrate the law of refraction, total internal reflection in water etc, as indicated earlier.

6. Conclusion

On the basis of developing simple and low cost devices for experimenting with the properties of light, the following conclusions are made:

- Simple and low cost devices can be made available and implemented in both primary and secondary school classrooms

- Basic and important properties of light can be observed in an ordinary classroom which is not a darkroom, using the simple devices proposed in the current study

- The experiments that can be carried out with the devices proposed in the current study include the straight propagation of light both in air and water, law of reflection, properties of refraction when light propagates from air to water and vice versa, critical angle, and total internal reflection.

Further experiments related to the properties of light will be developed in the near future.

1) law of refraction

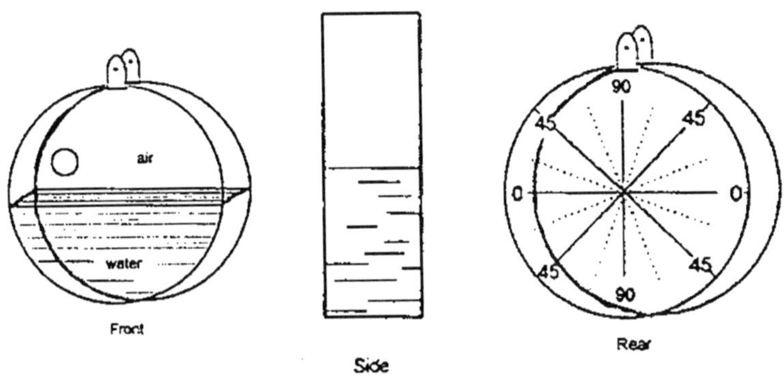

2) law of reflection

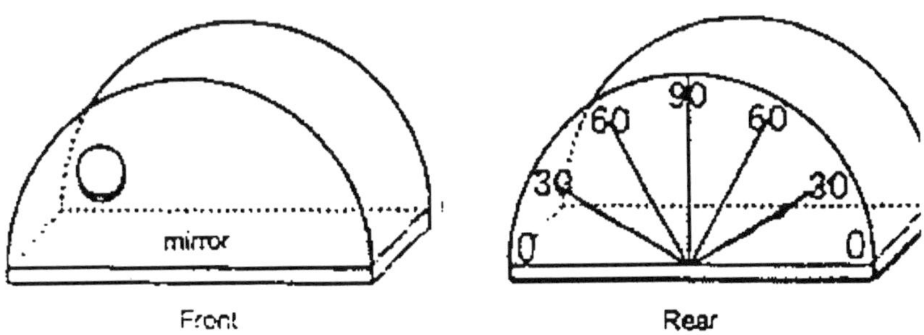

Figure 1 Diagram of the equipment

References

1. Hewitt, P. G. (1993). *Conceptual physics.* Harper Collins College Publishers.

2. Highsmith, P. (1975). *Physics, energy and our world.* W. B. Saunders Company.

3. Herbert, D. & Ruchlis, H. (1968). *Mr. Wizard's 400 experiments in science.* Book-lab, Inc.

4. Ministry of Education, Korea (1990). *Teachers guides from grade 1 to 6.* Korean textbook printing company.

SCIENCE FRICTION

A.KAZACHKOV, YU.POPKOV, E.SAVCHENKO, L.DIMITROVA & I.ANIKEEVA

Department of Physics, Kharkov National University, 4, Svobody Sq., Kharkov, 61077, Ukraine

It is almost impossible to exaggerate the role of friction phenomena in the history of human civilization. In an educational context, a variety of physical effects, innumerable technical devices and processes, and even some amazing toys – friction-reducing or friction-based - provide excellent topics for discussion with students learning physics. Accordingly, the current paper presents a series of class demonstrations and hands-on activities for high school and university students, with the particular intent that these topics can become introductory research studies. The proposed activities and experiments on sliding and rolling friction, triboelectricity, estimating the friction coefficient, and the unexpected (and frictionally explained) behavior of spinning tops, were found to stimulate students without exception, and certain students even became involved in co-authoring the current presentation. In particular, the proposed activities gave students an understanding of the physical nature of real-life phenomena, while appealing to their everyday experience and cultural background. The didactic aspects and educational impact of "science frictional" investigations are presented and analyzed.

1. Introduction

In the history of science and engineering, two approaches to frictional phenomena have prevailed:

Minimization of frictional force between surfaces and bodies in relative motion. In this sense, the invention of the wheel was among the utmost achievements of the human mind, since friction diminishes drastically when sliding motion is substituted with rolling. As such, sections 4 and 5 in the current paper demonstrate the difference between the two motions, and the friction-caused transfer from sliding to rolling.

Attempts to make frictional cohesion between contacting surfaces as strong as possible. Sections 2 and 3 present exciting examples of unexpectedly strong frictional connections that can be good topics for class demonstrations, followed by explanation and calculative analysis.

2. Frictional Self-Blocking

If two sufficiently thick books are interleaved, they cannot be pulled apart even by the strongest person. Although the frictional nature of this effect is obvious, student explanations of this mechanism are often errant. In fact, neither the square of the surfaces in contact nor the total weight of the books matters here. The point is that the thickness of the interleaved section is twice that of the individual book, therefore, the pulling force of each leaf (except for the central one, to be exact) has a component, normal to the plane of the pages.

The resulting frictional force F_{fr} must be proportional to this force of compression and thus to the pulling one. As such, the more one pulls, the stronger the frictional force that prevents the separation of the books! An expression for F_{fr} obtained by rough estimation [1], was shown to be in a good agreement with the exact formula when the number of interleaved sheets was over 50. Note, that the frictional force and leaf self-blocking increase rapidly with the number N of overleaved sheets ($F_{fr} \sim N^2$). The different geometries of proposed interleaving allow this model to be proved experimentally and the obtained formula to be checked without measuring any forces.

3. Euler's Superfriction

A rope wrapped around a cylinder (or any other object), is also strongly self-pressed to the surface of contact when pulled at the ends [2]. The expression $F_1 = F_2 e^{\mu\theta}$, connecting the pulling forces F_1 and F_2, the coefficient μ of the friction between the cylinder and the rope, and the angle of wrapping θ, was first obtained by Leonard Euler. Consequently, when assisted by static friction, a small force pulling on a rope can actually hold a huge weight tied at the another end. The current inquiry revealed that although very few people are aware of Euler's formula, literally everyone takes advantage of this superfriction, for example, holding sewed buttons on a coat, tying boot-laces, and the amazing ability of one person to pull a giant ship to shore using a rope properly wrapped around a mooring block.

Persuasive demonstrations are proposed to show that the value of Euler's friction force is independent of either the cylinder diameter or the length of the rope wrapped, yet changes dramatically with a variation in the wrapping angle θ.

4. Skidding: What's Wrong with The Brakes?

The problem of a skidding car has already been discussed in detail elsewhere ([3] probl. 26, [4] §74). Essentially, skidding is caused when the direction of the kinetic friction force strictly opposes that of the velocity of the relative motion of the rubbing surfaces. In the proposed demonstration, a toy cart, thread bobbin, and ordinary coin on an inclined plane, are used to represent a car skidding on an icy road. Moreover, the difference between sliding and rolling friction is also clearly demonstrated in this experiment (Fig. 1).

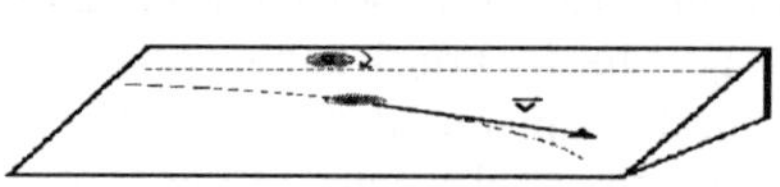

Fig. 1. Coins rolling and sliding with friction on an incline.

Calculating the trajectories of the bodies moving on an incline with friction is an excellent problem for students majoring in physics. It is common knowledge that the maximum value of the static friction force coefficient μ_s can be easily measured on an incline. Skidding-simulating experiments allow the value of the coefficient μ_k of the kinetic sliding friction to be derived from an analysis of the moving object trajectory.

5. Tops That Spin, Tops That Fall, Tops That Roll

People are used to the fantastic stability of spinning tops, yet wait till the rotation of the top slows and finally falls down. For most of the students involved in the current study, the inverse direction of the fallen top rotation (Fig. 2) was a big surprise.

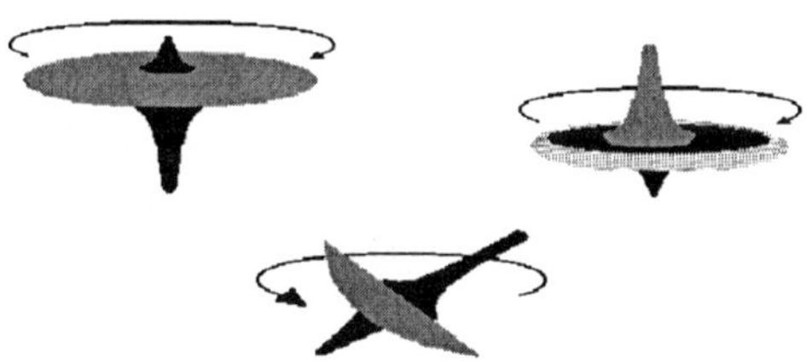

Fig. 2. Tops that spin and roll after they fall down.

At first had difficulty in explaining the behavior of the familiar toy, yet under proper experimental conditions, the friction-induced transfer from the fallen top's initial sliding to a rolling motion, became much more understandable. A physically similar, though much less fascinating, problem of a rotating cylinder placed on a horizontal surface with friction has also been previously analyzed ([4] §72).

A set of experimental problems related to describing the phenomena proved to be really amazing and useful for students. Tops of different shape were explored. Visualization (see Fig. 3) and interpretation of the trajectories of the sliding/rolling was a special challenge, requiring real imagination and insight.

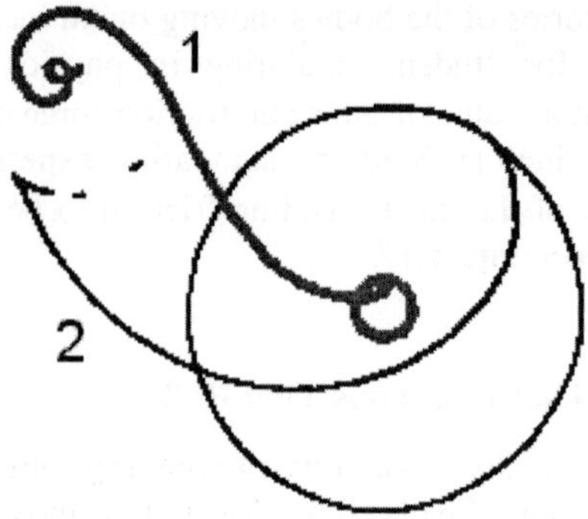

Fig. 3. Lipstick-visualized trajectories of fallen top's tip (1) and rim (2) motion.

6. Conclusion

When used in teaching practice, the proposed examples of friction-natured effects were found to have a strong educational impact. In addition, the various forms of friction livened up the atmosphere of the physics classes, making students more enthusiastic to master the subject.

Acknowledgements

Our interest in friction-blocking phenomena was mainly due to Prof. M.Euler, Kiel, plus the ideas of Dr. C.Ucke, Munich were followed as regards the use of spinning tops. Special thanks to D.Kazachkova for her assistance in designing the demonstrations and figures.

References

1. A. Muller, Paper Science, in: G. Born, H. Harreis, H. Litschke, . Treitz (eds.), *Hands-on Experiments in Physics Education.*, Duisburg, 1999.

2. A. A. Silin, *Friction and Its Role in Engineering*, Nauka, Moscow, 1976.

3. L. G. Aslamazov, I. Sh. Slobodetsky, *Physics Problems*, Nauka, Moscow, 1981.

4. S. P. Strelkov, Mechanics, Nauka, Moscow, 1976.

List of Participants

Name	Institution	E-mail address
Ahn, Ji-hye	Chongju National Education University	anjel99@hanmail.net
Akizo, Kobayashi	Dept of Science Education, Niigata University	kobayasi@ed.niigatau.ac.jp
Bentillo, Eulalia Nilo	University of the Philippines	enb@ismed.upd.edu.ph
Bong, Man Young	Kwang-ju Electronic Technical High School	photons@hanmail.net
Byon, Sunye	Dept of Physics Education, Seoul National University	sbyon@hotmail.com
Byun, Jeeyoung	Seokgok Middle School	00bjy00@hanmail.net
Cambaliza, Maria Obiminda	Department of Physics, Ateneo De Manila University	cambaliz@admu.edu.ph
Cha, Dongheon	Dept. of Physics Education, Pusan University	beastche@hanmail.net
Chang, Eunjin	Kyungpook National University	cute-jjang@hanmail.net
Cheon, Jo Hyun	Buyeong Girls' High School	cjh1024@hanmail.net
Chiaverina, Christopher C.	Physics Department, New Trier High School	fizzforfun@aol.com
Cho, Juhee	Chonnam National University	physics00@hanmail.net
Cho, O-gun	Dept. of Physics Education, Pusan University	puog34@chollian.net
Cho, Sook Kyoung	Pohang University of Science and Technology	skcho@postech.ac.kr
Cho, Yeo Jeong	Taegu University	chossaem@hanmail.net
Choe, Dong-keun	Dept. of Physics Education, Pusan National Univ.	rino@hyowon.pusan.ac.kr
Choe, Heejin	Chonnam National University	chjele@hanmir.com
Choe, Hongsoo	Kyungsang National University	
Choe, Young-joon	Division of Science Education, Taegu University	yjchoe@biho.taegu.ac.kr
Choi, Eun Hee	Munsan middle school	ce2356@hanmail.net
Choi, Hanyong	Korea National University of Education	
Choi, Hyukjoon	Korea National University of Education	rimb@unitel.co.kr
Choi, Hyun Il	Chongju National University of Education	sam7811@naver.com
Choi, Jaehyeok	Dept. of Physics Education, Seoul National University	myjaep@chollian.net
Choi, Kyou Lee	Korea Academy of Gifted Education	judy45@email.com
Choi, Kyunghee	Ewha Womans University	khchoi@mm.ewha.ac.kr
Choi, Myung-Hee	Chongju National University of Education	responce@hanmail.net
Choi, Seung-A	Korea National University of Education	dohyu@unitel.co.kr
Choi, Sungyoun	Ewha Womans University	otrumpt@yahoo.com
Choi, Won Suk	Jinpyeong Middle School	nettrek@chollian.net
Choi, Young Suk	Chongju National University of Education	c99ys@hanmail.net

Chung, Byunghoon	Chongju National University of Education	bhhosw95@sugok.chongju-e.ac.kr
Ferdinande, Hendrik	Ghent University	hendrik.ferdinande@rug.ac.be
Gunstone, Richard Francis	Faculty of Education, Monash University	dick.gunstone@education.monash.edu.au
Ha, Eunsun	Seoul National University	sun76@sun.ac.kr
Han, Changming	Department of Physics, Northeast Normal University	hanchangming@163.net
Hann, Ann-Chin	Inchon National Univ. of Education	ahann@hananet.net
Haruka, Onishi	Kagawa University	s01g117@stmail.ed.kagawa-u.ac.jp
Hayashi, Hirotaka	Ishin High School	h_hayasi@tcp-ip.or.jp
Hayashi, Toshio	Department of Physics, Kagawa University	hayashi@ed.kagawau.ac.jp
Heo, In-ha	Chonnam National University	75hades@hanmail.net
Hong, Sookhee	Ewha Womans University	scihong@lycos.co.kr
Hong, Young-Duck	Chonnam National University	ghddudejr@hanmail.net
Hwang, Sung-Won	Dept. of Physics Education, Seoul National University	solbdung@chollian.net
Hyodo, Toshio	Dept. of Basic Science, University of Tokyo	hyodo@phys.c.utokyo.ac.jp
Hyun, Bongsoo	Kyungpook National University	kevin9871@hanmail.net
Iida, Yoji	Sakuradai High School	YHE03553@nifty.ne.jp
Im, Sungmin	Seoul National University	ismphs@hanmail.net
Jang, KyoungAe	Physics Education, Seoul National University	kajang@donga.com
Jang, Se Jung	Korea National University of Education	jjssjj@cc.knue.ac.kr
Jeon, Jinku	Chongju National University of Education	jjeonjjinku@hanmail.net
Jeon, Seongug	Kwangju Soonil middle school	sim4@netian.com
Jeon, Woosoo	Gongju National University of Education	jeonwsoo@yahoo.co.kr
Jeong, Mi Suk	Sugok Middle School	mp-sp@hanmail.net
Jeong, Mira	Jeong-pyeong Elementary school	jmira11@hanmail.net
Jeong, Seong Oh	Pusan Science High School	jso77@yahoo.co.kr
Jeong, Sukyoung	Dept. of Physics Education, Chungbuk National Univ.	case01@hanmail.net
Jo, Bong Geun	Kyungpook National University	cham1410@yahoo.co.kr
Jo, Kwanghee John	Shingu Middle School	mylux@snu.ac.kr
Joo, Hyun-Na	Jiwon Middle School	gray7711@hanmir.com
Jossem, E. Leonard	The Ohio State Univ.	jossem@mps.ohio-state.edu
Juanich, Maria Jade Salera	Kagawa University	mjuanich@yahoo.com
Jung, Hyungsik	Seoul National University	ssdrim@dreamwiz.com
Jung, Hyun-Ju	Korea National University of Education	hzjung@hanmail.net

Jung, Su In	Song Jung Middle School	captin110@hanmail.net
Jung, Yong-Jae	Dept. of Physics Education, Seoul National Univ.	csn1pjy@hanmail.net
Kamishina, Yoshio	Faculty of Education, Shimane University	kamisina@edu.shimaneu.ac.jp
Kang, Eun Joo	Chil-bo Elementry School in Su-won	hellen_k@hanmail.net
Kang, Eunhyung	Onam Junior High School	ehkrosa@hanmail.net
Kang, Seung Hoon	Korea Academy of Gifted Education	kang21@edunet4u.net
Kawada, Hideo	Meiwa High School	s-cat@mtf.biglobe.ne.jp
Kawakatsu, Hiroshi	Department of Science Education, Kagawa University	kawakatu@ed.kagawau.ac.jp
Kazachkov, Alexander	Physical Department, Kharkov National University	kazachkov@ilt.kharkov.ua
Kim, Beomjung	Dept. of Physics Education, Chungbuk National Univ.	kbj1114@netian.com
Kim, Boemki	Korea National University of Education	kimbk@cc.knue.ac.kr
Kim, Changha	Pusan University	quity@chollian.net
Kim, Daishik	Chungbuk National University	nucleus@chungbuk.ac.kr
Kim, Dok Il	Chonnam Technical High School	twin96@dreamwiz.com
Kim, Dongsook	Chungbuk National University	snakii77@hotmail.com
Kim, Duk Young	Gumdang Middle School	kdukyoung@hanmail.net
Kim, Eun Sik	Chonnam National University	777dmstlrl@hanmail.net
Kim, Eunsook	Korea Institute of Curriculum and Evaluation	eunsooklm@hanmail.net
Kim, Heekyong	Seoul National University	science7@dreamwiz.com
Kim, Hyoungseok	Physics Education, Seoul National University	khsks@hitel.net
Kim, Hyundong	Korea Academy of Gifted Education	khd2@unitel.co.kr
Kim, Hyun-Young	Chongju National University of Education	anata99@hanmail.net
Kim, Ikgyun	Chungbuk National University	igsci@cbucc.chungbuk.ac.kr
Kim, Jaegyung	Taegu University	kjg@biho.taegu.ac.kr
Kim, Jin	Chonnam National University	
Kim, Joowan	Kyungpook National University	long95@hanmail.net
Kim, Jung Hwan	Division of Science Education, Taegu University	fightjh@orgio.net
Kim, Jung-Kuk	Dept. of Physics Education, Pusan National Univ.	
Kim, Jun-Sup	Oisa Elementary School	edu4kid@hanmail.net
Kim, Mani	Korea National University of Education	jjk2000@ibis.co.kr
Kim, Myeong Ki	Kyongsan Elementary School	4627083@hanmail.net
Kim, Myounghwan	Kim Institute for the Korean Gifted	mhkim@thinkid.com
Kim, Sohee	Seoul National University	ksh0703@hotmail.com
Kim, Sungyeon	Ewha Womans University	kimsy19@hanmail.net

Kim, Tae Sun	Korea National University of Education	S032238@hitel.net
Kim, Tae Young	Division of Science Education, Taegu University	adrine@hanmail.net
Kim, Won Suk	Pusan National University	baechu99@yahoo.co.kr
Kim, Woosoo	Chonnam National University	kws8942@hanmail.net
Kim, Yeounsoo	Korea National University of Education	windy98@blue.knue.ac.kr
Kim, Yongpyo	Cheongju Foreign Language High School	22dragon22@hanmail.net
Kim, Young-Min	Pusan National University	minkiyo@hyowon.pusan.ac.kr
Kim, Youngsam	Imae High School	kim03ys@netian.com
Klykov, Dmitry Yurievich	Sternberg Astronomical Institute, Moscow State Univ.	dyuk@sai.msu.ru
Koh, Jae-Kul	Kyungpook National University	
Koo, Dukgil	Seoul National University of Education	
Kwon, Jaesool	Korea National University of Education	jskwon@cc.knue.ac.kr
Kwon, Kee Yahm	Chimsan Middle School	yahm@netsgo.com
Kwon, Sung Gi	Taegu National University of Education	kwonsg@taegu-e.ac.kr
Lee, Dongkyoo	Taegu University	dklee@taegu.ac.kr
Lee, Gang Gil	Chonnam National University	karisma34@hanmail.net
Lee, Gyeongweon	Dept. of Physics Education, Pusan University	egw20@chollian.net
Lee, Heebok	Kongju National Univ.	heebok@kongju.ac.kr
Lee, Hong Jun	Chungbuk National University	metallica@just.chungbuk.ac.kr
Lee, I Mook	Yeo Cheon Middle School	imugyn@dreamwiz.com
Lee, Jeong-Ah	Kwangju Moon-heung Middle School	
Lee, Jeongwon	Seoul National University	hansawee@chollian.net
Lee, Keum-Hwi	ASPEN & KPS	khl@moak.chonbuk.ac.kr
Lee, Myungsook	Joongwon Junior High School	heifetz87@hanmail.net
Lee, Nee-Kyeong	Korea Institute of Curriculum and Evaluation	melee66@yahoo.com
Lee, Sanggyo	Chungbuk National University	apocalip@hanmir.com
Lee, Seok-Tae	Cheongju Foreign Language High School	bdtkm@dreamwiz.com
Lee, Soo Jin	Chonnam High School	sj1218@netian.com
Lee, Sung Yi	Pusan National University	lsungy94@hanmail.net
Lee, Sungmuk	Seoul National University	sungmuk@snu,ac.kr
Lee, Sun-Kyung	Chongju National University of Education	sklee@cje.ac.kr
Lee, Youngsu	Chongju National University of Education	scout111@hanamil.net
Ma, Sang Rin	Sugok Elementary School	msiajs@hanmail.net

Markova, Svetlana Nicolaevna	Sternberg Astronomical Institute, Moscow State Univ.	snm@sai.msu.ru
Masuko, Hiroshi	Azabu High School	hiromasuko@nifty.com
Mazzolini, Alex Peter	Swinburne University of Technology	amazzolini@swin.edu.au
Moon, Seong Sook	Korea National University of Education	giyeob@hanmail.net
Nemoto, Kazuaki	Hitachi First Sr. High School	rainbowman@ma4.justnet.ne.jp
Ogborn, Jon. Micheal	Institute of Physics, London	jon.ogborn@iop.org
Oh, Chul-Han	Kyungpook National University	choh@knu.ac.kr
Omori, Yasuhiro	Kagawa University	y-omori@mail.netwave.or.jp
Paik, Seoung-Hey	Korea National University of Education	shpaik@cc.knue.ac.kr
Park, Cheonin		chun1230@hanmail.net
Park, Churo	Seoul National University	churo@nownuri.net
Park, Hackyoo	Woosuk University	parkhakgu@woosuk.ac.kr
Park, Hyundeok	Korea National University of Education	
Park, Jongwon	Chonnam National University	jwpark94@chonnam.ac.kr
Park, Jong-Yoon	Ewha Womans University	jypark@ewha.ac.kr
Park, Mi-Jin	Pusan National University	alwls90@hanmail.net
Park, Sang Tae	Institute of Science Education, Kongju National Univ.	stpark@kongju.ac.kr
Park, Sangwoo	Seoul National University	pk1966@chollian.net
Park, Sujung	Sintanjinyong Elementary School	carrotsj@hanmail.net
Park, Sung-Jae	Physics Education, Seoul National University	paksj@snu.ac.kr
Park, Yunebae	Kyungpook National University	ypark@knu.ac.kr
Redish, Edward F.	Department of Physics, University of Maryland	redish@physics.umd.edu
Rodrigues, Alice Margariga	Center for Educational Research, University of Lisbon	amlbr@fc.ul.pt
Roh, So-young	Ha-an Middle School	mond11@netian.com
Rossing, Thomas	University of Northern Illinois	rossing@physics.niu.edu
Ryu, Sang-Min	Dept. of Physics Education, Pusan National Univ.	ryu4u@hanmail.net
Sahm, Juergen. A.	Technische Universitate Berlin	sahm@physik.tu-berlin.de
Seo, Eun-Suk	Jisan Middle School, Taegu	silqroad@hanmail.net
Seo, Sangoh	Korea National University of Education	younggae@blue.knue.ac.kr
Seong, Ye-Myoung	Myuncheon Elementary School	kong7242@hanmail.net
Shigeho, Maeda	Kambe High School	fenix@catnip.freemail.ne.jp
Shim, Jae-Gyu	Seoul National University	imshim@dreamwiz.com

Shim, So Jin	Taegu University	tkfkdqkek@hanmail.net
Shim, Yu Jin	Chongju National University of Education	spiny01@hanmail.net
Shimizu, Yuuki	Graduate School of Science, Nagoya University	yuuki@ihas.nagoyau.ac.jp
Shin, Ju Kyoung	Seoul National University	riza7@hanmail.net
Sin, Hyunok	Sintanjin Elementary School	hyuni3@hanmail.net
Solomon, Joan	Dept. of Basic Science, The Open University	j.h.solomon@open.ac.uk
Son, Yun Seop	Chungbuk National University	life2325@hanmail.net
Song, Jinwoong	Seoul National University	jwsong@snu.ac.kr
Song, Young Woog	Korea National University of Education	songywg@hanmail.net
Sugimoto, Norihiro	Sakuradai High School	EZX03202@nifty.ne.jp
Suh, Jungah	Dongma Junior High School	jahelen@hanmail.net
Sung, Kyungjin		peskj@mail.pen.go.kr
Suzuki, Yoshifumi	Kyushu Institute of Technology	ysuzuki@e-lab.kyutech.ac.jp
Takahashi, Wako	Komatsugawa Daini Junior High School	wako-t@kt.rim.or.jp
Takemoto, Keiichi	Kagawa Prefectural Education Center	takemoto-k@niji.or.jp
Takeshi, Ogata	Department of Science Education, Niigata University	atm@pa2.so-net.ne.jp
Takikawa, Yoji	International Christian University High School	takikawa@sannet.ne.jp
Talisayon, Vivien	University of Philippines	director@ismed.upd.edu
Tanaka, Masaki	Matsue Educational Center	tanaka@mec.dandan.ed.jp
Tibell, Gunnar	Uppsala University	gtibell@tsl.uu.se
Toda, Ichiro	Toyama Daiichi High School	itoda@edu.toyamau.ac.jp
Toda, Tsugiteru	Nagoya City Wakamiya Commercial High School	tsu-toda@tcp-ip.or.jp
Tokuya, Inoue	Tokyo Gakugei University	tokuya@yk.rim.or.jp
Trowbridge, Leslie Walter	Dept of Earth Science, Univ. of Northern Colorado	lwtrowbridge@winwins.com
Tsutsui, Masataka	Physics Education Society of Japan	masa777@ma4.justnet.ne.jp
Vicentini, Matilde	Dept. of Physics, Univ. Le Sepienzo	matilde.vicentini@roma1.infn.it
Wang, Zuyuan	Southwest Jiaotong University	cpyan@sc.hnomeway.com.cn
Yamaguchi, Kishichiro	University of the Ryukyus	yamaguci@edu.uryukyu.ac.jp
Yamamoto, Hisamori	Kita High School	hisamori@muh.biglobe.ne.jp
Yamaoka, Yoshiro	Kasugainishi High School	IGOYOSHI@synnet.or.jp
Yan, Yanlai	Shaaghai Jiaotong University	ylyan@online.sh.cn
Yang, Dong-Seok	Dept. of Science Education, Chungbuk National Univ.	

Yang, Hyung Jin	Korea University	yangh@tiger.korea.ac.kr
Yi, Sang Mi	Kyungpook National University	ysm0805@orgio.net
Yoo, Junehee	Korea Institute of Curriculum and Evaluation	jhyoo@kice.re.kr
Yoon, Chan Kyong	Gwangju Science High School	chan29@orgio.net
Yoon, Eun-jeong	Kyungpook National University	shining78@hanmail.net
Yoon, Hye Jung	Chong Ju National University of Education	yhj0915@hanmail.net
Yoon, Jin	Kanghyon Middle School	bjsdream@chollian.net
Youn, Joungsoon	Chongju National University of Education	arang25@hanmail.net
Yu, Byeong Ha	Namsin Elementary School	ybhhdh@hanmail.net
Yu, Kyeongsoo	Cheongju Foreign Language High School	you9405@hanmail.net
Yuk, Keun-Cheul	Kongju National University	gdyuk@kongju.ac.kr
Yum, Suengyul	Chongju National University of Education	jinroy@netian.com
Zhao, Kai Hua	Dept. Physics, Peking University	khzhao@pku.edu.cn

Author Index